An Introduction to
FISH MIGRATION

An Introduction to
FISH MIGRATION

Editors

Pedro Morais

CIIMAR - Centro Interdisciplinar de Investigação Marinha e Ambiental
Universidade do Porto
Porto
Portugal

and

CIMA - Centro de Investigação Marinha e Ambiental
Universidade do Algarve
Faro
Portugal

Françoise Daverat

IRSTEA - Institut National de Recherché en Sciences et
Technologies pour l'Environnement et l'Agriculture
Cestas
France

CRC Press
Taylor & Francis Group
Boca Raton London New York

CRC Press is an imprint of the
Taylor & Francis Group, an **informa** business
A SCIENCE PUBLISHERS BOOK

Cover acknowledgement

·   Left-hand side figure: reproduced by permission of Dr. Russell F. Thurow

·   Centre figure: reproduced by permission of the first editor, Dr. Pedro Morais

·   Right-hand side figure: reproduced by permission of Dr. João Encarnação

CRC Press
Taylor & Francis Group
6000 Broken Sound Parkway NW, Suite 300
Boca Raton, FL 33487-2742

First issued in paperback 2020

© 2016 by Taylor & Francis Group, LLC
CRC Press is an imprint of Taylor & Francis Group, an Informa business

No claim to original U.S. Government works

ISBN-13: 978-1-4987-1873-8 (hbk)
ISBN-13: 978-0-367-78306-8 (pbk)

Library of Congress Cataloging-in-Publication Data

Names: Morais, Pedro (Pedro Miguel) | Daverat, Françoise, 1975-
Title: An introduction to fish migration / Pedro Morais and Françoise Daverat.
Description: Boca Raton : Taylor & Francis, 2016. | "A CRC title." | Includes bibliographical references and index.
Identifiers: LCCN 2015041284 | ISBN 9781498718738 (hardcover : alk. paper)
Subjects: LCSH: Fishes--Migration.
Classification: LCC QL639.5 .M67 2016 | DDC 597.156/8--dc23
LC record available at http://lccn.loc.gov/2015041284

Visit the Taylor & Francis Web site at
http://www.taylorandfrancis.com

and the CRC Press Web site at
http://www.crcpress.com

Preface

The massive quantity and diversity of scientific research papers dealing with several aspects of fish migration hinder undergraduate and graduate students to find an easy introduction to the subject. Furthermore, the access to some of the seminal books and papers is also difficult. Therefore, when we planned the structure of this book, we aimed at covering the aspects we considered essential for undergraduate and graduate students to be introduced to the study of fish migrations, and to provide college professors and scientists with another complementary tool for their work. Indeed, as far as we are aware, this is the first book to cover all fish migratory tactics, i.e., potadromy, diadromy and oceanodromy, since the 1980's.

The book is divided in three parts. It was clear for us that the first part ought to revise the history of fish migration research, the definitions and concepts, as well as the challenges that fish migration scientists must consider in the future. In the second part, the book focuses on the general patterns and processes of each life history, but without focusing on any particular species or family, and it also briefly explores their conservation status. The most common techniques used to study fish migration are described in the third part of this book, and examples on the use of each technique are described for each migratory tactics, if suited for the technique and available in the literature.

We also claim that this book is not an end on itself. We are convinced that the momentum that exists in fish migration research, either on basic or more complex aspects of fish migration, plus the advances in technology and the growing perception on the importance to protect migratory fish species and their habitat, will force a continuous revision of this book through the years to come.

Finally, and above all, we would like to thank all authors for their outstanding contributions, and also for the time that they invested in this endeavor in detriment of personal and professional commitments. The reviewers' contributions were also relevant, their suggestions undoubtedly improved the quality of this book.

September 2015   **Pedro Morais**
CIIMAR - Centro Interdisciplinar de Investigação Marinha e Ambiental
Universidade do Porto
Portugal

and

CIMA - Centro de Investigação Marinha e Ambiental
Universidade do Algarve
Portugal

Françoise Daverat
IRSTEA - Institut National de Recherché en Sciences et Technologies pour
l'Environnement et l'Agriculture, France

Contents

An Overview on Fish Migration Research

PART 1

An Overview on Fish
Migration Research

CHAPTER 1

History of Fish Migration Research

Pedro Morais[1,2,*] and *Françoise Daverat*[3]

David Starr Jordan published 'The history of Ichthyology' in 1902, and dated the origin of Ichthyology back to Aristotle (383–322 BC) with the publication of 'The history of animals' (Aristotle *ca.* 350 BC). However, Jordan (1902) and others after him (Wheeler 1987) elected Petri Artedi (1705–1734) as the father of Ichthyology. Wheeler (1987) enumerated a series of facts highlighting the significance of Artedi's work to ichthyology: (i) revision of previous literature on fishes, (ii) clarification of fish species synonyms, (iii) presentation of a synopsis on how to deal with genera, (iv) accurate description of more than 70 European fish species, (v) introduction of a polynomial classification system. This classification system is considered Artedi's most important achievement, because it is the precursor of the binomial system proposed and used consistently by Carolus Linnæus after the 10th edition of 'Systema Naturæ' published in 1758 (Wheeler 1987). Artedi's book '*Ichthyologia cluding opera omnia the piscibus*' was published posthumously in 1738, after his premature death, which includes two unpublished works '*Bibliotheca Ichthyologica*' and '*Philosophia Ichthyologia*' (Wheeler 1987). Carolus Linnæus was the editor of Artedi's seminal book and acquired almost all of his knowledge on fishes from Artedi's work (Jordan 1902).

However, before Artedi's book, a few seminal books were published during four years in the 16th century on Mediterranean Sea fishes, namely the books of Pierre Bellon in 1553, Salviani in 1554, and of Guillaume Rondelet in 1556 (Jordan 1902). As a matter of interest, the identification (or misidentification) and distribution of sturgeons in Europe and northern Africa were at the center of a heated debate during the 16th century, as the one led by the Portuguese Renaissance man André de Resende with the publication of '*De Antiquitatibus Lusitaniae*' in 1593 (Resende 1593;

[1] CIMA-Centro de Investigação Marinha e Ambiental, Universidade do Algarve, Portugal.
[2] CIIMAR-Centro Interdisciplinar de Investigação Marinha e Ambiental, Universidade do Porto, Portugal.
[3] IRSTEA-Institut National de Recherché en Sciences et Technologies pour l'Environnement et l'Agriculture, France.
* Corresponding author: pmorais@ualg.pt

Rosado Fernandes 1991, 1996). In the late 17th century, the book '*Historia Piscium*' was published in 1686, a work that was initiated by Francis Willughby and concluded by John Ray, and was considered by Georges Cuvier as the precursor of modern ichthyology (Kusukawa 2000). In 1777, Francesco Cetti published the 'Storia naturale della Sardegna', where he described a putative migratory route for the Atlantic bluefin tuna into the Mediterranean, and before their arrival to spawn around Sicily, and which had a tremendous economic importance in Sicily at that time (Cetti 2000).

In Jordan's review, there is no mention of studies on migratory fishes (Jordan 1902), since during the 19th century most ichthyologists were naturalists dedicated to classify and list the species existing in their countries or in certain areas of interest. However, these lists might contain special references to migratory fish, as it was in the case of sturgeon in Portuguese rivers (Pimentel 1894 in Almaça and Elvira 2000; Baldaque da Silva 1895 in Almaça and Elvira 2000), which are now of paramount importance to understand the long-term fluctuations of populations and the influence of anthropogenic impacts. Despite the apparent absence of relevant studies on migratory fish during the 19th century, there were already a few ichthyologists carrying out influential works on anatomy, compared anatomy and palaeontology during the 19th century (Lee 1893, 1894; Jordan 1902). Also, at this time, Australia, Canada, Norway, United Kingdom and the United States were the only countries to possess a comprehensive literature on commercial fisheries, other than faunistic lists and anatomical studies (Herdman and Dawson 1902; Jordan 1901, 1902).

In 1916, and regarding fish migration studies, Alexander Meek (1865–1949) stated the following: "*During the last two decades especially, the problems of migration have been investigated with gradually increasing energy, and the results has been an accumulation of valuable knowledge, which, I hope, is already sufficient to bring out and to establish general principles*" (Meek 1916, p. vii). At the end of the 1940's, George S. Myers affirmed that Meek's book was the most influential work on fish migration up to then (Myers 1949). So, it is often claimed that the study of fish migrations was officially established as a discipline of ichthyology with the publication of 'The migrations of fish' by Alexander Meek (Meek 1916). However, it is interesting to mention that Aristotle (*ca.* 350 BC) already recognized the existence of fish migration, namely into the Euxine Sea (the archaic name of Black Sea) for the purposes of reproduction and feeding.

Meek (1916) made an incredible effort to compile information about several migratory fish species and families, and described the patterns of migration and distribution whenever the available information allowed it. It is interesting to note that he did not restrict his review to cover only the most emblematic migratory species, as salmons or eels. Indeed, during the early 20th century, Meek (1916) demonstrated that there was already a clear perception of the importance of tracking and studying oceanodromous species, as elasmobranchs and tunas. It was also clear for Meek (1916) that tides, ocean circulation and phylogeny could explain species life history patterns, or at least clarify parts of it. The three aspects that were just mentioned are still debated and the core of many research projects during the 21st century (e.g., Parenti 2008; Robins et al. 2012; Trancart et al. 2012), thus revealing Meek's vision.

Meek (1916) also had the same concern on migratory species conservation as scientists and managers have nowadays, focusing mainly on pollution and fishery management. If the conservation of species and biodiversity would not be a sufficient reason to protect migratory species, he also highlighted their economic importance to reinforce his point of view. As an example, we quote three passages illustrating some of Meek's concerns and recommendations:

> *"The history of some of our rivers bears eloquent testimony to the fact that contamination by sewage and trade effluents may reach such a degree as to totally prevent anadromous fishes like the salmon ascending to the spawning grounds. ... The Thyne is clearly approaching the limit of contamination, for every year reports are made as to a heavy death-rate of spawning migrants in the river opposite Newcastle. ... The growth of the cities and towns on the banks and the multiplication of industries have converted the tidal portion of the river into an open sewer"* (p. 123);

> *"While, therefore, the summer phase of such a school as that of the Northumberland plaice could be regulated by the Fisheries Committee of that county, it is clear that the regulation of Fisheries cannot be done in a parochial manner, but requires not merely National, but International, legislation to make it effective"* (p. 397);

> *"The evidence we already possess indicates that some of the species become winter migrants to the coast or shallow water during the years of immaturity, and they retain the habit of assembling in shallow water when they reach maturity. Even from an economic standpoint it is desirable to obtain a full knowledge of the distribution and migrations of the young of the important family Gadidae"* (p. 245).

After the publication of 'The migrations of fish' (Meek 1916), there were a few other books published in English literature during the 20th and 21st centuries covering several aspects of fish migration. Among these books, we highlight those of Roule (1933), Hasler (1966), Harden Jones (1968), Nakamura (1969), McCleave et al. (1984), McKeown (1984), McDowall (1988) and Lucas and Baras (2001).

In 1933, Louis Roule published 'Fishes. Their journeys and migrations', which was described by Robert M. McDowall as a description of fish migration using very "colourful terms" (McDowall 1988). This book is not as comprehensive as the one of Meek (1916), nor influential. Roule (1933) only presents, in his "colourful terms", the migration patterns of a few emblematic species, among which are included the European eel *Anguilla anguilla*, the Atlantic salmon *Salmo salar* and the shads *Alosa alosa* and *Alosa fallax*. We would classify Roule's (1933) book as a science dissemination book, where he colorfully described some aspects of migratory fish's life histories. The following passage is illustrative of the dual characteristic of Roule's (1933) book:

> *"Yet the journey always take the same course. The salmon arrive at intervals, each apparently its own master, yet they all behave in exactly the same way and follow the same road. What do they find, what signpost like that upon our roads, that*

they manifest a preference, so definite and so lasting? The direct and immediate reason for their choice certainly does not reside in the fish alone. Where would he find, within himself, one strong enough to make him adopt a certain course of action? Many of the salmon, coming for the first time, cannot have any idea where their journey will take them. Even those who have been there before can hardly recognize the things and places they have seen on a previous occasion" (Roule 1933, pp. 48–49).

It is easy to recognize two fundamental questions presented by Roule (1933), homing and orientation. Just these two topics generate a profusion of works and discussions on fishes from several habitats, along several stages of their life cycle and with distinct life history strategies, until the present times (e.g., Huntsman 1937; Hasler et al. 1958; Hasler 1960; Rommel Jr. and McLeave 1972; Scholz et al. 1976; Quinn 1980; Quinn and Brannon 1982; Levin et al. 1992; James et al. 2007; Skov et al. 2010; Leonard et al. 2012). The physiological aspects involved in diadromous migrations were also a topic of interest for many decades now, and the lifetime work of Maurice Fontaine on migratory fish physiology, and particularly his insights on the endocrine processes underlying fish migrations, were influential (Fontaine 1954).

Later, in 1966, Arthur D. Hasler published 'Underwater guideposts: homing of salmon', a thorough review on salmon migration strategies, generally taking into account his experience with Pacific salmon species (Hasler 1966). Hasler (1966) based his general description of salmon's life cycle considering two major hypothesis to explain salmon's homing: (a) odor cues that allows salmons to find their way home within the natal river basin, (b) sun-compass (sun azimuth and altitude) which guide salmons toward the natal river basin after completing their oceanic life phase. However, Hasler (1966) recognizes that these two hypothesis cannot explain every aspect of salmon's migration, and suggests that other cues might aid in navigation, as stellar and inertial cues.

Just two years later, in 1968, Frederick Robert Harden Jones published the next major publication on fish migration, which was classified by Parrish (1968) as *"one of the most important post-war additions to the literature on fish biology and will probably remain the most complete work on the subject of fish migration for many years to come"*. Since Meek's book, there was a tremendous development in the study of fish migration, which was summarized in Harden Jones' book 'Fish migration' (Harden Jones 1968). Harden Jones updated the knowledge on the life history patterns of some migratory species, all of them already presented in the books of Meek (1916) and Roule (1933) (i.e., salmon, trout, eel, herring, cod and plaice), and he also describes the mechanisms used by fish during their migrations and a few useful methods to study fish migration. So, at the end of the 1960's, the recognized and/or hypothesized mechanisms used by fish to start, orientate and end a migration journey included the reaction to some of the following external stimuli- chemical cues, water current, magnetic/electrical cues, temperature, celestial cues (solar or stellar) and topography. Harden Jones (1968) also revised a series of techniques used to infer fish migration or fish stock origin, which were obviously fewer than today, some of which were already abandoned for this particular purpose and replaced by other techniques. Harden Jones (1968) mentioned several techniques to infer fish migration, like the use of scales and otoliths as 'certificates of origin' through the comparative analyses of their microstructure and shape. Nowadays, the microchemical composition of these structures receives far more attention to study these topics (e.g., Borcherding et al. 2008; Elsdon et al. 2008; Daverat and Martin 2016), than microstructure or shape analyses (Quinn et al. 1999; Begg and Brown 2000). At that time, morphometric analyses were done using few morphological traits, for example Harden Jones (1968) refers to a study that used three morphological traits to distinguish between three stocks of landlocked sockeye salmon. Nowadays, with the advances of geometric morphometric software and new statistical techniques, scientists compare dozens of variables (e.g., landmarks and contours, besides just meristic and morphometric variables) to develop such types of studies (Saborido-Rey and Nedreaas 2000; Sequeira et al. 2011; Turki-Missaoui et al. 2011; Trella et al. 2013). Among the restricted number of techniques available at the end of the 1960's, fish tagging and fish marking were one of the most used techniques to infer fish migration, ever since the end of the 19th century (Harden Jones 1968). The main concept of these techniques are maintained till today, however numerous technological leaps took tagging into a completely new realm (e.g., acoustic and radio tags, pop-up satellite archival tags) (see Chapters 12 and 13 for more details) (Begout et al. 2016; Schaefer and Fuller 2016).

In 1969, Hiroshi Nakamura published a book dedicated solely to tunas, entitled 'Tuna: distribution and migration' (Nakamura 1969). This book, because of its specificity, complements perfectly the broader scoped book of Harden Jones (1968), by analyzing in detail the life histories of these oceanodromous species, and is still a reference book almost 50 years later (e.g., Couto et al. 2016).

In 1984, James D. McCleave, Geoffrey P. Arnold, Julian J. Dodson and William H. Neill edited a book entitled 'Mechanisms of migration in fishes' (McCleave et al. 1984), with contributions of some conference delegates that met in Italy in 1982. The editors distinguished the processes occurring during fish migration in different environments—ocean, coastal areas and estuaries, rivers. The book was divided in four sections, 'Migration in the open ocean', 'Migration in coastal and estuarine waters', 'Migration in rivers' and another entitled 'Special topics', which presented works on fish learning, orientation, behavior, bioenergetics. Most contributions were based on a species case study, though the topics covered were still investigated in the 2010's. Here are some examples of these contributions that might be of interest to many of today's researchers: 'Advection, diffusion, and drift migrations of larval fish' (Power 1984), 'Behavioral enviroregulation's role in fish migration' (Neill 1984), 'Influence of stock origin on salmon migratory behaviour' (Brannon 1984), 'The orientation of fish and the vertical stratification at fine- and micro-structure scales' (Westerberg 1984), 'Patterns, mechanisms and approaches to the study of migrations of estuarine-dependent fish larvae and juveniles' (Miller et al. 1984), 'Fish migration by selective tidal stream transport: first results with a computer simulation model for the European continental shelf' (Arnold and Cook 1984), 'Migration in coral reef fishes: ecological significance and orientation mechanisms' (Ogden and Quinn 1984), 'Mechanisms of fish migration in rivers' (Northcote 1984), 'Bioenergetic considerations in fish migration' (Weihs 1984).

Also in 1984, Brian A. McKeown published a book entitled 'Fish migration', which is the first, to our knowledge, covering potamodromous, diadromous and oceanodromous fish migrations (McKeown 1984). He also included a small chapter entitled 'Littoral migrations', and used some examples that in the face of current knowledge are not regarded as migrations, but rather as tidal, daily or ontogenetic movements, or are included in one of the mentioned migration strategies. The main focus of McKeown's book is on the processes involved during migration, rather than on updating the knowledge on the migration strategies of the most emblematic migratory species (McKeown 1984). Thus, McKeown (1984) focused his attention in four main chapters: 'Orientation', 'Bioenergetics', 'Physiology', 'Ecology and evolution'. Undoubtedly, the chapter 'Physiology' is the most comprehensive, which clearly reveals McKeown's expertise (e.g., McKeown and Peter 1976; McKeown et al. 1980).

Three years later, in 1987, the American Fisheries Society edited a book 'Common strategies of anadromous and catadromous fishes', with contributions of many scientists that participated in a conference the previous year (Dadswell et al. 1987). This book starts with two influential papers, 'The occurrence and distribution of diadromy among fishes' by McDowall (1987) and 'Evolution of diadromy in fishes' by Gross (1987). Both scientists revisited these topics (McDowall 1988; Gross et al. 1988; McDowall 1997), and these papers are still among the most well-known in fish migration research. In this book, we can find several papers on fish life history patterns (e.g., Boreman and Lewis 1987; Bruton et al. 1987; Loesch 1987), migration tactics (e.g., McCleave and Wippelhauser 1987; Healey and Groot 1987; Weihs 1987), physiology (e.g., Dutil et al. 1987; McCormick and Saunders 1987; McEnroe and Cech Jr. 1987), recruitment (e.g., Chadwick 1987; Moriarty 1987; Rothschild and DiNardo 1987) and reproduction (e.g., Kedney et al. 1987; Maurice et al. 1987), yet the opening papers of McDowall (1987) and Gross (1987) are those that clearly stand out, even today.

Four years after McKeown's book (1984), Robert M. McDowall published a book entitled 'Diadromy in fishes'. Although McDowall (1988) does not cover the life histories of oceanodromous and potamodromous fishes, his book is undisputedly a stepping-stone for fish migration research. The potential of McDowall's book was recognized at that time by Harden Jones (1989) and confirmed by the vast literature that still cites McDowall's book. McDowall's (1988) book is very stimulating, and he presents a series of interesting hypotheses. McDowall (1988) suggests that there is a relationship between the frequency of anadromous and catadromous species along a latitudinal gradient with the productivity of marine habitats. So, according to McDowall (1988), there are more anadromous species at higher latitudes because marine habitats are more productive there, while the frequency of catadromous species increases towards the tropics because aquatic productivity is higher in freshwater habitats than in marine habitats. This hypothesis was also published in 1988 by Mart R. Gross, Ronald M. Coleman and Robert M. McDowall (Gross et al. 1988). However, some breaches to this hypothesis were detected by Bloom and Lovejoy (2014), which suggest that predation, competition and even geological history may be, at least, as relevant as productivity in explaining the origins of diadromy. McDowall (1988) also tried to elucidate other patterns and proposed that diadromy is predominant in phylogenetically primitive families (e.g., lampreys, sturgeons, anguillid eels, several salmoniform families), that diadromy is widespread in some divergent phylogenetic lineages within families (e.g., shads, herrings, mullets, gobies), and that diadromy is intermittent or a minor characteristic in some other families (e.g., flounders, sculpins, scorpionfishes). McDowall (1988) also highlighted the fact that diadromy might be facultative in some species, i.e., polytypic species—populations that comprise migratory and non-migratory stocks. This phenomenon might also be a result of a landlocked process, either natural or due to anthropogenic influence. Those landlocked populations that are able to succeed might reacquire the ability to migrate once the opportunity for migration is restored, even if the population is landlocked for some generations. McDowall (1988) hypothesized that it is likely that landlocked populations of diadromous species might be at the base of some speciation processes, but he also made it clear that this speciation mechanism had to be evaluated with phylogenetic analysis prior to establishing any relationship between similar species. In 2001, McDowall (2001) revisited this topic, and added that the dispersive capacity of diadromous species, together with the facultative life history stages that some diadromous species have, could lead to diversification and speciation in diadromous fishes.

Finally, and already in the 21st century, Martyn C. Lucas and Etienne Baras published an outstanding book 'Migration of freshwater fishes' (Lucas and Baras 2001). In this book, they revised and updated the fish migration concepts, the types of migration associated with freshwater fish (i.e., feeding, refuge-seeking, spawning), the stimulus that trigger fish migration, the cues used by fish during migration, the methods used to study fish migration and some aspects regarding migratory freshwater fish conservation. Additionally, Lucas and Baras (2001) also summarized the main migration strategies of species belonging to 57 orders and families of fishes that migrate within freshwater habitats, or that might use freshwater habitats during a specific period of their life.

References

Almaça CF and Elvira B (2000) Past and present distribution of *Acipenser sturio* L., 1758 on the Iberian Peninsula. Boletín Instituto Español de Oceanografía 16: 11–16.

Aristotle (circa 350 BC) A History of Animals. Book VIII. Translated by D'ArcyWentworth Thompson. http://classics.mit.edu/Aristotle/history_anim.8.viii.html. Accessed December 28, 2014.

Arnold GP and Cook PH (1984) Fish migration by selective tidal stream transport: first results with a computer simulation model for the European continental shelf. pp. 227–261. *In*: McCleave JM, Arnold GP, Dodson JJ and Neill WH (eds.). Mechanisms of Migration in Fishes. Plenum Press, New York and London.

Begg GA and Brown RW (2000) Stock identification of haddock *Melanogrammus aeglefinus* on Georges Bank based on otolith shape analysis. Transactions of the American Fisheries Society 129: 935–945.

Begout M-L, Bau F, Acau A and Acolas M-L (2016) Methodologies for investigating diadromous fish movements: conventional, PIT, acoustic and radio tagging and tracking. pp. 214–250. *In*: Morais P and Daverat F (eds.). An Introduction to Fish Migration. CRC Press, Boca Raton, FL, USA (this book).

Boreman J and Lewis RR (1987) Atlantic coastal migration of stripped bass. pp. 331–339. *In*: Dadswell MJ, Klauda RJ, Moffitt CM, Saunders RL, Rulifson RA and Cooper JE (eds.). Common Strategies of Anadromous and Catadromous Fishes. American Fisheries Society Symposium I.

Brannon EL (1984) Influence of stock origin on salmon migratory behaviour. pp. 103–111. *In*: McCleave JM, Arnold GP, Dodson JJ and Neill WH (eds.). Mechanisms of Migration in Fishes. Plenum Press, New York and London.

Bruton MN, Bok AH and Davies MTT (1987) Life history styles of diadromous fishes in inland waters of Southern Africa. pp. 104–121. *In*: Dadswell MJ, Klauda RJ, Moffitt CM, Saunders RL, Rulifson RA and Cooper JE (eds.). Common Strategies of Anadromous and Catadromous Fishes. American Fisheries Society Symposium I.

Cetti F (2000) Storia naturale di Sardegna: a cura di Antonello Mattone e Piero Sanna. Ilisso. 452p. Available at www.sardegnacultura.it/documenti/7_49_20060407114902.pdf. Accessed October 5, 2014.

Chadwick EMP (1987) Causes of variable recruitment in a small Atlantic salmon stock. pp. 390–401. *In*: Dadswell MJ, Klauda RJ, Moffitt CM, Saunders RL, Rulifson RA and Cooper JE (eds.). Common Strategies of Anadromous and Catadromous Fishes. American Fisheries Society Symposium I.

Couto A, Baptista M, Furtado M, Sousa LL and Queiroz N (2016) Oceanodromous fish migrations. pp. 123–146. *In*: Morais P and Daverat F (eds.). An Introduction to Fish Migration. CRC Press, Boca Raton, FL, USA (this book).

Dadswell MJ, Klauda RJ, Moffitt CM, Saunders RL, Rulifson RA and Cooper JE (1987) Common strategies of anadromous and catadromous fishes. American Fisheries Society Symposium I. 561p.

Daverat F and Martin J (2016) Microchemical and schlerochronological analyses used to infer fish migration. pp. 149–168. *In*: Morais P and Daverat F (eds.). An Introduction to Fish Migration. CRC Press, Boca Raton, FL, USA (this book).

Dutil J-D, Besner M and McCormick SD (1987) Osmoregulatory and ionregulatory changes and associated mortalities during the transition of maturing American eels to a marine environment. pp. 175–190. *In*: Dadswell MJ, Klauda RJ, Moffitt CM, Saunders RL, Rulifson RA and Cooper JE (eds.). Common Strategies of Anadromous and Catadromous Fishes. American Fisheries Society Symposium I.

Elsdon TS, Wells BK, Campana SE, Gillanders BM, Jones CM, Limburg KE, Secor D, Thorrold SR and Walther BD (2008) Otolith chemistry to describe movements and life-history parameters of fishes: hypotheses, assumptions, limitations and inferences. Oceanography and Marine Biology: An Annual Review 46: 297–330.

Fontaine M (1954) Du determinisme physiologique des migrations. Biological Reviews 29: 390–418.

Gross MR (1987) Evolution of diadromy in fishes. pp. 14–25. *In*: Dadswell MJ, Klauda RJ, Moffitt CM, Saunders RL, Rulifson RA and Cooper JE (eds.). Common Strategies of Anadromous and Catadromous Fishes. American Fisheries Society Symposium I.

Gross MR, Coleman RM and McDowall RM (1988) Aquatic productivity and the evolution of diadromous fish migration. Science 239: 1291–1293.

Harden Jones FR (1968) Fish Migration. Edward Arnold (Publishers) Ltd., London. 325p.

Harden Jones FR (1989) There and back again. Nature 340: 276.

Hasler AD (1960) Guideposts of migrating fishes. Science 132: 785–792.

Hasler AD (1966) Underwater Guideposts: Homing of Salmon. The University of Wisconsin Press, Madison, Milwaukee, and London. 155p.

Hasler AD, Horrall RM, Wisby WJ and Braemer W (1958) Sun-orientation and homing fishes. Limnology and Oceanography 3: 353–61.

Healey MC and Groot C (1987) Marine migration and orientation of an ocean-type chinook and sockeye salmon. pp. 298–312. *In*: Dadswell MJ, Klauda RJ, Moffitt CM, Saunders RL, Rulifson RA and Cooper JE (eds.). Common Strategies of Anadromous and Catadromous Fishes. American Fisheries Society Symposium I.

Herdman WA and Dawson RA (1902) Fish and Fisheries of the Irish Sea and Especially of the Lancashire and Western Sea-Fisheries District. George Philip & Son Ltd., London. 98p.

Huntsman AG (1937) Migration and homing of Pacific salmon. Science 86: 55–56.

IATTC (Inter-American Tropical Tuna Commission) (2014) http://www.iattc.org. Accessed May 23rd 2014.

James NC, Cowley PD, Whitfield AK and Kaiser H (2007) Choice chamber experiments to test the attraction of postflexion *Rhabdosargus holubi* larvae to water of estuarine and riverine origin. Estuarine, Coastal and Shelf Science 77: 143–149.

Jordan DS (1901) The fish fauna of Japan, with observations on the geographical distribution of fishes. Science XIV: 545–567.

Jordan DS (1902) The history of Ichthyology. Science XVI: 427–456.

Kedney GI, Boulé V and Fitzgerald GJ (1987) The reproductive ecology of threespine sticklebacks breeding in fresh and brackish water. pp. 151–161. *In*: Dadswell MJ, Klauda RJ, Moffitt CM, Saunders RL, Rulifson RA and Cooper JE (eds.). Common Strategies of Anadromous and Catadromous Fishes. American Fisheries Society Symposium I.

Kusukawa S (2000) The Historia Piscium (1686) Notes and Records of the Royal Society of London 54: 179–197.

Lee FS (1893) A study of the sense of equilibrium in fishes. I. Journal of Physiology 15: 311–348.

Lee FS (1894) A study of the sense of equilibrium in fishes. Part II. Journal of Physiology 15: 192–210.

Leonard G, Maie T, Moody KN, Schrank GD, Blob RW and Schoenfuss HL (2012) Finding paradise: cues directing the migration of the waterfall climbing Hawaiian gobioid *Sicyopterus stimpsoni*. Journal of Fish Biology 81: 903–920.

Levin LE, Belmonte P and Gonzalez O (1992) Sun-compass orientation in the characid *Cheirodon pulcher*. Environmental Biology of Fishes 35: 321–325.

Loesch JG (1987) Overview of life history aspects of anadromous alewife and blueback herring in freshwater habitats. pp. 89–103. *In*: Dadswell MJ, Klauda RJ, Moffitt CM, Saunders RL, Rulifson RA and Cooper JE (eds.). Common Strategies of Anadromous and Catadromous Fishes. American Fisheries Society Symposium I.

Lucas MC and Baras E (2001) Migration of Freshwater Fishes. Blackwell Science, Great Britain. 420p.

Maurice KR, Blye RW and Harmon PL (1987) Increasing spawning by American shad coincident with improved dissolved oxygen in the tidal Delaware river. pp. 79–88. *In*: Dadswell MJ, Klauda RJ, Moffitt CM, Saunders RL, Rulifson RA and Cooper JE (eds.). Common Strategies of Anadromous and Catadromous Fishes. American Fisheries Society Symposium I.

McAllister DE, Craig JF, Davidson N, Delany S and Seddon M (2001) Biodiversity Impacts of Large Dams. UNEP & United Nations Foundation & IUCN. 68p.

McCleave JD and Wippelhauser GS (1987) Behavioral aspects of selective tidal stream transport in juvenile American eels. pp. 138–150. *In*: Dadswell MJ, Klauda RJ, Moffitt CM, Saunders RL, Rulifson RA and Cooper JE (eds.). Common Strategies of Anadromous and Catadromous Fishes. American Fisheries Society Symposium I.

McCleave JD, Arnold GP, Dodson JJ and Neill WH (eds.) (1984) Mechanisms of Migration in Fishes. NATO conference series. IV Marine Series, Vol. 14. Plenum Press, New York and London. 574p.

McCormick SD and Saunders RL (1987) Preparatory physiological adaptations for marine life of salmonids: osmoregulation, growth, and metabolism. pp. 211–229. *In*: Dadswell MJ, Klauda RJ, Moffitt CM, Saunders RL, Rulifson RA and Cooper JE (eds.). Common Strategies of Anadromous and Catadromous Fishes. American Fisheries Society Symposium I.

McDowall RM (1987) The occurrence and distribution of diadromy among fishes. pp. 1–13. *In*: Dadswell MJ, Klauda RJ, Moffitt CM, Saunders RL, Rulifson RA and Cooper JE (eds.). Common Strategies of Anadromous and Catadromous Fishes. American Fisheries Society Symposium I.

McDowall RM (1988) Diadromy in Fishes. Timber Press, Portland, Oregon. 308p.

McDowall RM (1997) The evolution of diadromy in fishes (revisited) and its place in phylogenetic analysis. Reviews in Fish Biology and Fisheries 7: 443–462.

McDowall RM (2001) Diadromy, diversity and divergence: implications for speciation processes in fishes. Fish and Fisheries 2: 278–285.

McEnroe M and Cech Jr. JJ (1987) Osmoregulation in white sturgeon: life history aspects. pp. 191–196. *In*: Dadswell MJ, Klauda RJ, Moffitt CM, Saunders RL, Rulifson RA and Cooper JE (eds.). Common Strategies of Anadromous and Catadromous Fishes. American Fisheries Society Symposium I.

McKeown BA (1984) Fish Migration. Croom-Helm, London. 224p.

McKeown BA and Peter RE (1976) The effects of photoperiod and temperature on the release of prolactin from the pituitary gland of the goldfish, *Carassius auratus* L. Canadian Journal of Zoology 54: 1960–1968.

McKeown BA, Jenks BG and Van Overbeeke AP (1980) Biosynthesis and release of prolactin from the pituitary gland of the rainbow trout, *Salmo gairdneri*. Comparative Biochemistry and Physiology 65: 705–709.

Meek A (1916) The Migrations of Fish. Edward Arnold, London. 427p. + xviii.

Miller JM, Reed JP and Pietrafesa LJ (1984) Patterns, mechanisms and approaches to the study of migrations of estuarine-dependent fish larvae and juveniles. pp. 209–225. *In*: McCleave JM, Arnold GP, Dodson JJ and Neill WH (eds.). Mechanisms of Migration in Fishes. Plenum Press, New York and London.

Moriarty C (1987) Factors influencing recruitment of the Atlantic species of anguillid eels. pp. 483–491. *In*: Dadswell MJ, Klauda RJ, Moffitt CM, Saunders RL, Rulifson RA and Cooper JE (eds.). Common Strategies of Anadromous and Catadromous Fishes. American Fisheries Society Symposium I.

Myers GS (1949) Usage of anadromous, catadromous and allied terms for migratory fishes. Copeia 1949: 89–97.

Nakamura H (1969) Tuna: Distribution and Migration. Fishing News Ltd., London. 76p.

Neill WH (1984) Behavioral enviroregulation's role in fish migration. pp. 61–66. *In*: McCleave JM, Arnold GP, Dodson JJ and Neill WH (eds.). Mechanisms of Migration in Fishes. Plenum Press, New York and London.

Northcote TG (1984) Mechanisms of fish migration in rivers. pp. 317–355. *In*: McCleave JM, Arnold GP, Dodson JJ and Neill WH (eds.). Mechanisms of Migration in Fishes. Plenum Press, New York and London.

Ogden JC and Quinn TP (1984) Migration in coral reef fishes: ecological significance and orientation mechanisms. pp. 293–308. *In*: McCleave JM, Arnold GP, Dodson JJ and Neill WH (eds.). Mechanisms of Migration in Fishes. Plenum Press, New York and London.

Parenti LR (2008) Life history patterns and biogeography: an interpretation of diadromy in fishes. Annals of the Missouri Botanical Garden 95: 232–247.

Parrish BB (1968) Fish migration. Nature 220: 1008–1009.

Power JH (1984) Advection, diffusion, and drift migrations. pp. 27–37. *In*: McCleave JM, Arnold GP, Dodson JJ and Neill WH (eds.). Mechanisms of Migration in Fishes. Plenum Press, New York and London.

Quinn TP (1980) Evidence for celestial and magnetic compass orientation in lake migrating sockeye salmon fry. Journal of Comparative Physiology 137: 243–248.

Quinn TP and Brannon EL (1982) The use of celestial and magnetic cues by orienting sockeye salmon smolts. Journal of Comparative Physiology 147: 547–52.

Quinn TP, Volk EC and Hendry AP (1999) Natural otolith microstructure patterns reveal precise homing to natal incubation sites by sockeye salmon (*Oncorhynchus nerka*). Canadian Journal of Zoology 77: 766–775.

Resende A (1593) *De Antiquitatibus Lusitaniae*. Matinus Burgensis academis typographus. Évora. 259p. Available at www.bdalentejo.net/BDAObra/BDADigital/Obra.aspx?ID=240#. Accessed November 23, 2014.

Robins PE, Neill SP and Giménez L (2012) A numerical study of marine larval dispersal in the presence of an axial convergent front. Estuarine, Coastal and Shelf Science 100: 172–185.

Rommel Jr. S and McLeave JD (1972) Oceanic electric fields: perception by American eels? Science 176: 1233–1235.

Rosado Fernandes RM (1991) André de Resende e o seu Asturjão africano (o *Angulo amazi* do *De Antiquitatibus Lusitaniae*). Humanitas XLIII-XLIV: 355–368. Available at www.uc.pt/fluc/eclassicos/publicacoes/ficheiros/humanitas43-44/21_Rosado_Fernandes.pdf. Accessed February 4, 2014.

Rosado Fernandes RM (1996) Asantiguidades de Lusitânia/André de Resende; introdução, tradução e comentário. Fundação Calouste de Gulbenkian. Lisboa. 660p. Available at www.bdalentejo.net/BDAObra/BDADigital/Obra.aspx?ID=265. Accessed October 17, 2014.

Roule L (1933) Fishes. Their Journeys and Migrations. George Routledge & Sons Ltd., London. 270p.

Saborido-Rey F and Nedreaas KH (2000) Geographic variation of *Sebastes mentella* in the Northeast Arctic derived from a morphometric approach. ICES Journal of Marine Science: Journal du Conseil 57: 965–975.

Schaefer KM and Fuller DW (2016) Archival and pop-up satellite archival tags: designs, attachments, data analyses, and applications in studies of large-scale movements of fish. pp. 251–289. *In*: Morais P and Daverat F (eds.). An Introduction to Fish Migration. CRC Press, Boca Raton, FL, USA (this book).

Scholz AT, Horrall RM, Cooper JC and Hasler AD (1976) Imprinting to chemical cues: the basis for home stream selection in salmon. Science 192: 1247–1249.

Sequeira V, Rodríguez-Mendoza R, Neves A, Paiva R, Saborido-Rey F and Gordo LS (2011) Using body geometric morphometrics to identify bluemouth, *Helicolenus dactylopterus* (Delaroche, 1809) populations in the Northeastern Atlantic. Hydrobiologia 669: 133–141.

Skov C, Aarestrup K, Baktoft H, Brodersen J, Brönmark C, Hansson L-A, Nielsen EE, Nielsen T and Nilsson PA (2010) Influences of environmental cues, migration history, and habitat familiarity on partial migration. Behavioral Ecology 21: 1140–1146.

Trancart T, Lambert P, Rochard E, Daverat F, Coustillas J and Roqueplo C (2012) Alternative flood tide transport tactics in catadromous species: *Anguilla anguilla*, *Liza ramada* and *Platichthys flesus*. Estuarine, Coastal and Shelf Science 99: 191–198.

Trella K, Podolska M, Nedreaas K and Janusz J (2013) Discrimination of the redfish (*Sebastes mentella*) stock components in the Irminger Sea and adjacent waters based on meristics, morphometry and biological characteristics. Journal of Applied Ichthyology 29(2): 341–351.

Turki-Missaoui O, M'Hetli M, Kraïem MM and Chriki A (2011) Morphological differentiation of introduced pikeperch (*Sander lucioperca* L., 1758) populations in Tunisian freshwaters. Journal of Applied Ichthyology 27: 1181–1189.

Weihs D (1984) Bioenergetic considerations in fish migration. pp. 487–508. *In*: McCleave JM, Arnold GP, Dodson JJ and Neill WH (eds.). Mechanisms of Migration in Fishes. Plenum Press, New York and London.

Weihs D (1987) Hydromechanics of fish migration in variable environments. pp. 254–261. *In*: Dadswell MJ, Klauda RJ, Moffitt CM, Saunders RL, Rulifson RA and Cooper JE (eds.). Common Strategies of Anadromous and Catadromous Fishes. American Fisheries Society Symposium I.

Westerberg E (1984) The orientation of fish and the vertical stratification at fine- and micro-structure scales. pp. 179–203. *In*: McCleave JM, Arnold GP, Dodson JJ and Neill WH (eds.). Mechanisms of Migration in Fishes. Plenum Press, New York and London.

Wheeler A (1987) Peter Artedi, founder of modern ichthyology. Proceedings of the Fifth Congress of European Ichthyologists (1985). pp. 3–10.

CHAPTER 2

Definitions and Concepts Related to Fish Migration

Pedro Morais[1,2,]* *and Françoise Daverat*[3]

The definition of migration is much more than a semantic issue, as pointed out by Hugh Dingle in his book 'Migration: the biology of life on the move' (Dingle 1996). Dingle (1996) and Dingle and Drake (2007) advocated that the definition ought to focus on individuals and on its behavioral aspects, even though migration can be described regarding its population outcomes. Therefore, taking this framework into account, Dingle (1996) considered that the definition proposed by J. S. Kennedy, in 1985, is the one that best describes migration across taxa. Thus, migration is defined as a persistent, undistracted and straightened-out movement, achieved through the animal's locomotory means or by actively seeking a transport medium (e.g., air or water currents), during which individuals remain undistracted by the resources they might find during migration by temporarily inhibiting 'station-keeping responses', and that might be repeated later in life (see J. S. Kennedy's definition in Dingle 1996, p. 25). For Dingle (1996), Kennedy's definition is the best definition of migration because (a) it emphasizes the organism's behavior and (b) it is predictive and not only descriptive, in the sense that it enumerates behavioral traits that a truly migratory animal should fulfill to be considered as such.

In the particular case of migratory fish, the lexicon and definitions associated with their study evolved during the 20th century. Essentially, there were two very distinct definitions, the one from Meek (1916) and then the one from Myers (1949), which had some refinements and ended in the definitions proposed by McDowall (1997).

The adjectives anadromous and catadromous were already used previously to Meek's book (1916), but he gave these adjectives a broader meaning than before. However, Meek's definitions were rather dubious and prone to confusion, because

---
[1] CIMA-Centro de Investigação Marinha e Ambiental, Universidade do Algarve, Portugal.
[2] CIIMAR-Centro Interdisciplinar de Investigação Marinha e Ambiental, Universidade do Porto, Portugal.
[3] IRSTEA-Institut National de Recherché en Sciences et Technologies pour l'Environnement et l'Agriculture, France.
* Corresponding author: pmorais@ualg.pt

anadromous and catadromous were used to define the directions of the migration, and not the migratory pattern as a whole. For example, and using Meek's own words, "*the migration of the salmon when it leaves the river for the sea is catadromous, and when it returns is anadromous*" (Meek 1916, pp. 18). Meek (1916) continued to develop his ideas and wrote "*...all degrees of anadromous migration from mid-ocean to the upper limits of streams may take place and corresponding catadromous migrations*" (pp. 18–19). However, Meek (1916) mentioned that "*the terms cannot be applied to indicate migrations with relation to current*" (pp. 19), so he also used the adjectives denatant and contranatant for this purpose – "*denatant, swimming or drifting or simply migrating with the current; contranatant, swimming or migrating against the current*" (pp. 19).

The definitions proposed by Meek (1916) for the adjectives anadromous and catadromous were not consensual, to the point that some American ichthyologists continued to use the previous narrower definitions (Myers 1949). Aware of this problem, Myers (1949) considered that "*the most practical system of classifying fish migrations would seem to be one based principally upon the physical type of route and direction followed, and this method is implicit in the already well accepted terms anadromous and catadromous*". Thus, Myers (1949) proposed the following definitions: diadromous – "*Truly migratory fishes which migrate between the sea and fresh water*"; anadromous – "*diadromous fishes which spend most of their lives in the sea and migrate to fresh water to breed*"; catadromous – "*diadromous fishes which spend most of their lives in fresh water and migrate to the sea to breed*"; amphidromous – "*diadromous fishes whose migration from fresh water to the sea, or vice-versa, is not for the purpose of breeding, but occurs regularly at some other definite stage of the life-cycle*"; potamodromous – "*truly migratory fishes whose migrations occur wholly within fresh water*"; oceanodromous – "*truly migratory fishes which live and migrate wholly in the sea*".

Myers' (1949) definitions were broadly accepted at that time, and are still accepted in its wider sense until today. Regarding diadromy in particular, McDowall (1988) considered that "*the terms anadromy, catadromy and amphidromy are exclusive, specialised forms of diadromy that seem to cover all possibilities*" (pp. 22). Despite this, McDowall (1988) mentioned the works of several scientists that complemented Myers' definitions, which were clearly summarized in a paper he published in 1997 (McDowall 1997). Thus, for McDowall (1997), diadromous migration must: a) occur regularly, b) be physiologically-mediated movements between two biomes, freshwater/brackish water and the sea, c) occur at predictable times, d) occur during a characteristic life history phase in each species, e) involve the majority of a species' population, f) be usually obligatory, and g) occur in opposite directions, i.e., one migration from freshwater/brackish water to the sea, and another in the opposite direction. Thus, based on this definition, McDowall (1997) proposed the following definitions for anadromy, catadromy and amphidromy:

Anadromy: "*diadromous fishes in which most feeding and growth are at sea prior to migration of fully grown, adult fish into fresh water to reproduce; either there is no subsequent feeding in fresh water, or any feeding is accompanied by little somatic growth; the principal feeding and growing biome (the sea) differs from the reproductive biome (fresh water)".*

Catadromy: *"diadromous fishes in which most feeding and growth are in fresh water prior to migration of fully grown, adult fish to sea to reproduce; there is either no subsequent feeding at sea, or any feeding is accompanied by little somatic growth; the principal feeding and growing biome (fresh water) differs from the reproductive biome (the sea)".*

Amphidromy: *"diadromous fishes in which there is migration of larval fish to sea soon after hatching, followed by early feeding and growth at sea, and then a migration of small postlarval to juvenile fish from the sea back into fresh water; there is further, prolonged feeding in fresh water during which most somatic growth from juvenile to adult stages occurs, as well as sexual maturation and reproduction; the principal feeding biome is the same as the reproductive biome (fresh water)".*

Regarding potamodromy and oceanodromy, McDowall (1988) considered that the essential difference between diadromous and non-diadromous migrations resides in the absence of osmoregulation during potamodromous and oceanodromous migrations. This observation is misleading, since potamodromous have to go under physiological (e.g., to store energy reserves – Brönmark et al. 2014) and also morphological adaptations prior to migration (see Chapter 4 for more details – Thurow 2016). The same holds true for oceanodromous fish, for example, only fish species with warm muscles, as tunas and some sharks, are capable of long-distance migrations due to the greater cruising speeds that this physiological adaptation concedes, and then take advantage of seasonal available resources in distant locations (Watanabe et al. 2015).

Migratory fish often exhibit a philopatric behavior, i.e., the return to their natal site, a term often called 'homing' or 'homing behavior', and observed for potamodromous (Rakowitz et al. 2009), anadromous (Dittman and Quinn 1996; Stepien and Faber 1998), catadromous (Hunter et al. 2003) and oceanodromous fish (Hueter et al. 2004; Jorgensen et al. 2009; Feldheim et al. 2013). However, it is important to highlight that philopatry is not a required behavior to classify a fish species as migratory (e.g., Waldman et al. 2008). Indeed, Dingle and Drake (2007) considered that the most emblematic examples of migration (across all taxa) may be the exception rather than the rule. This statement probably holds true for migratory fish species as well, as an increasing number of studies are showing that some of them display alternative life history strategies (or life history plasticity), and not a unique or predominant life history (examples: anadromous fish – brown trout *Salmo trutta* (Limburg et al. 2001), European eel *Anguilla anguilla*, Japanese eel *Anguilla japonica*, American eel *Anguilla rostrata* (Daverat et al. 2006), steelhead *Oncorhynchus mykiss* (Moore et al. 2014); amphidromous fish – Inanga *Galaxias maculatus* (Chapman et al. 2006), *Rhinogobius* spp. (Tsunagawa and Arai 2008); catadromous fish – European flounder *Platichthys flesus* (Morais et al. 2011; Daverat et al. 2012); potamodromous fish – oopu nakea *Awaous stamineus* (Hogan et al. 2014), roach *Rutilus rutilus* (Brodersen et al. 2014); oceanodromous fish – spiny dogfish *Squalus acanthias* (Campana et al. 2009)). Probably, the first cases of alternative life histories in fish that were noticed where those from landlocked diadromous fish populations. Interestingly, these landlocked populations might mimic migratory behavior in the new landlocked environment (Näslund 1992; Barriga et al. 2007) and they might undertake diadromous migrations once the barriers are eliminated (McDowall 1988).

Life history plasticity is defined as the *"ability of a single genotype to produce multiple phenotypes in response to variation in the environment"* (Pfennig et al. 2010), and it can be displayed by a population in two distinct ways. All individuals within a population exhibit a single life history strategy different from the most common for the species, or assumed as so (Morais et al. 2011), and this locally predominant life history might even differ within the same metapopulation (Morais, unpubl. data), or along the species distribution range (Daverat et al. 2012). There is also the hypothesis that individuals within a population display different life history strategies, with individuals forming resident and migratory contingents (Tzeng et al. 2003; Nims and Walther 2014; Gahagan et al. 2015; Gillanders et al. 2015). The simultaneous existence of contingents within a population is classified as partial migration (Jonsson and Jonsson 1993; Secor 1999). Partial migration is probably caused by a trade-off between maximizing fitness (growth, reproduction) with the resources available locally throughout the year and the risks associated with being resident (intraspecific competition, predation), and the energetic and physiological demands associated with undertaking a migratory journey to minimize predation and intraspecific competition to explore distant available resources (Chapman et al. 2012). Partial migration and fish life history plasticity should also be analyzed taking into consideration the conditional strategy hypothesis, which states that genetically monomorphic individuals decide on a tactic, depending on their status (size, sex, age) or condition (energy reserves), to acquire higher fitness, and since each tactic has unequal fitness, the one that produces better fitness will dominate (Gross 1996). Complementarily, fish personality traits should also be included among the status factor, and boldness in particular, since it can influence the migratory behavior of an individual (Chapman et al. 2011b). The existence of contingents within a population, or the existence of a portfolio of life history strategies displayed by a species along their distribution range, probably results from a successful evolutionary process to maximize individual fitness and consequently to enable a population with increased resilience and stability (Geffen 2009; Perrier et al. 2014; Gahagan et al. 2015), and probably providing species with an extraordinary adaptative and evolutionary tool that enables them to persist in varied and unpredictable environments (Pfennig et al. 2010; Chapman et al. 2011a; Daverat et al. 2012; Kerr and Secor 2012; Dodson et al. 2013; Crozier and Hutchings 2014; Valladares et al. 2014). Thus, it is clear that fish life history plasticity is a complex phenomenon and a powerful evolutionary driver, which must be studied taking into consideration different spatial and time scales to encompass the influence of evolution and local adaptations, and that can enlighten us on the patterns, processes, mechanisms and evolution of migratory fish.

References

Barriga JP, Battini MA and Cussac VE (2007) Annual dynamics variation of a landlocked *Galaxias maculatus* (Jenyns 1842) population in a Northern Patagonian river: occurrence of juvenile upstream migration. Journal of Applied Ichthyology 23: 128–135.

Brodersen J, Chapman BB, Nilsson PA, Skov C, Hansson L-A and Brönmark C (2014) Fixed and flexible: coexistence of obligate and facultative migratory strategies in a freshwater fish. PLOS ONE 9: e90294.

Brönmark C, Hulthén K, Nilsson PA, Skov C, Hansson LA, Brodersen J and Chapman BB (2014) There and back again: migration in freshwater fishes. Canadian Journal of Zoology 92: 467–479.

Campana SE, Joyce W and Kulka DW (2009) Growth and reproduction of spiny dogfish off the eastern coast of Canada, including inferences on stock structure. pp. 195–207. *In*: Gallucci VF, McFarlane GA and Bargmann GG (eds.). Biology and Management of Dogfish Sharks. American Fisheries Society, Bethesda, Maryland, USA.

Chapman A, Morgan DL, Beatty SJ and Gill HS (2006) Variation in life history of land-locked lacustrine and riverine populations of *Galaxias maculatus* (Jenyns 1842) in Western Australia. Environmental Biology of Fishes 77: 21–37.

Chapman BB, Brönmark C, Nilsson J-Å and Hansson L-A (2011a) The ecology and evolution of partial migration. Oikos 120: 1764–1775.

Chapman BB, Hulthén K, Blomqvist DR, Hansson L-A, Nilsson J-A, Brodersen J, Nilsson PA, Skov C and Brönmark C (2011b) To boldly go: individual differences in boldness influence migratory tendency. Ecology Letters 14: 871–876.

Chapman BB, Hulthén K, Brodersen J, Nilsson PA, Skov C, Hansson L-A and Brönmark C (2012) Partial migration in fishes: causes and consequences. Journal of Fish Biology 81: 456–478.

Crozier LG and Hutchings JA (2014) Plastic and evolutionary responses to climate change in fish. Evolutionary Applications 7: 68–87.

Daverat F, Limburg KE, Thibault I, Shiao J-C, Dodson JJ, Caron F, Tzeng W-N, Iizuka Y and Wickström H (2006) Phenotypic plasticity of habitat use by three temperate eel species, *Anguilla anguilla*, *A. japonica* and *A. rostrata*. Marine Ecology Progress Series 308: 231–241.

Daverat F, Morais P, Dias E, Babaluk J, Martin J, Eon M, Fablet R, Pécheyran C and Antunes C (2012) Plasticity of European flounder life history patterns discloses alternatives to catadromy. Marine Ecology Progress Series 465: 267–280.

Dingle H (1996) Migration: The Biology of Life on the Move. Oxford University Press, New York. 474p.

Dingle H and Drake VA (2007) What is migration? BioScience 57: 113–121.

Dittman A and Quinn T (1996) Homing in Pacific salmon: mechanisms and ecological basis. Journal of Experimental Biology 199: 83–91.

Dodson JJ, Aubin-Horth N, Thériault V and Páez DJ (2013) The evolutionary ecology of alternative migratory tactics in salmonid fishes. Biological Reviews 88: 602–625.

Feldheim KA, Gruber SH, DiBattista JD, Babcock EA, Kessel ST, Hendry AP, Pikitch EK, Ashley MV and Chapman DD (2013) Two decades of genetic profiling yields first evidence of natal philopatry and long-term fidelity to parturition sites in sharks. Molecular Ecology 23: 110–117.

Gahagan BI, Fox DA and Secor DH (2015) Partial migration of striped bass: revisiting the contingent hypothesis. Marine Ecology Progress Series 525: 185–197.

Geffen AJ (2009) Advances in herring biology: from simple to complex, coping with plasticity and adaptability. ICES Journal of Marine Science 66: 1688–1695.

Gillanders BM, Izzo C, Doubleday ZA and Ye Q (2015) Partial migration: growth varies between resident and migratory fish. Biology Letters 11: 20140850.

Gross MR (1996) Alternative reproductive strategies and tactics: diversity within sexes. Trends in Ecology and Evolution 11: 92–98.

Hogan JD, Blum MJ, Gilliam JF, Bickford N and McIntyre PB (2014) Consequences of alternative dispersal strategies in a putatively amphidromous fish. Ecology 95: 2397–2408.

Hueter RE, Heupel MR, Heist EJ and Keeney DB (2004) Evidence of philopatry in sharks and implications for the management of shark fisheries. Journal of Northwest Atlantic Fishery Science 35: 239–237.

Hunter E, Metcalfe JD and Reynolds JD (2003) Migration route and spawning area fidelity by North Sea plaice. Proceedings of the Royal Society B: Biological Sciences 270: 2097–2103.

Jonsson B and Jonsson N (1993) Partial migration: niche shift versus sexual maturation in fishes. Reviews in Fish Biology and Fisheries 3: 348–365.

Jorgensen SJ, Reeb CA, Chapple TK, Anderson S, Perle C, Van Sommeran SR, Fritz-Cope C, Brown AC, Klimley AP and Block BA (2009) Philopatry and migration of Pacific white sharks. Proceedings of the Royal Society B: Biological Sciences 277: 679–688.

Kerr LA and Secor DH (2012) Partial migration across populations of white perch (*Moronea mericana*): a flexible life history strategy in a variable estuarine environment. Estuaries and Coasts 35: 227–236.

Limburg KE, Landergren P, Westin L, Elfman M and Kristiansson P (2001) Flexible modes of anadromy in Baltic sea trout: making the most of marginal spawning streams. Journal of Fish Biology 59: 682–695.

McDowall RM (1988) Diadromy in Fishes. Timber Press, Portland, Oregon. 308p.

McDowall RM (1997) The evolution of diadromy in fishes (revisited) and its place in phylogenetic analysis. Reviews in Fish Biology and Fisheries 7: 443–462.

Meek A (1916) The Migrations of Fish. Edward Arnold, London. 427p. + xviii.

Moore JW, Yeakel JD, Peard D, Lough J and Beere M (2014) Life-history diversity and its importance to population stability and persistence of a migratory fish: steelhead in two large North American watersheds. Journal of Animal Ecology 83: 1035–1046.

Morais P, Dias E, Babaluk J and Antunes C (2011) The migration patterns of the European flounder *Platichthys flesus* (Linnaeus, 1758) (Pleuronectidae, Pisces) at the southern limit of its distribution range: Ecological implications and fishery management. Journal of Sea Research 65: 235–246.

Myers GS (1949) Usage of anadromous, catadromous and allied terms for migratory fishes. Copeia 1949: 89–97.

Näslund I (1992) Upstream migratory behaviour in landlocked Arctic charr. Environmental Biology of Fishes 33: 265–274.

Nims MK and Walther BD (2014) Contingents of Southern flounder from subtropical estuaries revealed by otolith chemistry. Transactions of the American Fisheries Society 143: 721–731.

Perrier C, Normandeau É, Dionne M, Richard A and Bernatchez L (2014) Alternative reproductive tactics increase effective population size and decrease inbreeding in wild Atlantic salmon. Evolutionary Applications 7: 1094–1106.

Pfennig DW, Wund MA, Snell-Rood EC, Cruickshank T, Schlichting CD and Moczek AP (2010) Phenotypic plasticity's impacts on diversification and speciation. Trends in Ecology & Evolution 25: 459–467.

Rakowitz G, Kubečka J, Fesl C and Keckeis H (2009) Intercalibration of hydroacoustic and mark–recapture methods for assessing the spawning population size of a threatened fish species. Journal of Fish Biology 75: 1356–1370.

Secor DH (1999) Specifying divergent migrations in the concept of stock: the contingent hypothesis. Fisheries Research 43: 13–34.

Stepien C and Faber JE (1998) Population genetic structure, phylogeography and spawning philopatry in walleye (*Stizostedion vitreum*) from mitochondrial DNA control region sequences. Molecular Ecology 7: 1757–1769.

Thurow R (2016) Life histories of potamodromous fishes. pp. 29–54. *In*: Morais P and Daverat F (eds.). An Introduction to Fish Migration. CRC Press, Boca Raton, FL (this book).

Tsunagawa T and Arai T (2008) Flexible migration of Japanese freshwater gobies *Rhinogobius* spp. as revealed by otolith Sr:Ca ratios. Journal of Fish Biology 73: 2421–2433.

Tzeng WN, Iizuka Y, Shiao JC, Yamada Y and Oka HP (2003) Identification and growth rates comparison of divergent migratory contingents of Japanese eel (*Anguilla japonica*). Aquaculture 216: 77–86.

Valladares F, Matesanz S, Guilhaumon F, Araújo MB, Balaguer L, Benito-Garzón M, Cornwell W, Gianoli E, van Kleunen M, Naya DE, Nicotra AB, Poorter H and Zavala MA (2014) The effects of phenotypic plasticity and local adaptation on forecasts of species range shifts under climate change. Ecology Letters 17: 1351–1364.

Waldman J, Grunwald C and Wirgin I (2008) Sea lamprey *Petromyzon marinus*: an exception to the rule of homing in anadromous fishes. Biology Letters 4: 659–662.

Watanabe YY, Goldman KJ, Caselle JE, Chapman DD and Papastamatiou YP (2015) Comparative analyses of animal-tracking data reveal ecological significance of endothermy in fishes. Proceedings of the National Academy of Sciences 112: 6104–6109.

CHAPTER 3

Trends and Challenges in Fish Migration Research

Pedro Morais[1,2,*] and *Françoise Daverat*[3]

Research on fish migration has always been a topic of great interest, and the number of papers published on fish migration, and registered in Scopus database, reflect that interest: 6624 papers published until August 2014. The number of papers published increased significantly since 1995. The average increase of papers published between 1990 and 1995 and between 2008 and 2013 was 518%, from 70.3 ± 5.9 papers to 434.7 ± 50.6 papers (Fig. 3.1A). The number of papers published before 1996 might be slightly underestimated, due to the reduced coverage of the used bibliographic database before that year. Though, the increasing trend is obvious after 1995 and certainly reveals the main trends that have occurred since then.

Most papers were published by US and Canadian scientists, which account for 26.6 and 9.7% of all country affiliations. The top-10 countries published over two thirds of all papers, 69.7% to be more precise (Fig. 3.1B). The number of papers published by country (overall top-20 countries) in 2013 are more correlated with nominal Gross Domestic Product, or GDP, than with GDP per capita, which explains the huge publishing discrepancy between continents. The contribution of the top-25 countries for the total number of papers published by continent is the following: North America – 36.3%, Europe – 35.4%, Asia – 9.1%, Oceania – 5.3%, Africa – 0.8%.

The distribution of fish migration papers by journals is more even, probably because there is no journal dedicated solely to fish migration research, and also due to the multidisciplinary topics and techniques involved in fish migration research. The top-10 publishing journals published 28.4% of all papers, of which Journal of Fish Biology (6.0%), Canadian Journal of Fisheries and Aquatic Sciences (3.2%) and Environmental Biology of Fishes (3.1%) are the top-three publishing journals (Fig. 3.1C).

---

[1] CIMA-Centro de Investigação Marinha e Ambiental, Universidade do Algarve, Portugal.
[2] CIIMAR-Centro Interdisciplinar de Investigação Marinha e Ambiental, Universidade do Porto, Portugal.
[3] IRSTEA-Institut National de Recherché en Sciences et Technologies pour l'Environnement et l'Agriculture, France.
* Corresponding author: pmorais@ualg.pt

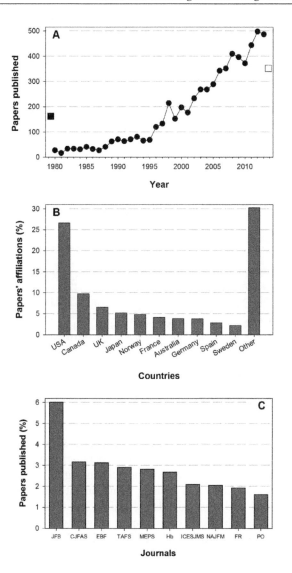

Figure 3.1. Evolution of the number of papers published on migratory fish from 1980 till 2013, and recorded in Scopus database (A). The total number of papers published, and recorded in Scopus database, between 1961 and 1979 is represented by a black square in Fig. 3.1A. The top-10 publishing countries (B) and journals (C) on fish migration during this period are also shown. Legend: JFB – Journal of Fish Biology, CJFAS – Canadian Journal of Fisheries and Aquatic Sciences, EBF – Environmental Biology of Fishes, TAFS – Transactions of the American Fisheries Society, MEPS – Marine Ecology Progress Series, Hb – Hydrobiology, ICESJMS – ICES Journal of Marine Sciences, NAJFM – North American Journal of Fisheries Management, FR – Fisheries Research, PO – PLOS ONE. This bibliographic research was done in Scopus, and the keywords used were the following: "fish migration" and anadromous, or anadromy, or catadromous, or catadromy, or amphidromous, or amphidromy, or diadromous, or diadromy, or oceanodromous, or oceanodromy, or potamodromous, or potamodromy, or plasticity, or contingents, or 'life history', or ocean, or sea, or river, or conservation, or fisheries, or management, or physiology, or phylogeny or cues, or methodology, or patterns, or processes, but excluding from the output the keywords 'zebra fish', 'zebrafish' and '*Danio rerio*'. Only papers published in English were considered in this analysis.

The main keywords used in papers about fish migration reflect the enormous bias between the number of papers published in different countries and continents. If such a thing would exist, the 'typical' fish migration paper would be on the physiological aspects of Salmonidae spawning migrations in North American rivers, and it would result from the cooperation between US and Canadian scientists (Fig. 3.2). Salmonid species are definitely the most studied species, namely the Pacific salmons and trouts *Oncorhynchus* spp. (e.g., *Oncorhynchus tshawytscha*, *Oncorhynchus mykiss*, *Oncorhynchus nerka*), the Atlantic salmon *Salmo salar* and the brown trout *Salmo trutta*. Other species are also among the main keywords used, as the European eel *Anguilla anguilla* or the Atlantic cod *Gadus morhua*. Cyprinidae and Clupeidae are also among the most studied families of migratory fish species (Fig. 3.2).

Figure 3.2. Wordle™ representation of the most commonly cited keywords in fish migration studies between 1980 and 2013, according to Scopus database.

Papers on the ecology of migratory fish species covered a series of topics through the years since 1980, like 'life history', 'population dynamics', 'population structure', 'habitat use' and 'growth, development and aging'. In terms of migratory fish's conservation, the keywords 'fishery management' and 'dams' are the most commonly cited, and in the case of 'dams' mainly due to the problems associated with the destruction and blocking of salmonids spawning habitats – although dams are not an exclusive problem of salmonids. Regarding, the techniques used to study fish migration, tagging, telemetry and otoliths have been the most frequently used since 1980 (Fig. 3.2).

As for future research topics, we are convinced that the number of studies linking climate change and migratory fish will gain increased relevance, since there are concerns that climate change may exceed the capacity of developmental, genetic, and demographic mechanisms that species have evolved to deal with environmental change (Ozgul et al. 2009; Chevin et al. 2010, 2013; Bernhardt and Leslie 2013). Therefore, much of the focus has to be put into migratory fish conservation and management (e.g., Lassalle and Rochard 2009; Hague et al. 2011; Koehn 2011; Finstad and Hein 2012; Beatty et al. 2014). Thus, climate change pressure on migratory fish populations will further enhance the need to investigate this topic, as it might be a triggering mechanism for the dispersal of populations, in order to escape unfavorable habitats and colonize new favorable habitats. Indeed, fish migration and fish life history plasticity are considered fundamental mechanisms explaining fish species distribution after glaciation events. Therefore, it is possible that at least some migratory fish may display resilience to climate change (Valladares et al. 2014), in opposition to resident or non-migratory fish species. It is important to recall that ecological resilience is dependent of rapid phenotypic change, which occurs through both ecological and evolutionary processes (Ozgul et al. 2009), and combines the resistance to increasingly frequent and severe disturbances and is expressed through the capacity for recovery and self-organization and the ability to adapt to new conditions (Bernhardt and Leslie 2013). This adaptive capacity is, in its turn, the expression of a combination of traits, namely phenotypic plasticity, species range shifts, and microevolution (Bernhardt and Leslie 2013; Chevin et al. 2013). Therefore, the complexity linking climate change and migratory fish calls for a transdisciplinary research effort to support wise conservation policies. In fact, conservation measures as assisted colonization, or displacement of endangered species, are being considered if populations cannot adapt fast enough to the rate of habitat favorability change or if they cannot disperse quickly enough.

As we celebrate the 100th anniversary of Alexander Meek's seminal book 'The migration of fish' (Meek 1916), the differences between the techniques and proxies currently used to study fish migration and those used in the early 1900's are tremendous. Certainly whatever the predictions that we might make on the future of fish migration research, all of them will be less ambitious than those the future holds for us. Yet, we are sure that technical advances will pave the way for reaching previously unattainable data, shedding light to novel insights and, above all, exploring new questions. For example, the increasing miniaturization of electronic components used in electronic tags, long-lasting and smaller batteries, combined with new physiological sensors and lower production costs will certainly continue to revolutionize fish tagging studies (Cooke et al. 2013). Also, newer and more sensitive analytical techniques, either when using fish otoliths or tissues as proxies to infer fish migration and habitat use, will evolve with technological progress and interdisciplinary approaches (Campana and Neilson 1985; Campana 2005). Indeed, it is incredible to recall the technological advances achieved regarding electronic tracking of fish movements and migration, since the rise of echo-sounders use during the 1930's and 1940's (Harden Jones 1968). Otoliths are another extraordinary example on the advances achieved to study fish migration since the late 1960's. Harden Jones (1968) states that otoliths were used as certificates of origin, in the sense that similar otolith shapes would correspond to specimens with equal origin, and that was basically it. The daily deposition of otolith growth rings was described by Giorgio Panella only in 1971 (Panella 1971), and otolith microchemistry only began to rise after the mid-1980's with the seminal works of Richard L. Radtke and colleagues and also of John M. Kalish (Radtke and Targett 1984; Kalish 1989; Radtke 1989; Radtke and Morales-Nin 1989). Yet so much has been accomplished since the mid-1980's.

As a concluding remark, and in our opinion, fish migration research will continue to evolve due to technological progress, but it also has to rely on cooperation between developed and developing countries to tighten the gaps on both fundamental research and conservation efforts. In this sense, one of the best examples is the work accomplished by the Inter-American Tropical Tuna Commission (IATTC), currently with 21 members and four cooperating non-members from North and South

America, Asia, Oceania, Africa and Europe (IATTC 2014). It is also clearer that fruitful conservation efforts require effective and long-term environmental awareness programs, targeting all segments of the population and involving stakeholders (e.g., fishermen, fisheries industry, dam managers, water agencies, regional and national legislators). The outcomes are not immediate, but they will certainly flourish in the future. One of the most emblematic examples on environmental awareness activities related with fish migration was the celebration of the 'World Fish Migration Day' on May 24th 2014, where 50,000 people participated in 273 events, from 53 countries in all continents (WFMD 2014). It is certainly due to public and media perception and pressure that, in some countries, it is possible to dismantle weirs and dams that hinder the migration of diadromous or potamodromous fish (McAllister et al. 2001; Arthington et al. 2003; Doyle et al. 2005; Brown et al. 2013). The same holds true for oceanodromous fishes, since it is only possible to ban or to put on halt certain targeted or non-targeted fisheries, to recover the stocks of over-exploited species, with the involvement of all parties, including the public and media (Tolotti et al. 2015). However, in some situations, economic pressure is so overwhelming that environmental awareness and conservation impetus are not sufficient to hinder the conservation impetus of migratory fish (Dugan et al. 2010; Tolotti et al. 2015). So, diadromous and potamodromous fish conservation also has to rely on newer and efficient technologies to harvest energy in hydroelectric power plants, in the sense that fewer of them will have less impact, because fish passages and fish hatcheries do not compensate the long-term impacts of dams and weirs (Brown et al. 2013). While in the case of oceanodromous species, it is essential to improve fishing gears to avoid the accidental capture of endangered and protected oceanodromous species and others (Tolotti et al. 2015). Also, learning from previous experiences, either positive or negative, is fundamental for developing wiser water and fishery management plans (Ferguson et al. 2011; Tolotti et al. 2015). Despite all economic pressures (fisheries, water development plans) that world migratory fishes are subject to, we are still confident that the future will be beneficial for migratory fish species if fundamental and applied research continues in developed countries, if it expands to developing countries in a consistent manner, but only if encompassed by public awareness programs and media pressure.

References

Arthington AH, Lorenzen K, Pusey BJ, Abell R, Halls AS, Winemiller KO, Arrington DA and Baran E (2003) River fisheries: ecological basis management and conservation. *In*: Welcomme RL and Petr T (eds.). Proceedings of the Second International Symposium on the Management of Large Rivers for Fisheries. Cambodia. FAO. pp. 21–60.

Beatty SJ, Morgan DL and Lymbery AJ (2014) Implications of climate change for potamodromous fishes. Global Change Biology 20: 1794–1807.

Bernhardt JR and Leslie HM (2013) Resilience to climate change in coastal marine ecosystems. Annual Review of Marine Science 5: 371–392.

Brown JJ, Limburg KE, Waldman JR, Stephenson K, Glenn EP, Juanes F and Jordaan A (2013) Fish and hydro power on the U.S. Atlantic coast: failed fisheries policies from half-way technologies. Conservation Letters 6: 280–286.

Campana SE (2005) Otolith science entering the 21st century. Marine & Freshwater Research 56: 485–495.

Campana SE and Neilson JD (1985) Microstructure of fish otoliths. Canadian Journal of Fisheries and Aquatic Sciences 42: 1014–1032.

Chevin L-M, Collins S and Lefèvre F (2013) Phenotypic plasticity and evolutionary demographic responses to climate change: taking theory out to the field. Functional Ecology 27: 967–979.

Chevin L-M, Lande R and Mace GM (2010) Adaptation, plasticity, and extinction in a changing environment: Towards a predictive theory. PLOS Biology 8: e1000357.

Cooke SJ, Midwood JD, Thiem JD, Klimley P, Lucas MC, Thorstad EB, Eiler J, Holbrook C and Ebner BC (2013) Tracking animals in freshwater with electronic tags: past, present and future. Animal Biotelemetry 1: 5.

Doyle MW, Stanley EH, Orr CH, Sellec AR, Sethib SA and Harbor JM (2005) Stream ecosystem response to small dam removal: lessons from the Heartland. Geomorphology 71: 227–244.

Dugan PJ, Barlow C, Agostinho AA, Baran E, Cada GF, Chen D, Cowx IG, Ferguson JW, Jutagate T, Mallen-Cooper M, Marmulla G, Nestler J, Petrere M, Welcomme RL and Winemiller KO (2010) Fish migration, dams, and loss of ecosystem services in the Mekong basin. AMBIO 39: 344–348.

Ferguson JW, Healey M, Dugan P and Barlow C (2011) Potential effects of dams on migratory fish in the Mekong river: Lessons from salmon in the Fraser and Columbia rivers. Environmental Management 47: 141–159.

Finstad AG and Hein CL (2012) Migrate or stay: terrestrial primary productivity and climate drive anadromy in Arctic char. Global Change Biology 18: 2487–2497.

Fontaine M (1954) Du determinisme physiologique des migrations. Biological Reviews 29: 390–418.

Hague MJ, Ferrari MR, Miller JR, Patterson DA, Russell GL, Farrell AP and Hinch SG (2011) Modelling the future hydroclimatology of the lower Fraser River and its impacts on the spawning migration survival of sockeye salmon. Global Change Biology 17: 87–98.

Harden Jones FR (1968) Fish Migration. Edward Arnold (Publishers) Ltd., London. 325p.

IATTC (Inter-American Tropical Tuna Commission) (2014) http://www.iattc.org. Accessed May 23rd 2014.

Kalish JM (1989) Otolith microchemistry: validation of the effects of physiology, age and environment on otolith composition. Journal of Experimental Marine Biology and Ecology 132: 151–178.

Koehn JD (2011) Climate change and Australian marine and freshwater environments, fishes and fisheries: introduction. Marine & Freshwater Research 62: 981–983.

Kusukawa S (2000) The Historia Piscium (1686). Notes and Records of the Royal Society of London 54: 179–197.

Lassalle G and Rochard E (2009) Impact of twenty-first century climate change on diadromous fish spread over Europe, North Africa and the Middle East. Global Change Biology 15: 1072–1089.

McAllister DE, Craig JF, Davidson N, Delany S and Seddon M (2001) Biodiversity impacts of large dams. UNEP & United Nations Foundation & IUCN. 68p.

Meek A (1916) The Migrations of Fish. Edward Arnold, London. 427p. + xviii.

Ozgul A, Tuljapurkar S, Benton TG, Pemberton JM, Clutton-Brock TH, Coulson T (2009) The dynamics of phenotypic change and the shrinking sheep of St. Kilda. Science 325: 464–467.

Panella G (1971) Fish otoliths: daily growth layers and periodical patterns. Science 173: 1124–1127.

Radtke RL (1989) Strontium-calcium concentration ratios in fish otoliths as environmental indicators. Comparative Biochemistry and Physiology Part A: Physiology 92: 189–193.

Radtke RL and Morales-Nin B (1989) Mediterranean juvenile bluefin tuna: life history patterns. Journal of Fish Biology 35: 485–496.

Radtke RL and Targett TE (1984) Rhythmic structural and chemical patterns in otoliths of the Antarctic fish *Notothenia larseni*: Their application to age determination. Polar Biology 3: 203–210.

Tolotti MT, Filmater JD, Bach P, Travassos P, Seret B and Dagorn L (2015) Banning is not enough: The complexities of oceanic shark management by tuna regional fisheries management organizations. Global Ecology and Conservation 4: 1–7.

Valladares F, Matesanz S, Guilhaumon F, Araújo MB, Balaguer L, Benito-Garzón M, Cornwell W, Gianoli E, van Kleunen M, Naya DE, Nicotra AB, Poorter H and Zavala MA (2014) The effects of phenotypic plasticity and local adaptation on forecasts of species range shifts under climate change. Ecology Letters 17: 1351–1364.

WFMD (World Fish Migration Day) (2014) World Fish Migration Day 2014-Report. World Fish Migration Platform. The Netherlands. 8p.

PART 2

Life Histories of Migratory Fishes

CHAPTER 4

Life Histories of Potamodromous Fishes

Russell F. Thurow

Definitions of potamodromy, potamodromous migrations and movements

Potamodromous fishes move and complete their life cycle entirely within freshwater. Myers (1949) proposed the term potamodromous to distinguish freshwater migratory fishes from diadromous fishes, which migrate between the sea and freshwater and oceanodromous fishes that migrate wholly within the sea. Diadromous fishes include anadromous, catadromous and amphidromous fishes (see Chapter 2, Morais and Daverat 2016). Despite its historical precedence, potamodromous has not been broadly accepted. Three other terms, 'non-anadromous', 'resident', and 'inland' are more commonly substituted in the fisheries literature. Unfortunately, these three terms have multiple definitions, as well as regional connotations which may confound their application to a broad geographic area (Gresswell et al. 1997). Consequently, potamodromous provides a more precise and more broadly applicable definition of fishes that remain wholly within freshwater.

Although potamodromous fishes are widespread among freshwater fish assemblages, the significance of potamodromy has received far less attention than diadromy (Northcote 1998). Unlike diadromy, no global analysis of potamodromous species has been undertaken, and it is limited by the difficulties in amassing information for inconspicuous and little-studied species, especially in the tropics (Flecker et al. 2010). Potamodromous fishes were included in a group-by-group review by Lucas and Baras (2001) of the migration and life cycle characteristics of species representative of families of fishes exhibiting migration in fresh and brackish water environments. Lucas and Baras (2001) explained that information was limited for some groups of freshwater fishes mostly found in tropical freshwater regions. This was partly because

USDA-Forest Service-Rocky Mountain Research Station, 322 East Front Street-Suite 401, Boise, Idaho 83702, USA.

of a paucity of information concerning spatial ecology at the species level; some of these groups (Cichlidae, Characiformes and Siluriformes) are very speciose, totaling over 5,000 species and representing nearly 50% of all fish species in freshwater (Lucas and Baras 2001). Flecker et al. (2010) reported that in both the tropics and temperate zone, potamodromy is likely the most common form of migration in stream fishes. Similarly, Lucas and Baras (2001) observed that in many large tropical rivers, more than 95% of the migratory fishes are potamodromous.

Migration and movements between biomes on a daily, seasonal, or annual basis, represents a fundamental aspect of the ecology of populations and individuals (Hobson 1999). Despite residing only in freshwater, for a variety of reasons, potamodromous fishes move and migrate various distances throughout their life cycle. As Dingle and Drake (2007) observed, our understanding of the movements of organisms has been hindered by imprecise and ambiguous terminology. As a result, it may be useful to begin by defining the terms movement and migration.

Movement may be defined as the act of changing locations or positions. In potamodromous fishes, these movements are most commonly associated with seeking essential resources (i.e., food) and they may be in response to other organisms (i.e., seeking cover from predators). For example, a potamodromous sculpin *Cottus* spp. residing in a pool may suddenly move and change 'position' to consume an aquatic insect larvae on the stream substrate. This same sculpin may change its 'location' in the same pool by moving beneath a boulder to escape an avian predator. Dingle (1996) observed that most movements occur within a relatively well defined area or home range. An organism travels or moves within its home range to acquire the resources it needs to survive. The size of the home range will tend to vary depending on the habitat and the size and movement abilities of the organism (Dingle 1996). Consequently, the sculpin in our example above, will have a much smaller home range compared to the average home range size (mean home range of 146 ha) for a 70–100 cm long muskellunge *Esox masquinongy* (Miller and Menzel 1986).

Movements and associated behaviors within a home range have also been termed 'station keeping' and perhaps the most prominent example is foraging (Kennedy 1985; Dingle 1996). Foraging is a repetitive and meandering movement that focuses on locating resources (food, cover, or mates). Foraging characteristically occurs on short timescales and small spatial scales within the home range (Dingle and Drake 2007). In our examples above, depending in part on food availability, the much larger and faster swimming muskellunge will potentially forage within a much larger area compared to the sculpin. A specialized form of foraging, that includes to-and-fro movements, is the diel vertical movement of fishes such as alewife *Alosa pseudoharengus* to consume zooplankton (Janssen and Brandt 1980). Another example of a 'station keeping' behavior is the territorial behavior that results in agonistic encounters between individuals. McNicol et al. (1985) reported frequent agonistic interactions between young-of-the-year brook trout *Salvelinus fontinalis*, in a second-order woodland stream.

Migration differs substantially from the often repetitive movements within home ranges described above. Home range movements are primarily in response to local resources. Migration is considered a more specialized type of movement, often, but not necessarily occurring at larger temporal and spatial scales. Dingle (1996) emphasized that migration differs from other types of movements both qualitatively and quantitatively. Unlike the sculpin, muskellunge, and alewife movements described above, Dingle (1996) observed that migration is not a proximate response to resources nor does it serve to keep an organism in its habitat. Rather, migration results in fishes moving from one habitat and relocating to another habitat outside their home range. Dingle (1996) observed that the most distinctive feature of migration is that migrants do not respond to sensory cues from resources (e.g., food or shelter) that would typically elicit responses. For example, during their annual migration to spawning sites, adult walleye *Stizostedion vitreum vitreum* may feed or temporarily seek shelter. However, unlike the above referenced foraging movement of a muskellunge within its home range, the presence of an abundant food source or suitable cover will not cause migrating walleye to stop. Their upstream migration ceases only when adult walleye arrive at their spawning location (i.e., Crowe 1962).

Migration involves two levels, the behavioral level that applies to individuals and the ecological level that applies to populations (Dingle and Drake 2007). Therefore, a broad conceptual understanding of migration encompasses both its mechanism and its function. Northcote (1978) distinguished migration from other types of fish movements and suggested four main features: (1) resulting in an alternation between two or more well-separated habitats; (2) occurring with regular periodicity (often seasonal) within the individual lifespan; (3) involving a large fraction of the population; and (4) being directed rather than a random wandering or passive drift. Decades later, Dingle and Drake (2007) suggested that migration represents four different but overlapping concepts which are very similar to those of Northcote (1978). The concepts of Dingle and Drake (2007) are also applicable to potamodromous fishes: (1) a type of locomotory activity that is notably persistent, undistracted, and straightened out (i.e., fall downstream migrations of adult westslope cutthroat trout *Salmo clarkia lewisii* to overwintering areas in large pools) (Bjornn and Mallet 1964); (2) a relocation of the animal that is on a much greater scale, and involves movement of much longer duration, than those arising in its normal daily activities (i.e., spring upstream migrations of mature walleye to spawning areas) (Crowe 1962); (3) a seasonal to-and-fro movement of populations between regions where conditions are alternately favorable or unfavorable (i.e., summer upstream migrations of brook trout seeking cold water refugia (Petty et al. 2012) followed by fall downstream migrations to more suitable overwintering habitats) (Chisholm et al. 1987); and (4) movements leading to redistribution or dispersal within a spatially extended population (i.e., downstream drift of newly hatched white sucker *Catostomus commersoni* larvae to areas with higher zooplankton production) (Corbett and Powles 1986). Dingle and Drake (2007) explained that migration of types 1 and 2 relate to individual organisms, while types 3 and 4 explicitly concern populations. Further, type 1 migration describes a process, whereas the other three migrations describe the outcomes (for individuals or populations) of migration by individuals.

Baker (1978) proposed that the sum of all migrations and movements during an organism's lifetime be termed a 'lifetime track'. An organism's lifetime track is essentially the time series of its successive locations throughout its lifetime

(Dingle and Drake 2007). Technological advances in tracking devices now allow biologists to more explicitly monitor a fishes' lifetime track. Hanson et al. (2007), for example, applied a whole-lake acoustic telemetry array to closely monitor the three-dimensional position of largemouth bass *Micropterus salmoides* across multiple temporal and spatial scales. Between November 2003 and April 2004, the authors simultaneously monitored 20 largemouth bass with transmitters (equipped with pressure and temperature sensors) at 15 second intervals with sub-meter accuracy. A fishes' lifetime track is influenced by its size, life history traits, ability to migrate, geographic range, and habitat. Dingle (1996) emphasized an important concept when he observed that the composite of movements, migrations, and stationary elements that form the lifetime track is determined by natural selection. The dynamics of fish populations are in turn influenced by the migrations and movements of the individual fishes it contains (Dingle 1996).

A strategy can be defined as a genetically determined life history type or behavior which has evolved because it maximizes fitness of individuals and populations (Gross 1987). Fitness can be defined as lifetime reproductive success. As Gross et al. (1988) summarized, the importance of food intake for growth, decreased mortality, increased fecundity, and improved breeding success is well documented. Gross et al. (1988) compared the distribution of diadromous fishes to global patterns in aquatic productivity and concluded that food availability is an important factor determining both where migratory fishes occur and their direction of movement.

Migration and movements are very widespread strategies in potamodromous fishes (Northcote 1978) and ultimately result in fish switching habitats. Salmonids, for example, change habitats many times during their growth and development, and each change within and across life stages involves migration (Thorpe 1988). Early studies of fish migration relied on external marks to track individuals between habitats in an effort to characterize timing and duration. As Lucas and Baras (2001) observed, the relative inadequacy of early techniques used to investigate the migration of freshwater fishes contributed to the idea that many freshwater fishes exhibit very little movement, which is now viewed as a misplaced paradigm (Gowan et al. 1994). Fisheries biology is moving from descriptive studies to more mechanistic approaches that strive to understand the ecological and evolutionary importance of migration. Cooke et al. (2008), for example, advanced understanding of sockeye salmon *Oncorhynchus nerka* migration through the integration of disciplines including physiology, behavior, functional genomics, and experimental biology.

Understanding habitat connectivity and the characteristics of essential habitats utilized by potamodromous species, throughout their often complex life histories, is essential to their effective conservation. Such knowledge can effectively be directed to conserve the habitats that are critical for various species life stages (e.g., Myers et al. 1987). In this chapter, we include a broad spatial and temporal spectrum of migrations and movements by potamodromous fishes; ranging from short-distance (~ 1–2 meter) diel movements of juvenile bull trout *Salvelinus confluentus* seeking winter concealment in interstitial areas of stream substrates (Thurow 1997), to very long distance (> 650 km) spawning migrations of adult Colorado pikeminnow *Ptychocheilus lucius* over several months (Irving and Modde 2000).

Taxonomic and biogeographic distribution of potamodromous fishes

Taxonomists currently list 33,592 identified fish species worldwide with 217 new species added in 2015 as of August 3, 2015 (Eschmeyer and Fong 2015). In 2006, 27,977 species of fish were identified and about 43% (11,952) were considered strictly freshwater species (Nelson 2006). If 40% of currently known fish species worldwide reside strictly in freshwater, then it is likely that more than 13,000 fish species meet our definition as potamodromous fishes. Flecker et al. (2010) observed that, collectively, potamodromous species can represent a substantial proportion of fish biomass even in the largest freshwater ecosystems. In South America, for example, potamodromous fishes are dominated by large pimelodid catfish and characins, many of commercial importance. In Africa, potamodromous species include characins, siluroids, cyprinids, and mormyrids that move from lakes to tributaries and upstream swamps to spawn. In Asia, among the best known potamodromous fishes are pangasiid catfish and cyprinids, such as some barbs, as well as members of the genus *Tor* that are known to ascend Himalayan streams (Welcomme 1985).

Ross (2013) reported that the freshwater fish fauna of North America is the most diverse and thoroughly researched temperate fish fauna in the world. As a result of the abundance of literature describing North American freshwater fishes, this chapter will focus on well-studied potamodromous fishes within North America.

Worldwide, potamodromous fishes represent at least 31 orders of fishes. Thirteen of those orders are not native to North America. These include: Atheriniformes, Ceratodontiformes, Characiformes, Gonorynchiformes, Gymnotiformes, Mugiliformes, Osteoglossiformes, Pleuronectiformes, Rajiformes, Scorpaeniformes, Synbranchiformes, Syngnathiformes, and Tetraodontiformes. North American potamodromous species are extremely diverse and represent 18 distinct orders: Lepisosteiformes (gars), Amiiformes (bowfin), Hiodontiformes (mooneye), Clupeiformes (alewife), Osmeriformes (smelt), Percopsiformes (trout-perch, cave fishes, and pirate perch), Acipenseriformes (sturgeons, and paddlefishes), Cypriniformes (minnows, carp, and suckers), Siluriformes (catfishes), Esociformes (pikes and pickerels), Salmoniformes (whitefish, trout, and salmon), Scorpaeniformes (sculpin); Perciformes (bass, sunfish, perch, cichlids, and drums), Gadiformes (burbot), Atheriniformes (silversides), Cyprinodontiformes (top minnows, killifish, and pupfish), Gasterosteiformes (sticklebacks), and Petromyzontiformes (lamprey) (Eschmeyer 2013).

Within North America, these 18 orders represent 29 families of fish that reside wholly within freshwater. An additional two families (Clupeidae and Osmeridae) were formerly anadromous, but have been introduced, as alewife and rainbow smelt *Osmerus esperlantus*, respectively, and are now landlocked within the freshwaters of the Great Lakes and other North American waters (Scott and Crossman 1973). Adding to this diverse list of potamodromous species is the potential for several anadromous species to develop potamodromous populations. Northcote (1997), for example, described four North American species of Pacific salmon that now have permanent freshwater residence. Three species, pink salmon *Oncorhynchus gorbuscha*, coho salmon *Oncorhynchus kisutch*, and Chinook salmon *Oncorhynchus tshawytscha* developed potamodromous populations after introductions to the Great Lakes (Scott and Crossman

1973). Sea lamprey *Petromyzon marinus* also established potamodromous forms after introductions to the Great Lakes (Clemens et al. 2010). Potamodromous populations develop naturally in landlocked sockeye or kokanee *Oncorhynchus nerka*, as well as in landlocked salmon or Ouananiche; the freshwater form of Atlantic salmon *Salmo salar*. Boucher (2004) reported that prior to 1868 landlocked salmon occurred naturally in four Maine River Basins. After extensive stocking, Maine supported one of the largest sport fisheries for landlocked salmon in the world with fisheries in 176 lakes and about 464 km of rivers and streams (Boucher 2004). Outside North America, other anadromous salmonid forms of *Oncorhynchus* including masu salmon *Oncorhynchus masou*, and Biwa trout *Oncorhynchus rhodurus* also develop potamodromous populations (Northcote 1997). Other formerly anadromous species such as white sturgeon *Acipenser transmontanus* have similarly developed potamodromous forms after becoming landlocked above dams and impoundments in the Columbia and Kootenai Rivers (Jager et al. 2001). Some landlocked fish populations may reacquire former life history strategies after being 'unlocked' (see Chapter 2).

Key characteristics of potamodromous fishes

The 18 orders of North American potamodromous fishes represent thousands of diverse fish species that exhibit a variety of life stages, life history strategies, and associated movements. Despite this diversity, all of the North American potamodromous fishes persisting in riverine or lacustrine environments share some common life stages and life history strategies.

Life stages

Schlosser's (1991) provided a useful synthesis of freshwater fish life stages. Fish vary dramatically in size and behavior, from embryo to larvae, then to juvenile and subsequently to sub-adult and to adult (Fig. 4.1). For more detailed descriptions of the life stages of a variety of North American fishes, see Scott and Crossman (1973). Life begins when fertilized eggs are either buried within substrates, broadcast over the surface of substrates, broadcast into the water column, or attached to plant material. Eggs mature after an incubation period lasting anywhere from a few days (i.e., Cypriniformes) to several months (i.e., Salmoniformes) when they hatch to produce a free embryo phase (Schlosser 1991). During the brief free embryo phase, potamodromous fishes rely on energy sources provided entirely by an egg yolk sac. For a thorough review of yolk sac absorption, see Heming and Buddington (1988). In some groups (i.e., Salmoniformes, Petromyzontiformes) the yolk-sac embryos remain hidden in the substrate and do not emerge until the yolk sac is completely absorbed (Schlosser 1991). In others (i.e., walleye) hatched embryos begin feeding before the yolk sac is absorbed, after which fry disperse into open water (Scott and Crossman 1973). As soon as the free embryo phase is complete, the fish begin feeding on external energy sources, and at this point they are termed larvae. The larval phase is variable in length and ends after completion of the axial skeleton and development of a fully formed organ system and fins (Schlosser 1991). When fully formed the

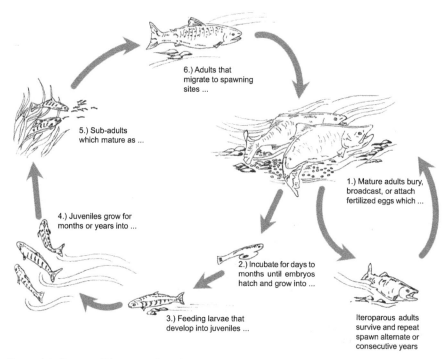

Figure 4.1. Common life stages of North American potamodromous fishes in riverine and lacustrine environments (revised from Thurow 1982).

larvae become juveniles (Schlosser 1991). During the juvenile stage, fish undergo a number of seasonally favorable periods with rapid growth, followed by seasonally unfavorable periods with reduced growth until sexual maturity is reached (Schlosser 1991). In North American temperate streams, these favorable and unfavorable periods frequently involve migration between summer and winter habitats. Depending on the species, the juvenile life stage may encompass months or years (Schlosser 1991). Juveniles ultimately develop into sub-adults, the life stage immediately prior to sexually mature adults. After sexual maturity is attained, adults complete spawning migrations to locate appropriate sites for egg deposition and re-initiation of the life cycle (Schlosser 1991). In iteroparous species, surviving adults return to repeat spawn in alternate or consecutive years (Schlosser 1991).

Life-history strategies

Within these general life stages, tremendous diversity occurs in the specific life-history characteristics of the different fish species. For example, substantial variation occurs in spawning migrations; seasonal occurrence of eggs, young, and adults; and feeding habitats. Consequently, no single life history definition is all inclusive and, as Northcote (1997) observed, one may wish to further define the life history forms of potamodromous fishes. Riverine reproductive migrations of salmonids have been applied to partition them into four main life history forms: fluvial (spawn and rear in large rivers and streams), fluvial-adfluvial (spawn in tributaries and rear in streams, rivers, and tributaries), lacustrine-adfluvial (spawn in lake tributaries and rear primarily in lakes), and allacustrine (spawn in lake outlets and rear primarily in lakes) (Varley and Gresswell 1988) (Fig. 4.2). These various life stages and life history types may be applied to begin describing the diverse types of migrations; including reasons for migrations, migration timing, and the ways fish migrate.

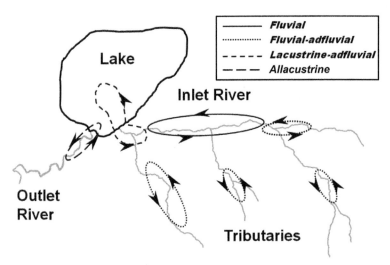

Figure 4.2. Potamodromous salmonid life history forms (adapted from Varley and Gresswell 1988). Ovals and arrows represent migration paths between spawning and rearing areas in rivers, tributaries, and lakes.

Types of migration

Potamodromous fishes exhibit complex life cycles and habitat-use patterns that are integrated with the diversity of their various life stages and associated body sizes (Northcote 1984; Schlosser 1991). Northcote (1984) explained that migratory behavior arises from spatial, seasonal, and ontogenetic separation of optimal habitats for growth, survival, and reproduction. Schlosser (1991) described the basic migrations of stream fishes among three types of habitat (feeding, overwintering, and spawning). Northcote (1997) examined riverine populations of 34 species of salmonids in detail, and he summarized potamodromy as a cyclic sequence of three types of migrations (trophic, refuge, and reproductive) between three respective habitats (feeding, overwintering, and spawning) (Fig. 4.3).

Migration is known to be an important tactic for thermoregulation of coldwater species (Petty et al. 2012). Consequently, refuge migrations by potamodromous fishes may be of two primary types; migrations by fish seeking overwinter refuge habitat and refuge migrations by fishes seeking cover or thermoregulation during non-winter periods. The three types of migration outlined by Schlosser (1991) and Northcote (1997) were adopted by me and I revised destination habitat types to include both

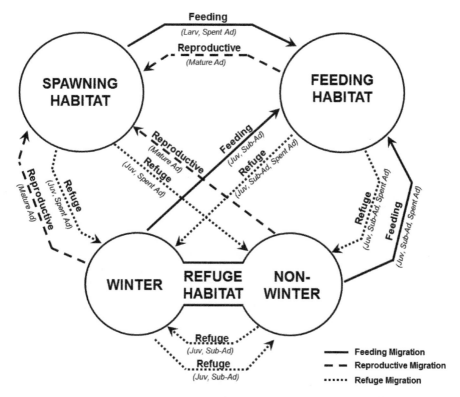

Figure 4.3. Generalized movements (Feeding, Reproductive, and Refuge Migrations) by North American potamodromous fishes with emphasis on patterns of migration between essential feeding, spawning, and refuge habitats (revised from Schlosser 1991 and Northcote 1997). Legend: Larvae-Larv, Juvenile-Juv, Sub-Adults-Sub-Ad, Adults-Ad.

overwintering refuge and non-winter refuge habitats. One can therefore summarize potamodromy as a cyclic sequence of migrations (feeding, refuge and reproduction) among four types of habitat (feeding, winter refuge, non-winter refuge and spawning) (Fig. 4.3).

After hatching, most potamodromous fishes migrate during each of the four subsequent mobile life stages (larvae, juvenile, sub-adult, and adult) (Figs. 4.1, 4.3). The diversity of destination habitats and movement patterns by potamodromous fishes in riverine and lacustrine systems reflect seasonal habitat preferences, as well as shifts in preferred habitats as fishes develop among life stages (Northcote 1984; Schlosser 1991). This broad temporal and spatial scale of movement can be illustrated by a series of examples from each of the four major life stages summarized in Table 4.1. After larval fish emerge, some may drift passively with stream or lake currents (Brown and Armstrong 1985). Others may begin feeding, even before their yolk sac is absorbed (i.e., walleye noted above).

After fish attain the juvenile stage, the diversity of movements may further increase. If there is an abundant food supply, if cover is adequate (i.e., water depth,

Table 4.1. Generalized migrations and movements by North American potamodromous fishes in riverine and lacustrine environments. These movements reflect seasonal habitat preferences as well as life stage related shifts in preferred habitats as fishes develop to maturity.

Life Stage	Type	Movement	Destination Habitat	References
I. Adult	Maturing adults (Pre-Spawning)	Feeding migration Refuge migration Refuge migration Diel summer movements Diel winter movements	Feeding/rearing Thermal refugia/staging Overwintering Feeding to resting and back Concealment out and back	Crowe 1962 Bjornn and Mallet 1964 Swanberg 1997 Janssen and Brandt 1980 Wang et al. 2007
	Gravid adults (Spawning)	Reproductive migration	Spawning	Irving and Modde 2000
	Spent adults (Post-Spawning) Iteroparous	Feeding migration Refuge migration Refuge migration Diel summer movements Diel winter movements	Feeding/rearing Thermal refugia/staging Overwintering Feeding to resting and back Concealment out and back	Swanberg 1997 Stevens and DuPont 2011
	Kelts (Post-spawning)	Passive dispersal and death	Downstream in riverine or drift in lacustrine	Brown and Mackay 1995
II. Embryos	Emerged with yolk sac	Diel movements Passive dispersal	Feeding/rearing Downstream in riverine or drift in lacustrine	Scott and Crossman 1973
III. Larvae	Free swimming larvae	Passive dispersal Feeding migration Refuge migration Refuge migration Diel summer movements Diel winter movements	Downstream in riverine or drift in lacustrine Feeding/rearing Thermal refugia/cover Overwintering Feeding to resting and back Concealment out and back	Tyus and McAda 1984 Corbett and Powles 1986

| IV. Juvenile | Diverse juvenile ages and sizes | Feeding migration
Refuge migration
Refuge migration
Diel summer movements
Diel winter movements | Feeding/rearing
Thermal refugia/cover
Overwintering
Feeding to resting and back
Concealment out and back | Wolfert 1963
McNicol et al. 1985
Chisholm et al. 1987
Thurow 1997 |
| V. Sub-adult | Sexually immature | Feeding migration
Refuge migration
Refuge migration
Diel summer movements
Diel winter movements
Pied Piper migrations | Feeding/rearing
Thermal refugia/staging
Overwintering
Feeding to resting and back
Concealment out and back
Unknown | Janssen and Brandt 1980
Schill et al. 1994
Petty 2012 |

overhead cover) and if other rearing conditions are favorable (i.e., suitable water temperatures), juveniles may establish a home range and move within that range until winter when they will complete a refuge migration to overwintering refuge habitat (Schlosser 1991). However, juveniles may migrate to feeding and or non-winter refuge habitats outside a home range if food or cover is lacking, or if water temperatures increase above the suitable range (Petty et al. 2012). The scale and number of movements are also influenced by the longevity of the life stages. Female lake sturgeon *Acipenser fulvescens*, for example, may not attain sexual maturity until an age of approximately 20 years (Threader and Broussaeu 1986). Consequently, juvenile lake sturgeon may complete numerous migrations to a series of feeding, non-winter refuge, and winter refuge habitats (Threader and Broussaeu 1986). As Northcote (1997) observed, preferred habitats are not necessarily the same ones occupied during the previous year migrations. The locations of preferred habitats may change as the fish grow, age, and have different habitat requirements (Schlosser 1991). In their final life stages, both sub-adult and adult fishes also complete annual migrations to a series of feeding and winter refuge habitats (Schlosser 1991), as well as non-winter refuge habitats (Petty et al. 2012).

Gerking (1959) defined homing as "the return to a place formerly occupied instead of going to other equally probable places". The effects of homing or fidelity to the same habitats are addressed below. Longer lived species, such as the lake sturgeon (maximum age of 154 years) (Scott and Crossman 1973), have the potential to complete hundreds of annual feeding and refuge migrations. To add to this complexity of movements, adults of all species also complete reproductive migrations to appropriate spawning habitats (Schlosser 1991). Since many potamodromous species, such as Yellowstone cutthroat trout *Oncorhynchus clarkii bouvieri*, are iteroparous, surviving adults complete spawning migrations over multiple years which further increases the complexity of migrations (Thurow et al. 1988). If repeat spawning adults return to the same spawning site, this 'homing' behavior will reduce the number of different spawning habitats an individual fish migrates to. If fish do not have high fidelity to the same spawning site, this will add further spatial complexity to spawning migrations.

As Northcote (1997) observed, the apparent simplicity of a table or figure used to summarize movements of potamodromous fishes is deceiving. Within each of the four mobile life stages, many types of movements are repeated and a large diversity of habitats may be utilized over very large temporal and spatial scales. Examples of feeding, refuge, and spawning migrations are provided below. Temporal and spatial scales of migration between these habitats are directly influenced by the species life histories and habitat requirements. Flathead catfish *Pylodictis olivaris*, for example, may not require expansive stretches of river in order to complete critical life stages, so they tend to have relatively smaller home ranges and exhibit more localized movements (Daugherty and Sutton 2005). Alternatively, alligator gars *Atractosteus spatula* and other long-lived, large-bodied, and highly mobile species, as sturgeon *Acipenser* spp. and paddlefish *Polydon* spp., tend to have large home ranges (Minns 1995) and complete extensive migrations because habitats used for reproduction, refuge, and feeding are dispersed. Diel movements or other movements within home ranges, which are of more restricted spatial and temporal scale, are addressed later.

Feeding migrations

Migrations to preferred feeding habitats, with either more abundant or more suitable prey, may be completed by all four mobile life stages (larvae, juveniles, sub-adults, and adults). Larval Colorado pikeminnow drifted from spawning areas in the lower Yampa River downstream to shallow, productive nursery habitats in the Green River (Tyus and McAda 1984). The authors observed that this larval life history strategy, which likely benefited the Colorado pikeminnow formerly, could be implicated in its decline in the lower Colorado River, where dams may have blocked migration routes and degraded nursery habitats. Similarly, newly hatched white sucker larvae drifted downstream on a feeding migration to areas with higher zooplankton production (Corbett and Powles 1986).

Juveniles of many species migrate to preferred feeding areas. In Lake Erie, yearling walleye migrated from a known nursery area and traveled primarily north toward the Western Basin during their first year, and in succeeding years moved progressively toward the extreme western end of the lake. Marked juvenile walleye migrated an average of 40 km from the nursery area and one juvenile migrated more than 320 km (Wolfert 1963).

Wang et al. (2007) monitored movements of adult walleye across Lake Erie and suggested these migrations may be a response to spatial patterns in prey abundance (soft-rayed prey preferred cooler temperatures). Knight et al. (1984) and Knight and Vondracek (1993) observed shifts in walleye diets according to the availability of prey, and confirmed that adult walleyes prefer to feed on soft-rayed fish (i.e., spottail shiner *Notropis hudsonius*, and clupeids (i.e., alewife) rather than on spiny-rayed fish (i.e., yellow perch *Perca flavescens*)). Clupeids and spiny-rayed fish are fast-growing forage fish that become invulnerable to walleye predation after one growing season, whereas smaller soft-rayed fish of all ages are easily caught and digested by walleyes (Knight and Vondracek 1993).

Surviving adults also migrate to feeding or refuge habitats after spawning. The distance that iteroparous forms migrate, their post-spawning condition, and the quality of the habitats they migrate to, likely influence their ability to survive and repeat spawn (Brown and Mackay 1995). Brown and Mackay (1995) reported cutthroat trout post-spawning mortality of less than 14%, but noted that other researchers have observed much higher (60%) post-spawning mortality rates. This relatively low spawning mortality may have been a result of shorter migrations to spawning areas in their watershed, compared with cutthroat trout migrations in the other studied basins (Brown and Mackay 1995).

Refuge migrations

Refuge migrations occur for a variety of purposes, including seasonal refuge from severe conditions, such as extreme low temperatures during winter or low water and dissolved oxygen deficit in floodplains during the dry season (Flecker et al. 2010). Refuge migrations may be of two primary types: (1) migrations by fish seeking overwinter refuge habitat; (2) migrations by fish seeking refuge habitats during non-winter periods.

Non-winter refuge migrations: As water temperature increase in spring, mature bull trout begin migrating from overwintering refuge habitats, at lower elevations, toward spawning areas, at higher elevations (Swanberg 1997). Spawning does not commence for several months, and between spring and late summer, adult bull trout migrate to and stage in thermally suitable refuge habitats for varying periods of time (Swanberg 1997). As water temperatures increased in mid-June, adult bull trout sometimes exhibited rapid upstream migrations to higher elevation habitats (Schill et al. 1994). Movements related to the seeking of thermal refugia have also been reported for other potamodromous species. Brook trout migrated upstream in summer while seeking cold water refugia (Petty et al. 2012). Stevens and Dupont (2011) observed westslope cutthroat trout, rainbow trout *Oncorhynchus mykiss*, and mountain whitefish *Prosopium williamsoni* moving into cooler Coeur d'Alene River side channels as main-river water temperatures increased. Movements and aggregations of salmonids seeking coldwater refugia in streams have also been documented for brook trout (Baird and Krueger 2003) and rainbow trout (Kaya et al. 1977; Ebersole et al. 2001; Sutton et al. 2007). Lake dwelling brook trout (Biro 1998) and lake trout *Salvelinus namaycush* (Snucins and Gunn 1995) have been similarly observed moving into cooler water areas as water temperatures rose in lakes.

Potamodromous fishes may also exhibit high fidelity to suitable summer refuge habitats. Individual smallmouth bass *Micropterus dolomieu* returned to the same 5 km reach of summer habitat (Langhurst and Schoenike 1990). Daugherty and Sutton (2005) similarly observed flathead catfish homing to summer habitats.

Winter refuge migrations: Overwinter ecology of stream-dwelling fishes is perhaps the least understood aspect of their life history. Many fishes occupy different habitats in winter than in summer. As water temperatures decline, fish move from summer habitat into suitable overwintering areas, often at much lower elevations (Bjornn and Mallet 1964). These lower elevation overwintering habitats may provide more benign conditions such as warmer water temperatures, less anchor ice, and more opportunities to escape predators. The distances fish move also seem to be influenced by the proximity of suitable overwintering habitat (Chapman and McLeod 1987). For example, at the onset of winter, stream-dwelling salmonids in the Intermountain West (northwestern USA) typically adopt two overwintering strategies, migration to more suitable overwinter habitats or concealment within their home range if the local habitat is suitable (Thurow 1997). Chapman and McLeod (1987) suggested juvenile salmonids seek overwintering areas in the most upstream locations near summer rearing areas. After locating suitable overwinter habitat, juvenile salmonids typically select areas of low water velocity and enter concealment cover (Edmundson et al. 1968; Cunjak 1988; Thurow 1997). In contrast, adult fishes often overwinter in deep water habitats. For example, adult westslope cutthroat trout migrated more than 100 km downstream to overwinter in large, deep pools (Bjornn and Mallet 1964). Similarly, as water temperatures declined below 16°C in autumn, smallmouth bass migrated long distances downstream (69–87 km) from summer habitats to overwintering habitats in deep pools (Langhurst and Schoenike 1990). Munther (1970) and Paragamian (1981) similarly observed adult smallmouth bass moving into deep pools as temperatures cooled, but neither described long-range movements. Langhurst and Schoenike (1990) suggest the paucity of deep pools near summer habitat may have been the reason for the long migration of smallmouth bass they documented. Fall migrations of bull trout to lower elevation overwintering habitats are also well documented. Many adult bull trout migrate, more than 100 km, to overwinter in deep pools in the lower portions of watersheds (Schill et al. 1994; Swanberg 1997; Hogan and Scarnecchia 2006). Brook trout (Chisholm et al. 1987) and channel catfish *Ictalurus punctatus* have also been observed migrating downstream in fall to more suitable overwintering habitats (Pellet et al. 1998).

Homing behavior in fishes is believed to facilitate development of population-specific adaptations to the habitat occupied (Leggett 1977). For example, alligator gar exhibited high fidelity to overwintering sites (Kluender 2011). This fidelity to high quality overwintering areas may optimize survival.

Spawning migrations

Seasonal migrations to spawning sites are very common in potamodromous fishes. Mature walleye complete upstream migrations to spring spawning areas (Crowe 1962), while fall spawning species, such as bull trout, also migrate to spawning areas (Swanberg 1997). Natal homing or natal philopatry is well documented in several potamodromous species, most commonly in salmonids (Hasler and Scholz 1983; Northcote 1984). Homing may also result in reproductive isolation producing fish stocks unique in behavior, energetics, and reproductive characteristics (Leggett 1977).

High levels of natal homing have been reported for coregoninies, thymallines, and salmonines (Northcote 1997). Although non-salmonids are less studied, homing to a previous spawning location has been reported in Colorado pikeminnow (Tyus 1985), walleye (Crowe 1962), longnose suckers *Catostomus catostomus* and white suckers (Geen et al. 1966), northern pike *Esox lucius* (Miller et al. 2001), muskellunge (Crossman 1990), channel catfish (Pellet et al. 1998), paddlefish *Polyodon spathula* (Firehammer and Scarnecchia 2007), razorback suckers *Xyrauchen texanus* (Tyus and Karp 1990), white bass *Morone chtysops* (Horral 1981), and alligator gar (Kluender 2011).

There are advantages to maintaining high levels of reproductive homing: eggs are deposited in suitable habitat and homing tends to balance the number of spawners with the reproductive capacity of the area (Northcote 1997). Despite the benefits of homing, some straying may also have a long term selective advantage; enabling species to invade new areas and repopulate old ones in the wake of stochastic events (Lindsey et al. 1959). Following the 1980 eruption of Mount St. Helens in Washington (USA), native cutthroat trout and sculpin, that found refugia in ice covered lakes or less-impacted tributaries, were able to recolonize streams where fish populations had been extirpated (Bisson et al. 2005).

In some potamodromous species, sub-adults complete a unique type of migration that may be associated with adult fish movements. These migrations could be termed 'Pied Piper' migrations, since immature, sub-adult fish appear to follow mature adults as they are migrating to spawning sites. Schill et al. (1994) reported sub-adult bull trout migrating upstream in Rapid River along with mature bull trout from April–July. In August, the sub-adult bull trout stopped migrating before reaching spawning sites and subsequently reversed their migration back downstream to other habitats (Schill et al. 1994). The authors reported that the likelihood of this behavior increased in bull trout smaller than 45 cm in length, suggesting the downstream movements may be associated with the seeking of summer thermal refugia or feeding habitats.

Diel and lesser scale movements

As described previously, movements within home ranges are primarily in response to local resources and typically consist of station-keeping behaviors such as foraging or agonistic behavior. Such movements also differ from migrations because of their more restricted spatial and temporal scales. Smaller and shorter-lived species, such as *Cottus* spp., tend to require smaller home ranges if all critical habitats are available locally in contrast to the species described above (i.e., gar and sturgeon) with much large home ranges and extensive migrations.

Diel movements represent a specialized type of movement within home ranges and may be associated with feeding or refuge movements. The extent and timing of diel vertical movements of adult alewives in Lake Michigan, for example, coincided with diel movements of mysis zooplankton (*Mysis relicta*). Both mysis and adult alewife concentrated at the bottom during the day and migrated upwards to the base of the thermocline at night, with their stomach contents indicating the alewife vertical movements were mechanistically linked to feeding behavior (Janssen and Brandt 1980). In winter, at water temperatures less than 2°C, juvenile bull trout exhibited diel behavioral movements. During the day, all bull trout were concealed in the substrate, while at night, some bull trout moved out of daytime concealment cover into the water column (Thurow 1997). At night, Thurow (1997) observed feeding and resting, primarily in pool and run habitats.

Benefits of potamodromy

Fish migration has been described as a life history syndrome involving energetic trade-offs between movements and energetic output (Schaffer and Elson 1975; Leggett 1977). Movements between habitats create both costs and benefits; costs include energy and physiological demands for osmoregulation, the energetic demands of swimming, and exposure to predators and disease (Gross 1987). The benefits of potamodromy could be organized into three categories: survival benefits to potamodromous fishes, benefits to humans, and lastly, benefits to the functioning of the entire ecosystem.

The benefits of potamodromous migratory behavior to individual fishes and populations were described by Northcote (1984) as arising from spatial, seasonal, and ontogenetic separation of optimal habitats for growth, survival, and reproduction. As described above in the section 'Types of Migration' and Table 4.1, published research suggests that migratory behavior allows potamodromous fishes to: (1) optimize growth by accessing more productive areas; (2) improve survival: perhaps via improved growth; increased overwinter survival; access to refugia from severe conditions such as drought, unsuitable temperatures or low oxygen concentration; and predator avoidance; (3) enhance reproductive fitness: perhaps via improved adult condition, increased fecundity, and access to optimal spawning habitat. Northcote (1997) additionally observed that North American potamodromous fishes have probably recolonized rivers and streams repeatedly over the past million years or more in the face of several glaciations, ice recessions, and interglacial periods. To do so, they may have evolved migratory behavioral patterns adapted to life in highly changeable and unpredictable systems (Northcote 1997). Consequently, migratory behavior also allows potamodromous fishes to: (4) recolonize previously extirpated habitats; (5) disperse to vacant habitats; and (6) maintain beneficial aspects of source/sink dynamics (Hanski and Gilpin 1991), even in very dynamic landscapes.

The importance of freshwater migratory species to humans has long been realized. Humans have exploited migratory freshwater fishes for thousands of years (Lucas and Baras 2001). Potamodromous fishes continue to support essential commercial and recreational fisheries across their world wide range. Revenga et al. (2000) reported that in 1997, inland fisheries landings accounted for 7.7 million metric tons. Taking into account the inland capture, fisheries are estimated to be underreported by two or three times, so the contribution to direct human consumption is likely to be at least twice as high (Revenga et al. 2000). The authors reported that at the global level, inland fisheries landings have been increasing since 1984 with most of this increase in Asia, Africa, and more moderately in Latin America. In North America, Europe, and the former Soviet Union, landings have declined, whereas in Oceania they have remained stable (Revenga et al. 2000).

Over the past two decades, there is increasing recognition that migratory species can also be major ecological drivers shaping both the structure and function of freshwater ecosystems (Flecker et al. 2010). Potamodromous fishes provide benefits to the entire ecosystem via a host of direct and indirect mechanisms as consumers, ecosystem engineers, modulators of biogeochemical processes, and transport vectors (Flecker et al. 2010). Consequently, the loss of key species can have widespread consequences in ecosystems (e.g., Hooper et al. 2005) and this has also led to growing interest in the roles species have in ecosystem function.

Flecker et al. (2010) provided a thorough description of the different processes by which potamodromous fishes subsidize streams and how these subsidies are linked to migration type. The authors described different types of fish migrations and considered their importance from the perspective of ecosystem subsidies. Material subsidies are the transfer of energy, nutrients, and other resources resulting in direct changes in resource pools within ecosystems. In contrast, process subsidies arise from feeding, spawning, or other activities of migratory species that directly affect process rates within recipient ecosystems. Although the presence of migratory individuals can modulate ecosystem functioning under both types of subsidy; the key difference is that material subsidies involve direct delivery of new material (e.g., fish carcasses, gametes), whereas process subsidies affect the dynamics and cycling of existing material (e.g., movement of substrate during spawning, parasitic hosts) (Flecker et al. 2010). For example, the physical and chemical effects of removing algae and periphyton by grazing and sediment-feeding fishes, such as prochilodontids, as well as seed dispersal by large-bodied frugivorous characins, represent potentially key process subsidies by migratory fishes in some large South American rivers (Flecker et al. 2010). Flecker et al. (2010) speculated that process subsidies are more widespread than material subsidies from migratory stream fishes, because they are independent of the type of migration patterns, life history, and distance traveled.

The potential for migratory fish to represent major material subsidies is largest when: (1) the biomass of migrants is high relative to ecosystem size; (2) the availability of nutrients and energy is low in the recipient ecosystem (i.e., oligotrophic); and (3) there is an effective mechanism for liberating nutrients and energy from migratory fishes and retaining those materials within the food web of the recipient ecosystem (Flecker et al. 2010). The authors note that the most efficient mechanisms for liberating nutrients generally involve: (1) local mortality of migrants in the recipient ecosystem due to programmed senescence in semelparous species; (2) local migrant mortality due to predation, parasitism, and disease in iteroparous species; or (3) excretion and gamete deposition by spawning fishes. Regardless of whether nutrients are re-released via decomposition of carcasses, excretion, or gamete release, a mechanism for the liberation and retention of nutrients and energy originating elsewhere is crucial for material subsidies to be significant. Although some of the best examples of material subsidies derived from migratory fishes have emerged from research on Pacific salmon, potamodromous fishes also have considerable potential to represent major material subsidies, especially when they display the requisite features of large migrant biomass and high local mortality or nutrient release in streams of comparatively low nutrient status (Flecker et al. 2010). The authors noted that perhaps the most likely potamodromous candidates for significant nutrient inputs to North American streams are the large, abundant, and widely distributed suckers and redhorses (Catostomidae). The authors cite research by Linderman et al. (2004) documenting that runs of longnose suckers *Catostomus catostomus* exceeded those of Pacific salmon in Alaska's George River. Although most catostomids are long-lived and iteroparous, Flecker et al. (2010) summarized research to illustrate high breeding mortality, as well as results in oligotrophic tributaries of Lake Michigan which indicate that spring migrations of white sucker and longnose sucker are closely associated with a time-lagged increase in dissolved phosphorus concentrations. Though they have not been studied in the context of material subsidies, substantial inputs of energy and nutrients to streams might also be provided by many other potamodromous North American fishes, including percids, salmonids, esocids, moronids, and osmerids (Flecker et al. 2010).

In addition to conveying material subsidies, migratory fishes can strongly affect stream ecosystem processes through their feeding and other activities (Flecker et al. 2010). The authors posit that in addition to migrant biomass, the potential for migratory fish to represent strong process subsidies is influenced by 'migrant interaction strength' and the degree to which a migratory species is functionally unique in a particular ecological setting. Flecker et al. (2010) noted that, by definition, strong interactors would be keystone species; their impacts on ecosystem structure and function would be substantial and disproportionately greater than would be predicted based on their biomass alone. For example, migratory fishes that are hosts of parasitic stages of mussel larvae are functionally unique, and even small numbers of fishes as hosts could be crucial to the dispersal and demography of mussel populations (Flecker et al. 2010).

Migratory fishes can influence important process subsidies in stream ecosystems through a diversity of mechanisms including functioning as: physical ecosystem engineers, chemical ecosystem engineers or modulators of nutrient cycles, seed dispersers, directly and indirectly as consumers, and vectors of contaminants and pathogens (Flecker et al. 2010). Two examples of process subsidies in North American potamodromous fishes are described below, for more information on other types of process subsidies, please see Flecker et al. (2010). Yellowstone cutthroat trout, for example, act as ecosystem engineers during spawning by constructing redds and altering the morphology of the stream bed, removing fine sediments, dislodging aquatic insects and potentially increasing drift, and coarsening the substrate (Thurow and King 1994). Fishes can modify their chemical environment by altering element cycles directly (e.g., Percidae excretion and egestion) or indirectly (e.g., reduced algal demand caused by Catostomidae feeding) (Flecker et al. 2010). A large migration of fish that stay and feed within the recipient local stream can therefore constitute both material (addition of carcasses and gametes) and process subsidies from an excretion standpoint (Flecker et al. 2010).

Morphological adaptations for migration

It is beyond the scope of this chapter to address the highly variable morphological adaptations of potamodromous fishes. These adaptations may influence their swimming performance during migration and movements. Readers are urged to review Videler (1993): Chapter 2 describes the structure of the muscles as swimming apparatus; Chapter 3 describes the body axis and fins; and Chapter 4 describes how body shape, skin, and other special adaptations affect swimming performance. On an interesting note, Portz and Tyus (2004) emphasized the importance of experimental observation for examining potential morphological adaptations for swimming. The authors reported that native Colorado River Basin humpback chub *Gila cypha* and razorback sucker possess a large nuchal hump. Portz and Tyus (2004) noted that although several authors have suggested the hump confers a hydrodynamic advantage to life in fast flow, this premise has not been confirmed with experimental work. Instead, Portz and Tyus (2004) argue that the large humps represent convergent evolution prompted by predation from sympatric Colorado pikeminnow, the top piscivore in the Colorado River system. Lack of jaw teeth and a relatively small jaw gape limit the maximum prey size that Colorado pikeminnow can consume and the large nuchal hump provides a deep body that is difficult or impossible to ingest (Portz and Tyus 2004).

Research needs

Our ability to conserve and restore potamodromous fishes will be enhanced by increased knowledge in several key research areas. Despite the rich history of excellent work that has been accomplished to date, additional research is needed to improve our future understanding of: metapopulation dynamics, detailed migratory behaviors, overwintering behaviors and habitats, and the effects of a changing climate.

Metapopulation-scale information is critical for understanding factors that influence fish population persistence. Hanski and Gilpin (1991) defined a metapopulation as a group of spatially disjunct populations linked by immigration and emigration. Consequently, some populations and their migrations may be disproportionately important for the survival of the species. Distinguishing between source and sink populations is fundamental to identifying populations essential for species persistence. As highlighted by Rosenfeld and Hatfield (2006), failing to distinguish source and sink populations may result in protection of sinks instead of sources, inappropriate identification of critical habitat, and underestimation of the probability of extinction. These authors listed three key information needs at the metapopulation scale: (1) determining the status of discrete populations as sources or sinks; (2) identifying corridors for dispersal and evaluating the probability of exchange between populations; and (3) assessing the probability of subpopulation persistence based on risk of extinction from combined natural and anthropogenic impacts.

Despite many decades of work, our understanding of the migratory behavior of many potamodromous species remains incomplete. Lee et al. (1997) focused a major portion of their comprehensive assessment of the distribution and status of fishes in the interior Columbia River Basin, on seven 'key' salmonids, in part, because these species were widely distributed and well understood. However, despite the rich history and many decades of excellent salmonid research, Northcote (1997) reported that our understanding of the migratory behavior of many salmonids remained incomplete. As Cooke et al. (2008) observed, given the complexity of migration and its role in a myriad of management and conservation situations, in addition to understanding migration timing and extent, we also need to understand the fundamental processes that enable some fish to migrate vast distances, the causes of mortality during migrations, and the factors that cause some fish to migrate and others not to. With few exceptions, the migratory behavior of most non-salmonid potamodromous fishes is even less well understood, especially in remote areas. As Flecker et al. (2010) observed, in the temperate zone, the ecological significance of material subsidies by potamodromous fishes is a ripe area for research. Migration studies would also benefit from being more broadly based; there is a need to focus on migration as a behavioral, ecological, and evolutionary phenomenon (Dingle and Drake 2007). Dingle and Drake (2007) also observed that since movements to exploit separated and ephemeral habitats transcend species and taxonomic groups, research should do likewise. New technologies and interdisciplinary approaches that integrate positional telemetry with other disciplines (e.g., stress physiology, functional genomics, oceanography, experimental biology) hold promise to enhance future research on fish migration and ultimately provide fisheries managers with the knowledge to better manage and conserve migratory fishes globally (Cooke et al. 2008).

Overwinter ecology of fishes is perhaps the least understood aspect of their life history, and the need for winter investigations has long been recognized (Hubbs and Trautman 1935). We have an incomplete understanding of winter habitat, the extent of winter movements, or how winter conditions regulate fish populations. For example, Chisholm et al. (1987) observed that, despite the array of winter habitat research on brook trout, no studies had focused on the extent of winter movement or the specific habitat features selected. Similarly, although several studies suggest that the abundance and quality of overwinter habitat may limit fish abundance (Bustard and Narver 1975; Campbell and Neuner 1985; McMahon and Hartman 1989), the role of winter conditions in regulating fish populations remains poorly understood. Identifying and describing overwinter habitat is an important step in maintaining critical habitats

(Thurow 1997) and conserving native fishes. Additional research is necessary to improve our understanding of the extent of winter movements, fish behaviors during winter, and the role of overwinter habitat in regulating potamodromous fish populations.

Potamodromous fishes are dependent on an abundant supply of water of a suitable temperature. Consequently, studies dedicated to estimate the impacts of climate change on potamodromous fishes will improve the management of habitats and species. Wenger et al. (2010), for example, observed that hydrologic regimes in the western United States have undergone substantial changes over the last half century, including trends toward earlier snowmelt runoff (Mote et al. 2005), reduced water yields and lower summer flows (Luce and Holden 2009), and increased or altered flood risk. Consequently, Isaac and Rieman (2013) observed that the question is not whether, but how fast, stream biotas are shifting or being extirpated by temperature increases associated with climate change. Although empirical evidence exists for shifts in the timing of migrations and spawning (Crozier et al. 2011), as well as poleward and upstream range expansions (Milner et al. 2011), little evidence exists of broadscale range contractions, despite the extensive changes predicted by numerous bioclimatic models. Better approaches are needed to document the response of stream biotas to climate change. Such information is fundamental to understanding if species responses are accurate predictions of the rate at which isotherms near thermally mediated species boundaries are shifting to higher elevations or latitudes (Isaac and Rieman 2013).

Conservation and restoration of potamodromous fishes

Potamodromous fishes are imperiled world-wide and more than 20% of the world's freshwater fish are extinct or have become threatened or endangered in recent decades (Revenga et al. 2000). Revenga et al. (2000) observed that globally, the greatest overall threat for the long-term sustainability of inland fishery resources is the loss of fishery habitat and the degradation of the terrestrial and aquatic environment; historical trends in commercial fisheries data for well-studied rivers show dramatic declines over the 20th century, mainly from habitat degradation, invasive species, and overharvesting. Liermann et al. (2012) assessed implications of dams for global freshwater fish diversity and reported that nearly 50% of the 397 freshwater ecoregions evaluated were obstructed by large- and medium-size dams, and approximately 27% faced additional obstruction. Threatened ecoregions were found on all continents (Liermann et al. 2012). In North America's interior Columbia River basin, Lee et al. (1997) reported that 45 of 88 native fish taxa were identified as threatened, sensitive, or of special concern by state or federal agencies or the American Fisheries Society. Eight of those species are anadromous which results in 37 of 80 potamodromous fish taxa identified as threatened, sensitive, or of special concern within the interior Columbia River Basin.

Identification and protection of critical habitat is central to the management of species at risk (Rosenfeld and Hatfield 2006). The rationale for protecting critical habitat is rooted in the observation that particular habitats are often disproportionately important to population limitation (Fausch et al. 2002), and therefore habitat protection can be prioritized. A general consensus of strategies, designed to conserve species and aquatic biological diversity, is that conservation and rehabilitation should focus first on the best remaining examples of aquatic biological integrity and diversity (Thurow et al. 1997). However, protection of critical habitats and population stronghold will not be sufficient; such reserves never will be large or well distributed enough to maintain biological diversity (Franklin 1993). Watershed rehabilitation and the development of more ecologically compatible land-use policies are also required (Thurow et al. 1997). As Rosenfeld and Hatfield (2006) observed, a key component of potamodromous fish species persistence is the management of habitat and human activities outside of critical habitat. Ultimately, conservation of potamodromous fishes will require a more integrated, broad-scale view of management than has been practiced historically. An assumed goal of ecosystem management is to maintain, or rehabilitate, the integrity of aquatic ecosystems and to provide for the long-term persistence of native (and in some cases desirable nonnative) fishes and other species (Grumbine 1994). Note that non-native species might be desirable if they fill an open niche such as some Pacific salmon species in the Great Lakes (Kohler and Courtenay 1986) or if they provide fisheries values in cases where habitats have been so severely degraded as to be unsuitable for native species restoration. As Thurow et al. (1997) observed, achieving the goal of ecosystem management will require the maintenance, or rehabilitation, of a network of well-connected, high-quality habitats that support a diverse assemblage of native species, the full expression of potential life histories and their movements, and the genetic diversity necessary for long-term persistence and adaptation in a variable environment. Ecosystem management, then, also implies using active management to reestablish more complete or natural structure, function, and processes whenever possible (Thurow et al. 1997). Lastly, effective conservation and restoration efforts for potamodromous fishes will also require an improved understanding of the effects of a changing climate (Isaac and Rieman 2013) as well as increasing knowledge of metapopulation dynamics, detailed migratory behaviors, and overwintering behaviors and habitats.

References

Baird OE and Krueger CC (2003) Behavioral thermoregulation of brook and rainbow trout: comparison of summer habitat use in an Adirondack River, New York. Transactions of the American Fisheries Society 132: 1194–1206.

Baker RR (1978) The Evolutionary Ecology of Animal Migration. Hodder and Stoughton, London.

Biro PA (1998) Staying cool: behavioral thermoregulation during summer by young-of-year brook trout in a lake. Transactions of the American Fisheries Society 127: 212–222.

Bisson PA, Crisafulli CM, Fransen BR, Lucas RE and Hawkins CP (2005) Responses of fish to the 1980 eruption of Mount St. Helens. pp. 163–181. *In*: Dale VH, Swanson FJ and Crisafulli CM (eds.). Ecological Responses to the 1980 Eruption of Mount St. Helens Springer, New York.

Bjornn TC and Mallet J (1964) Movements of planted and wild trout in an Idaho river system. Transactions of the American Fisheries Society 93: 70–76.

Boucher DP (2004) Landlocked Salmon Management Plan. Maine Department of Inland Fisheries and Wildlife, Division of Fisheries and Hatcheries, Augusta, Maine.

Brown AV and Armstrong ML (1985) Propensity to drift downstream among various species of fish. Journal of Freshwater Ecology 3: 3–17.

Brown RS and Mackay WC (1995) Spawning ecology of cutthroat trout (*Oncorhynchus clarki*) in the Ram River, Alberta. Canadian Journal of Fisheries and Aquatic Sciences 52: 983–992.

Bustard DR and Narver DW (1975) Aspects of the winter ecology of juvenile coho salmon (*Oncorhynchus kisutch*) and steelhead trout (*Salmo gairdneri*). Journal of the Fisheries Research Board of Canada 32: 667–680.

Campbell RF and Neuner JH (1985) Seasonal and diurnal shifts in habitat utilization by resident rainbow trout in western Washington Cascade Mountain streams. pp. 39–48. *In*: Olson FW, White RG and Hamre RH (eds.). Symposium on Small Hydropower and Fisheries. American Fisheries Society, Bethesda, Maryland.

Chapman DW and McLeod KP (1987) Development of criteria for fine sediment in the Northern Rockies ecoregion. U.S. Environmental Protection Agency, Seattle, Washington.

Chisholm IM, Hubert WA and Wesche TA (1987) Winter stream conditions and use of habitat by brook trout in high-elevation Wyoming streams. Transactions of the American Fisheries Society 116: 176–184.

Clemens BJ, Binder TR, Docker MF, Moser ML and Sower SA (2010) Similarities, differences, and unknowns in biology and management of three parasitic lampreys of North America. Fisheries 35: 580–594.

Cooke SJ, Hinch SJ, Farrell AP, Patterson DA, Miller-Saunders K, Welch DW, Donaldson MR et al. (2008) Developing a mechanistic understanding of fish migrations by linking telemetry with physiology, behavior, genomics and experimental biology: an interdisciplinary case study on adult Fraser River sockeye salmon. Fisheries 33: 321–339.

Corbett BW and Powles PM (1986) Spawning and larva drift of sympatric walleyes and white suckers in an Ontario stream. Transactions of the American Fisheries Society 115: 41–46.

Crossman EJ (1990) Reproductive homing in muskellunge (*Esox masquinongy*). Canadian Journal of Fisheries and Aquatic Sciences 47: 1803–1812.

Crowe WR (1962) Homing behavior in walleyes. Transactions of the American Fisheries Society 91: 350–354.

Crozier LG, Scheuerell MD and Zabel RW (2011) Using time series analysis to characterize evolutionary and plastic responses to environmental change: a case study of a shift toward earlier migration date in sockeye salmon. The American Naturalist 178: 755–773.

Cunjak RA (1988) Behavior and microhabitat of young Atlantic salmon (*Salmo salar*) during winter. Canadian Journal of Fisheries and Aquatic Sciences 45: 2156–2160.

Daugherty DJ and Sutton TM (2005) Seasonal movement patterns, habitat use, and home range of flathead catfish in the lower St. Joseph River, Michigan. North American Journal of Fisheries Management 25: 256–269.

Dingle H (1996) Migration: The Biology of Life on the Move. Oxford University Press.

Dingle H and Drake VA (2007) What is migration? Bioscience 57: 113–121.

Ebersole JL, Liss WJ and Frissell CA (2001) Relationship between stream temperature, thermal refugia and rainbow trout (*Oncorhynchus mykiss*) abundance in arid-land streams in the northwestern United States. Ecology of Freshwater Fish 10: 1–10.

Edmundson E, Everest FE and Chapman DW (1968) Permanence of station in juvenile chinook salmon and steelhead trout. Journal of the Fisheries Research Board of Canada 25: 1453–1464.

Eschmeyer WN (ed.) (2013) Genera, Species, References. Electronic version accessed November 1st 2014 (http://research.calacademy.org/research/ichthyology/catalog/fishcatmain.asp).

Eschmeyer WN and Fong JD (2015) Species by Family/Subfamily in the Catalog of Fishes. Updated 2 Aug 2015. Electronic version accessed Aug 30, 2015 (http://researcharchive.calacademy.org/research/ichthyology/catalog/SpeciesByFamily.asp).

Fausch KD, Torgersen CE, Baxter CV and Li HW (2002) Landscapes to riverscapes: bridging the gap between research and conservation of stream fishes. BioScience 52: 483–498.

Firehammer JA and Scarnecchia DL (2007) The influence of discharge on duration, ascent distance, and fidelity of the spawning migration for paddlefish of the Yellowstone-Sakakawea stock, Montana and North Dakota, USA. Environmental Biology of Fishes 78: 23–36.

Flecker AS, McIntyre PB, Moore JW, Anderson JT, Taylor BW and Hall RO, Jr. (2010) Migratory fishes as material and process subsidies in riverine ecosystems. American Fisheries Society Symposium 73: 559–592.

Franklin JF (1993) Preserving biodiversity: species, ecosystems, or landscapes? Ecological Applications 3: 202–205.

Geen GH, Northcote TG, Hartman GF and Lindsey CC (1966) Life histories of two species of catostomid fishes in Sixteen mile Lake, British Columbia, with particular reference to inlet stream spawning. Journal of the Fisheries Board of Canada 23: 1761–1788.

Gerking SD (1959) The restricted movement of fish populations. Biological Reviews 34: 221–242.

Gowan C, Young MK, Fausch KD and Riley SC (1994) Restricted movement in resident stream salmonids: a paradigm lost? Canadian Journal of Fisheries and Aquatic Sciences 51: 2626–2637.

Gresswell RE, Kershner J, Dunham JB and Gresswell RE (1997) Ecology and management of potamodromous salmonids: introduction to ecology and management of potamodromous salmonids. North American Journal of Fisheries Management 17: 1027–1028.

Gross MR (1987) Evolution of diadromy in fishes. American Fisheries Society Symposium 1: 14–25.

Gross MR, Coleman RM and McDowall RM (1988) Aquatic productivity and the evolution of diadromous fish migration. Science 239: 1291–1293.

Grumbine RE (1994) What is ecosystem management? Conservation Biology 8: 27–38.

Hanski I and Gilpin M (1991) Metapopulation dynamics: brief history and conceptual domain. Biological Journal of the Linnean Society 42: 3–16.

Hanson KC, Cooke SJ, Suski CD, Niezgoda G, Phelan FJS, Tinline R and Philipp DP (2007) Assessment of largemouth bass (*Micropterus salmoides*) behaviour and activity at multiple spatial and temporal scales utilizing a whole-lake telemetry array. Hydrobiologia 582: 243–256.

Hasler AD and Scholz AT (1983) Olfactory Imprinting and Homing in Salmon: Investigations into the Mechanism of Imprinting Process. Zoophysiology. Springer-Verlag, New York.

Heming TA and Buddington RK (1988) Yolk absorption in embryonic and larval fishes. Fish Physiology 11(part A): 407–446.

Hobson KA (1999) Tracing origins and migration of wildlife using stable isotopes: a review. Oecologia 120: 314–326.

Hogan DM and Scarnecchia DL (2006) Distinct fluvial and adfluvial migration patterns of a relict charr (*Salvelinus confluentus*) stock in a mountainous watershed, Idaho, USA. Ecology of Freshwater Fish 15: 376–387.

Hooper DU, Chapin FS, Ewel JJ, Hector A, Inchausti P, Lavorel S, Lawton JH, Lodge DM, Loreau M, Naeem S, Schmid B, Setala H, Symstad AJ, Vandermeer J and Wardle DA (2005) Effects of biodiversity on ecosystem functioning: A consensus of current knowledge. Ecological Monographs 75: 3–35.

Horrall RM (1981) Behavioral stock-isolating mechanisms in Great Lakes fishes with special reference to homing and site imprinting. Canadian Journal of Fisheries and Aquatic Sciences 38: 1481–1496.

Hubbs CL and Trautman MB (1935) The need for investigating fish condition in winter. Transactions of the American. Fisheries Society 65: 51–56.

Irving DB and Modde T (2000) Home-range fidelity and use of historic habitat by adult Colorado pikeminnow (*Ptychocheilus lucius*) in the White River, Colorado and Utah. Western North American Naturalist 60: 16–25.

Isaak DJ and Rieman BE (2013) Stream isotherm shifts from climate change and implications for distributions of ectothermic organisms. Global Change Biology 19: 742–751.

Jager HI, Chandler JA, Lepla KB and Van Winkle W (2001) A theoretical study of river fragmentation by dams and its effects on white sturgeon populations. Environmental Biology of Fishes 60: 347–361.

Janssen J and Brandt SB (1980) Feeding ecology and vertical migration of adult alewives (*Alosa pseudoharengus*) in Lake Michigan. Canadian Journal of Fisheries and Aquatic Sciences 37: 177–184.

Kaya CM, Kaeding LR and Burkhalter DE (1977) Use of a cold-water refuge by rainbow and brown trout in a geothermally heated stream. The Progressive Fish-Culturist 39: 37–39.

Kennedy JS (1985) Migration: Behavioral and ecological. pp. 5–26. *In*: Rankin MA (ed.). Migration: Mechanisms and Adaptive Significance. Contributions in Marine Science 27 (suppl.). Austin: Marine Science Institute, University of Texas.

Kluender ER (2011) Seasonal Habitat Use of a Leviathan, Alligator Gar, at Multiple Spatial Scales in a River-floodplain Ecosystem (Doctoral dissertation, University of Central Arkansas. Department of Biology).

Knight RL and Vondracek B (1993) Changes in prey fish populations in western Lake Erie, 1969–88, as related to walleye, *Stizostedion vitreum*, predation. Canadian Journal of Fisheries and Aquatic Sciences 50: 1289–1298.

Knight RL, Margraf FJ and Carline RF (1984) Piscivory by walleye and yellow perch in western Lake Erie. Transactions of the American Fisheries Society 113: 677–693.

Kohler CC and Courtenay WR (1986) American Fisheries Society position on introductions of aquatic species. Fisheries 11: 39–42.

Langhurst RW and Schoenike DL (1990) Seasonal migration of smallmouth bass in the Embarrass and Wolf Rivers, Wisconsin. North American Journal of Fisheries Management 10: 224–227.

Lee DC, Sedell JR, Rieman BE, Thurow RF and Williams JE (1997) Broadscale assessment of aquatic species and habitats. An assessment of ecosystem components in the interior Columbia Basin and portions of the Klamath and Great basins. Volume 3, Chapter 4. USDA Forest Service General Technical Report PNW-GTR-405.

Leggett WC (1977) The ecology of fish migrations. Annual Review of Ecology and Systematics 8: 285–308.

Liermann CR, Nilsson C, Robertson J and Ng RY (2012) Implications of dam obstruction for global freshwater fish diversity. BioScience 62: 539–548.

Linderman JC, Jr., Molyneaux DB, Folletti DL and Cannon DJ (2004) George River salmon studies. Alaska Department of Fish and Game, Regional Information Report No. 3A04–17, Juneau.

Lindsey CC, Northcote TG and Hartman GF (1959) Homing of rainbow trout to inlet and outlet spawning streams at Loon Lake, British Columbia. Journal of the Fisheries Board of Canada 16: 695–719.

Lucas MC and Baras E (2001) Migration of Freshwater Fishes. Blackwell Scientific Publications, Oxford, UK.

Luce CH and Holden ZA (2009) Declining annual streamflow distributions in the Pacific Northwest United States, 1948–2006, Geophys. Res. Lett., 36, L16401, doi:10.1029/2009GL039407.

McMahon TE and Hartman GF (1989) Influence of cover complexity and current velocity on winter habitat use by juvenile coho salmon (*Oncorhynchus kisutch*). Canadian Journal of Fisheries and Aquatic Sciences 46: 1551–1557.

McNicol RE, Scherer E and Murkin EJ (1985) Quantitative field investigations of feeding and territorial behaviour of young-of-the-year brook charr (*Salvelinus fontinalis*). Environmental Biology of Fishes 12: 219–229.

Miller LM, Kallemeyn L and Senanan W (2001) Spawning-site and natal-site fidelity by northern pike in a large lake: mark–recapture and genetic evidence. Transactions of the American Fisheries Society 130: 307–316.

Miller ML and Menzel BW (1986) Movements, homing, and home range of muskellunge (*Esox masquinongy*), in West Okoboji Lake, Iowa. Environmental Biology of Fishes 16: 243–255.

Milner AM, Robertson AL, Brown LE, Sønderland SH, McDermott M and Veal AJ (2011) Evolution of a stream ecosystem in recently deglaciated terrain. Ecology 92: 1924–1935.

Minns CK (1995) Allometry of home range size in lake and river fishes. Canadian Journal of Fisheries and Aquatic Sciences 52: 1499–1508.

Morais P and Daverat F (2016) Definitions and concepts related with fish migration. pp. 14–19. *In*: Morais P and Daverat F (eds.). An Introduction to Fish Migration. CRC Press, Boca Raton, FL, USA.

Mote PW, Hamlet AF, Clark MP and Lettenmaier DP (2005) Declining mountain snowpack in western North America. Bull. Am. Meteorol. Soc. 86: 39–49.

Munther GL (1970) Movement and distribution of smallmouth bass in the middle Snake River. Transactions of the American Fisheries Society 99: 44–53.

Myers GS (1949) Usage of anadromous, catadromous and allied terms for migratory fishes. Copeia 1949: 89–97.

Myers JP, Morrison RIG, Antas PZ, Harrington BA, Lovejoy TE, Sallaberry M and Tarak A (1987) Conservation strategy for migratory species. American Scientist 75: 18–26.

Nelson JS (2006) Fishes of the World. 4th Edition. John Wiley & Sons, New York.

Northcote TG (1978) Migratory strategies and production in freshwater fishes. pp. 326–359. *In*: Gerking SD (ed.). Ecology of Freshwater Fish Production. Blackwell Scientific Publications, Oxford, UK.

Northcote TG (1984) Mechanisms of fish migration in rivers. pp. 317–355. *In*: McCleve JD, Arnold GP, Dodson JJ and Neill WH (eds.). Mechanisms of Migration in Fishes. Plenum Press, New York.

Northcote TG (1997) Potamodromy in Salmonidae living and moving in the fast lane. North American Journal of Fisheries Management 17: 1029–1045.

Northcote TG (1998) Migratory behaviour of fish and its significance to movement through riverine fish passage facilities. pp. 3–18. *In*: Jungwirth M, Schmutz S and Weiss S (eds.). Fish Migration and Fish Bypasses. Fishing New Books, Blackwell Scientific Publications, Oxford, UK.

Paragamian VL (1981) Some habitat characteristics that affect abundance and winter survival of smallmouth bass in the Maquoketa River, Iowa. pp. 45–53. *In*: Krumhoz LA (ed.). Warmwater Streams Symposium. Southern Division, American Fisheries Society, Bethesda, Maryland.

Pellett DT, Van Dyck GJ and Adams JV (1998) Seasonal migration and homing of channel catfish in the Lower Wisconsin River, Wisconsin. North American Journal of Fisheries Management 18: 85–95.

Petty JT, Hansbarger JL, Huntsman BM and Mazik PM (2012) Brook trout movement in response to temperature, flow, and thermal refugia within a complex Appalachian riverscape. Transactions of the American Fisheries Society 141: 1060–1073.

Portz D and Tyus H (2004) Fish humps in two Colorado River fishes: a morphological response to cyprinid predation? Environmental Biology of Fishes 71: 233–245.

Revenga C, Brunner J, Henninger N, Kassem K and Payne R (2000) Pilot Analysis of Global Ecosystems. Freshwater Systems. World Resources Institute, Washington, DC.

Rosenfeld JS and Hatfield T (2006) Information needs for assessing critical habitat of freshwater fish. Canadian Journal of Fisheries and Aquatic Sciences 63: 683–698.

Ross ST (2013) Ecology of North American Freshwater Fishes. University of California Press.

Schaffer WM and Elson PF (1975) The adaptive significance of variations in life history among local populations of Atlantic salmon in North America. Ecology 577–590.

Schill D, Thurow R and Kline P (1994) Seasonal movement and spawning mortality of fluvial bull trout in Rapid River, Idaho. Progress Report. Project: F-73-R-15, Job2. Idaho Department of Fish and Game. Boise.

Schlosser IJ (1991) Stream fish ecology: a landscape perspective. BioScience 41: 704–712.

Scott WB and Crossman EJ (1973) Freshwater fishes of Canada. Fisheries Research Board of Canada Bulletin 184.

Snucins EJ and Gunn JM (1995) Coping with a warm environment: behavioral thermoregulation by lake trout. Transactions of the American Fisheries Society 124: 118–123.

Stevens BS and DuPont JM (2011) Summer use of side-channel thermal refugia by salmonids in the North Fork Coeur d'Alene River, Idaho. North American Journal of Fisheries Management 31: 683–692.

Sutton RJ, Deas ML, Tanaka SK, Soto T and Corum RA (2007) Salmonid observations at a Klamath River thermal refuge under various hydrological and meteorological conditions. River Research and Applications 23: 775–785.

Swanberg TR (1997) Movements of and habitat use by fluvial bull trout in the Blackfoot River, Montana. Transactions of the American Fisheries Society 126: 735–746.

Thorpe JE (1988) Salmon migration. Science Progress 72: 345–370.

Threader RW and Broussaeu CS (1986) Biology and management of the lake sturgeon in the Moose River, Ontario. North American Journal of Fisheries Management 6: 383–390.

Thurow RF (1997) Habitat utilization and diel behavior of juvenile bull trout (*Salvelinus confluentus*) at the onset of winter. Ecology of Freshwater Fish 6: 1–7.

Thurow RF and King JG (1994) Attributes of Yellowstone cutthroat trout redds in a tributary of the Snake River, Idaho. Transactions of the American Fisheries Society 123: 37–50.

Thurow RF, Corsi CE and Moore VK (1988) Status, ecology, and management of Yellowstone cutthroat trout in the upper Snake River drainage, Idaho. American Fisheries Society Symposium 4: 25–36.

Thurow RF, Lee DC and Rieman BE (1997) Distribution and status of seven native salmonids in the interior Columbia River basin and portions of the Klamath River and Great basins. North American Journal of Fisheries Management 17: 1094–1110.

Tyus HM (1985) Homing behavior noted for Colorado squawfish. Copeia 213–215.

Tyus HM and McAda CW (1984) Migration, movements and habitat preferences of Colorado squawfish, *Ptychocheilus lucius*, in the Green, White and Yampa rivers, Colorado and Utah. The Southwestern Naturalist 289–299.

Tyus HM and Karp CA (1990) Spawning and movements of razorback sucker (*Xyrauchen texanus*), in the Green River basin of Colorado and Utah. The Southwestern Naturalist 427–433.

Varley JD and Gresswell RE (1988) Ecology, status, and management of the Yellowstone cutthroat trout. American Fisheries Society Symposium 4: 13–24.

Videler JJ (1993) Fish Swimming, Vol. 10. Chapman and Hall, London.

Wang HY, Rutherford ES, Cook HA, Einhouse DW, Haas RC, Johnson TB and Turner MW (2007) Movement of walleyes in Lakes Erie and St. Clair inferred from tag return and fisheries data. Transactions of the American Fisheries Society 136: 539–551.

Welcomme RL (1985) River fisheries. FAO Fisheries Technical Paper 262.

Wenger SJ, Luce CH, Hamlet AF, Isaak DJ and Neville HM (2010) Macroscale hydrologic modeling of ecologically relevant flow metrics. Water Resources Research 46.

Wolfert DR (1963) The movements of walleyes tagged as yearlings in Lake Erie. Transactions of the American Fisheries 92: 414–420.

CHAPTER 5

Life Histories of Anadromous Fishes

*Marie-Laure Acolas** and *Patrick Lambert*

Introduction, anadromous fishes: who are they?

Anadromous fishes are those among diadromous fishes (truly migratory fishes which migrate between the sea and freshwater) who spend most of their lives in the sea and migrate to freshwater to breed (Myers 1949). These movements between habitats occur at a specific timing in the ontogeny and concern either part or all individuals of a population (Fontaine 1975). But who are these anadromous fish? According to several authors, the number of anadromous species is not yet fixed and is probably underestimated, and estimates varied between 109 (McDowall 1988) and 175 anadromous species (Riede 2004), which corresponds to 23 and 31 families around the world, respectively. Considering the 33,059 fish species described around the world (Eschmeyer and Fong 2014), anadromous species represents about 0.5% of the whole fish species. They occur among the class of Actinopterygii (among 28 family over 491) and Cephalaspidorphii (among the three families of the order Petromyzontiformes: Petromyzontidae, Geotriidae and Mordaciidae). No anadromous species are mentioned in the literature in the following classes: Myxini, Elasmobranchii, Holocephali and Sarcopterygii.

Anadromous fishes represent more than 20% of the total number of species in the orders Acipenseriformes (71.4%), Petromyzontiformes (23.9%) and Salmoniformes (21%) (Fig. 5.1), more than 5% in the orders Osmeriformes (8.7%), Clupeiformes (9.8%), Elopiformes (11.1%, which corresponds to one species only) and Gasterosteiformes (6.9%, which corresponds to two species only) (Fig. 5.1). Among the other orders, the anadromous species seem more anecdotic and it often corresponds to only one species and in a particular environment. For example, the

IRSTEA—National Research Institute of Science and Technology for Environment and Agriculture, Aquatic Ecosystems and Global Changes research unit, Diadromous migratory fish team, 50 avenue de Verdun, 33612 Gazinet Cestas, France.
* Corresponding author: marie-laure.acolas@irstea.fr

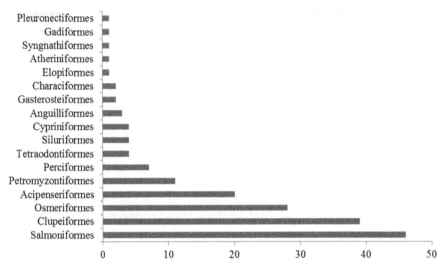

Figure 5.1. Number of species described as anadromous in 17 fish orders (based on data from Riede (2004) thanks to queries made on Fishbase).

common perch (*Perca fluviatilis*) was described as anadromous in the freshwater part of the Baltic Sea (Nesbø et al. 1998) which is probably questionable, and is reported as anadromous in the Groms database (www.groms.de).

Anadromous species are present in numerous parts of the world (Fig. 5.2), but they seem to be mainly observed in the Northern Hemisphere (Fig. 5.3). The orders Salmoniformes, Clupeiformes, Osmeriformes, Acipenseriformes and Petromyzontiformes gather the most anadromous species, 82% of all anadromous species, which occur mainly in the Baltic Sea, in North Atlantic and in North Pacific Oceans (Fig. 5.3). All orders occur in South and East Europe-Mediterranean, Adriatic, Black, Aral and Caspian Seas, except the Osmeriformes. All orders occur in Asia (China, Japan and Okhotsk Seas), except the Petromyzontiformes, and in northern seas (Arctic, Barents and Bering Seas), except the Clupeiformes. None of these orders occur in South Atlantic, Indian or west Pacific Oceans, except Clupeiformes and a few Petromyzontiformes species (Southwest Atlantic and Southwest Pacific Oceans). None of these five orders (Salmoniformes, Clupeiformes, Osmeriformes, Acipenseriformes and Petromyzontiformes) has been mentioned in the Red Sea. However, the differences between the two hemispheres may be partly biased by the fewer and more recent studies developed at the Southern Hemisphere, which can be very rare or nonexistent for some orders.

Why do they migrate?

The migration behavior of all migratory species is supposed to have entangled genetic and environmental components (Pulido 2007). However, the characterization of a gene or of a gene pool specific for anadromy is still a scientific challenge (Amstutz et al. 2006; Bruford 2006; Dodson et al. 2013), since many factors seem to

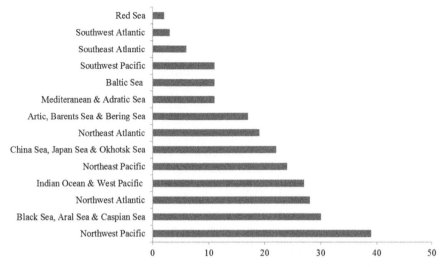

Figure 5.2. Number of anadromous species reported in different seas. The same species can occur in different areas.

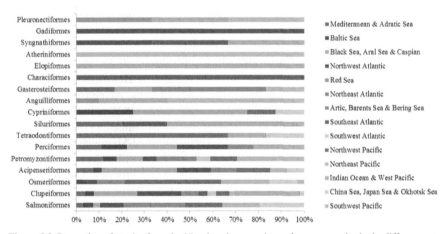

Figure 5.3. Proportion of species from the 17 orders that contain anadromous species in the different parts of the world's oceans and seas.

be involved in triggering migration. Some triggers may be endogenous, linked to fish size or growth (Bohlin et al. 1996; Acolas et al. 2012; Dodson et al. 2013), metabolism (Forseth et al. 1999) or sex (Rundio et al. 2012; Ohms et al. 2014). Migration triggers may also be exogenous, and linked to environmental conditions (Yako et al. 2002; Byrne et al. 2004), latitude (Jonsson et al. 2001; Ohms et al. 2014), altitude (Bohlin et al. 2001), density dependence (Morita et al. 2000; Marco-Rius et al. 2013), food abundance (Olsson et al. 2006) or even to climate variability (Otero et al. 2014). Moreover, the effects of these factors may interact and vary between populations (Jonsson and Jonsson 1993; Thorpe et al. 1998), and even among individuals of the same population (Olsson and Greenberg 2004). Even over endogenous and exogenous factors, evolution mechanisms might act on the persistence of the different migratory tactics when they coexist (Hendry et al. 2004). According to Gross (1987), a migratory life history could persist in a population if the migrant's fitness is higher than non-migrant's fitness for the same period of time, when taking into account certain migration costs, as mortality, later age at reproduction and energy expenditure. This benefit/cost ratio, also called trade-off, was mainly defined theoretically (Hendry and Stearns 2004) and very few empirical studies are available to support the founding assumptions, because these studies are difficult to implement (Bohlin et al. 2001; Kinnison et al. 2001). Sahashi and Morita (2013) studied how migration costs influence which members migrate in 10 populations of anadromous salmonids (white-spotted charr *Salvelinus leucomaenis* and masu salmon *Oncorhynchus masou*). They compared size at maturity for resident males, which can be considered to be the threshold trait that determines the migratory tactics used within a population, and they highlighted that size at maturity for resident males was smaller in fish located further from the sea, where migration costs are presumably higher. They suggest that migration costs are a significant convergent selective force on migratory tactics and life-history traits in nature. Despite the benefit/cost ratio involved when deciding to engage in a migration, or not, the migratory behavior of a species represents an undeniable advantage to colonize new places and to adapt to variable environmental conditions (Thorpe et al. 1998).

The origin of anadromy is still under debate. Gross (1987) suggested that anadromous species had a freshwater origin, through an intermediate stage of amphidromy. His main argument was based on a higher proportion of anadromous species associated with a higher productivity in sea water than in freshwater at high latitudes (Gross et al. 1988). In this context, in cold temperate and subpolar regions, freshwater fishes are more likely to cross into the sea as facultative wanderers and then to evolve into amphidromous and then anadromous species. Later, McDowall (1997) debated this argument since few freshwater fishes were known to be facultative marine wanderers and there were both amphidromous and anadromous species of a probable marine ancestry that reproduce in freshwaters. From phylogenetic studies, he concluded that each of the various forms of diadromy clearly has multiple origins throughout the diversity of fishes, and the migratory patterns that occur have taxonomic group-specific idiosyncrasies. In 2008, the same author questioned the explanation of overrepresentation of anadromy in northern high latitudes by a high primary productivity, and proposed biogeography history with a post-Pleistocene invasion of far northern new suitable freshwater habitats and species' thermal preference to explain the northern distribution of anadromous fish (McDowall 2008). For some species (e.g., Clupeiformes) predation, competition and even geological history may be at least as important as productivity (Bloom and Lovejoy 2014). A common origin among anadromous fish has not been demonstrated, and we can assume that anadromy corresponds to an adaptation of a species in its evolutionary history, in a particular environment and it was independent of a marine or a freshwater origin. Then, anadromy can be seen as a life history tactic.

Within the same population different tactics may occur with strictly migratory individuals (growth at sea) and resident individuals (growth in freshwater), a phenomenon that has been called as partial migration (Chapman et al. 2012) or contingent behaviors (Zlokovitz et al. 2003; Kerr and Secor 2010). In salmonids, within the same population, there are a gradient of tactics from residency to anadromy, a topic that has been widely discussed (Jonsson and Jonsson 1993; McDowall 1988; Quinn and Myers 2004; Cucherousset et al. 2005; Dodson et al. 2013), especially within the frame of species evolution (Hendry et al. 2004). Partial migration has been also described in other orders, yet not extensively, such as in Gasterosteiformes for the three-spined stickleback *Gasterosteus aculeatus* (Kitano et al. 2012), among Perciformes for White perch *Morone americana* (Kerr et al. 2009; Kerr and Secor 2012) or among Acipenceriformes for shortnose sturgeon *Acipenser brevirostrum* (Dionne et al. 2013). The existence of partial migrations have been recognized mainly due to studies using biotelemetry (Chapman et al. 2011; Cucherousset et al. 2005; Neuenfeld et al. 2007), or otolith microchemistry of trace element and stable isotopes (Honda et al. 2012; Kerr and Secor 2012). Quinn and Myers (2004) concluded, from a review on Pacific salmon and trout migratory behavior and relying on Rounsefell (1958) work, that "*anadromy is not a single trait with two conditions (anadromous or non-anadromous)… it reflects a suite of life history traits that are expressed as points along continua for each species and population*".

When do they migrate?

For each anadromous species, migrations occur at a precise timing in the ontogeny (Fig. 5.4). It is a synchronous movement among the same population. Moreover larvae or juveniles leave freshwater at a specific period of the year. For some species,

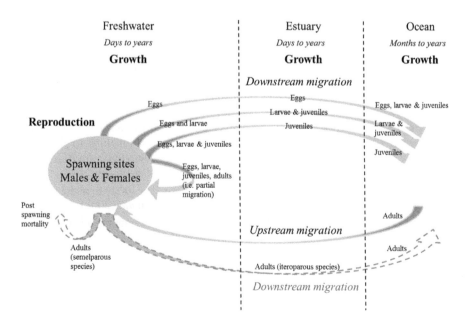

Figure 5.4. A general representation of the biological cycle of anadromous species.

juveniles spent some time in the estuary before reaching the sea, ranging from 11 days for allis shad *Alosa alosa* (Lochet et al. 2009) from two–seven years for the European sturgeon *Acipenser sturio* (Acolas et al. 2011a). Juveniles can spend between a few months and a few years at sea and then return to freshwater to spawn at a specific age and size (Fig. 5.4). Depending on the species, the freshwater stay can be highly variable for the young stages, for example sea lamprey *Petromyzon marinus* larvae can hide in the gravel for years (three to eight years) (Beamish and Potter 1975; Quintella et al. 2003), while the eggs and larvae of striped bass *Morone saxatilis* drift rapidly towards the sea (Dunning et al. 2009). In other cases, as for salmonids, the alevins emerge from gravel and juveniles can spend between one and two years in freshwater, and then they progressively colonize riverine habitats before reaching the sea (Baglinière and Maisse 1999). However, in some populations, where partial migration occurs, some individuals can perform all their biological cycle in freshwater and establish a resident population (Fig. 5.4) (Klemetsen et al. 2003).

Migration corresponds to a phenological event. In regard to the timing of migrations onset, it might vary along a latitude gradient for the same species, mostly due to temperature differences (Peer and Miller 2013; Otero et al. 2014). For example, the spawning migration of allis shad populations along the North East Atlantic coast starts in December for the Moroccan populations and in August for a few French populations. For all populations, the temperature ranges between 8°C and 13°C at the onset of the migration and between 16°C and 22°C at the end of the period. Other factors, such as tide intensity and river discharge also trigger adult migration (Mennesson-Boisneau et al. 2000). Besides the age and size at maturity of allis shad, populations are influenced by latitudinal gradient, being higher towards lower latitudes (Lassalle et al. 2008).

How do they migrate?

As explained above, whatever the life stage, downstream and upstream migrations can be triggered by exogenous (e.g., water discharge, temperature, photoperiod) (e.g., Yako et al. 2002) and endogenous factors (e.g., genetic, physiology) (e.g., Dodson et al. 2013). The interactions between these factors have been demonstrated for some species, mainly for salmonids (Jonsson and Jonsson 1993; Thorpe et al. 1998), and the mechanisms underlying migration are complex and probably species-specific (Glebe and Leggett 1981). Anadromy involves long distance migrations that implies high energy expenditure and requires great behavioral and physiological performances (Cooke et al. 2011). Some species perform long migrations, for example, the beluga sturgeon *Huso huso* commonly carries out a run of more than 1000 km in the Danube and Volga Basins (Vecsei et al. 2002), the American shad *Alosa sapidissima* can travel 2000 km (Maltais et al. 2010), and Pacific salmons migrate between 3700 and 7400 km at sea (Royce 1968). On the contrary, some species perform very short migration, for example the delta smelt *Hypomesus transpacificus* migrates less than 100 km (Sommer et al. 2011).

The cues and strategies used during migration, as well as the morphological and physiological adaptations used to enhance the migration performances of anadromous species also change through ontogeny, and are succinctly described next for each development stage, namely for larvae, juveniles and adults.

The downstream movement of eggs and larvae in fluvial habitats is linked more to dispersal rules than to a specific orientation mechanism, so the term 'larval drift' is often used (Bell 2009; Borcherding et al. 2014). The water current is mainly used to disperse, and the dispersal-timing is linked to the larvae's development characteristics. For example, salmonids eggs are buried in the gravel to avoid dispersal before reaching a development stage that allows them to counteract currents and choose a suitable habitat (Heggenes and Traaen 1988; Bardonnet et al. 2006). In the case of Acipenseridae, their eggs are dense and become rapidly adhesive to cause rapid sinking into the bottom without drifting (Jatteau 1998), while sea lampreys lay their eggs in a nest and ammocoetes larvae hatch a few weeks later, emerging from the nest, and drift downstream (Almeida and Quintella 2013). These ammocoetes larvae, which are filter-feeders, have the morphological adaptability to bury in the sediment where they can stay for several years before metamorphosis and initiate the downstream migration (Almeida and Quintella 2013). Some species' larvae, at an early development stage, are able to withstand the physiological challenges of migration towards the sea: in the case of the Baltic Sea whitefish *Coregonus lavaretus*, this is due to an early physiological adaption to salt water (Jokikokko et al. 2012), similar to pink salmon *Oncorhynchus gorbuscha*, that can enter sea water soon after emergence (Sackville et al. 2012). However, for other anadromous species' larvae, natural mortality increases during early downstream migration caused by a river flow increase, as seen in striped bass *Morone saxatilis* (Dunning et al. 2009). In Alosinae, downstream migration towards the sea can occur in shoals before their first winter (Aprahamian et al. 2002), a behavioral strategy usually used by pelagic species at sea to decrease predation risk (Grobis et al. 2013).

The migration of juveniles from freshwater towards the sea occurs when fish are physiologically ready to osmoregulate and support salt water (Hoar 1988). Osmoregulation has been studied extensively, especially in salmonids, which during the smoltification process face both morphological and behavioral changes (McCormick et al. 2013; Moore et al. 2013). The salmon parrs become smolt and their color changes, they lose their territorial behavior, and they also show negative rheotaxis and begin schooling (Moore et al. 2013). For other species, there are no morphological changes, but individuals have to reach a certain size and/or age to be able to physiologically support salt water (Allen et al. 2011). In addition, exogenous stimuli (e.g., temperature, water velocity, photoperiod) are also needed to trigger the downstream movement to provide juveniles with the best environmental conditions possible (Yako et al. 2002; Iafrate and Oliveira 2008).

In some families, a metamorphosis happens before juveniles initiate migration, such as for lampreys (Almeida and Quintella 2013; Silva et al. 2013). When sea lampreys become juveniles, they acquire a similar shape as the adult after metamorphosis, which constitutes a physiological and morphological preparation for their journey towards the sea (Reis-Santos et al. 2008). The juveniles are named macrophtalmia and this parasitic phase lasts up to two years (Silva et al. 2013). Then, juveniles can migrate downstream, triggered by an increase of river discharge, but they also have the possibility to hook onto other species to feed and to migrate at the same time, which enables them to save energy (Silva et al. 2013).

The downstream migration of juveniles can also have a diel rhythm for some species. For example, the Atlantic salmon *Salmo salar* presents mainly a nocturnal downstream migration pattern both in freshwater and in estuarine environments (Moore et al. 1995). In the case of sea lamprey *Petromyzon marinus*, macrophtalmia migrate downstream during the night and burrow during daylight (Almeida and Quintella 2013). These diel rhythm behaviors, migration during the night and rest during the day, are probably a strategy to decrease predation risk (Reebs 2002).

The time spent by juveniles in the estuary is also species-specific. For sturgeons, it can last several years with some movements between the sea and the estuary (Dadswell 2006; Rochard et al. 2001). Among salmonids, the estuary might just be a transition habitat between freshwater and saltwater for some species (e.g., steelhead trout *Oncorhynchus mykiss* spend less than one day in the estuary (Romer et al. 2013)), while other species actively use the estuary as a feeding area for a few months (e.g., one to six months for sub-yearling ocean-type chum salmon *Oncorhynchus keta* (Weitkamp et al. 2014)). Besides, in the estuary, juveniles might use the ebb tide during migration, as a strategy to limit energy expenditure (Moore et al. 1995; Taverny et al. 2002).

Both for juveniles and adults, the mechanisms of orientation are poorly known (Blinder et al. 2011), however migratory fish are suspected to use a series of orientation mechanisms to find suitable habitats, namely solar cues (Hasler et al. 1958), water currents (Thomson et al. 1994), olfactory cues (Dittman and Quinn 1996) or a geomagnetic orientation thanks to magnetite particles associated with the lateral line (Moore et al. 1990). At sea, some authors propose that migratory fish use infrasound patterns thanks to the sensitivity of the otolith organ (Sand and Karlsen 2000), and differences in the temperature and salinity of oceanic currents are also mentioned because it could create a vertical map with a unique odor which would lead individuals to their natal stream (Blinder et al. 2011). Homing, i.e., the return of the adult to their natal river for spawning, is supposed to allow fish to spawn in areas that have recently provided successful spawning habitat (Cury 1994) and permits the evolution of local adaptations (Taylor 1991). It is suspected to be a common life history trait among anadromous fish, leading to the suggestion that anadromy and homing may have co-evolved (McDowall 2001). For salmonids, homing of both sexes is particularly well known. They might use olfactory cues to home in their natal streams with high accuracy (Dittman and Quinn 1996; Ueda 2012), or even geomagnetic cues (Putman et al. 2013). Generally, these imprinting mechanisms occur during early life stages, and this information of their natal stream allows them to return as adults. For salmonids, this imprinting is supposed to take place during the parr/smolt transformation when plasma thyroxine level increase, which could favor the olfactory-imprinting (Dittman et al. 1996). For other anadromous species homing studies are scarce, but there are some studies on shads (Dodson and Leggett 1974; Waters et al. 2000; Walther et al. 2008) and sturgeons (Waldman et al. 2002) which found that imprinting and recognition of the natal stream might occur. This 'natal homing' ability is not the rule and straying might occur in a population, which can be viewed, if it concerns a small proportion of the population, as an opportunity to colonize unexploited habitats and to increase genetic resilience (Keefer and Caudill 2014). Lampreys may also be an exception to the 'rule of natal homing' in anadromous fishes (Waldman et al. 2008), since they use olfactory cues to find a suitable spawning habitat but without needing to return to the natal stream. For sea lamprey, the selection of a spawning stream is mediated by the release of a pheromone by stream-resident larvae which functions as an instinctively recognized indicator of habitat suitability (Sorensen et al. 2003).

Adult spawners need to have high swimming performances to reach their spawning site, since they swim against the current (Hinch and Rand 2000; Makiguchi et al. 2007). They must undergo, once again, through physiological modifications (e.g., osmoregulation, hormonal level) to prepare for their spawning migration, which they do without feeding and thus depleting their energy reserves (Kacem et al. 2013). Morphological modifications associated with sexual dimorphism might also occur during the spawning migration for some species: for example, the lower jaw of male salmons increases because it is used for fighting (Blair et al. 1993; Kinnison et al. 2003). Competition between males for access to nests and females has been observed between large anadromous Atlantic salmon males (combatants) and small resident males, which mature as parr (sneakers), and depending on the winner, it leads to small but significant differences on offspring quality, such as physiological performances linked to muscle metabolic capacities and development (Morasse et al. 2008). Moreover, male salmonids might also display an agonistic behavior after eggs are laid in the nest to protect them from cannibalism, as observed for the brown trout *Salmo trutta* (Tentelier et al. 2011).

Spawners also need specific environmental conditions to reproduce effectively, and water temperature and river discharge are particularly important for successful upstream migration (Acolas et al. 2006; Yi et al. 2010). For some species, such as allis shad, the upstream reproductive migration can even be stopped if the environmental conditions are not suitable: the upstream migration is stopped below 10–11°C (Acolas et al. 2006; Mennesson-Boisneau et al. 2000) and the reproduction activity itself is inhibited for temperatures below 14°C at the latitude of Brittany (France) (Acolas et al. 2006); yet these thresholds may vary with latitude (Cassou-Leins et al. 2000). For species that built a nest, as lampreys and salmonids, the spawners select the most appropriate substrates for their progeny (Gardner et al. 2012). Lampreys grab and move the cobbles with their mouth to dispose them adequately to protect the eggs, while female salmonids dig and cover the nest with their tail to protect the clutch (Tsuda et al. 2006). Finally, as for many fish species, once the spawning sites are reached (Fig. 5.4), some anadromous species will die after their first and unique reproduction (semelparous species), while others will be able to carry out several reproductions in their life (iteroparous species) (Crespi and Teo 2002; Hasselman et al. 2013).

Insights on anadromous species conservation

Why anadromous species are in peril?

Anadromous fish travel between rivers and the ocean, through transitional water areas of different size, so they are vulnerable to many threats associated to these habitats. The main threats on habitat that can affect migratory fish are:

- Habitat fragmentation and modification of river's hydraulic regime (Rolls et al. 2014). For anadromous fish, the presence of dams hinders the access of adults to spawning grounds (Zhou et al. 2014), and it increases the mortality risk of juveniles during downstream migration through entrainment against water intake screens of hydro-electrical power plants (Williams et al. 2001; Keefer et al. 2012). The large number of dams constructed in rivers, mainly since the mid-20th century, has led to river habitat loss and modification of population dynamics for many anadromous species (Locke et al. 2003; Morita et al. 2009; Hall et al. 2011; Junge et al. 2014). For example, for American shad, about 4000 km of an original 11200 km of spawning habitat have been lost due to dams (Limburg et al. 2003). To remediate these problems, fish ladders have been built and can be successful for some species, such as salmonids (Bangsgaard et al. 2014), but they are not adapted to all species, such as shads (Brown et al. 2013). For species that need to access spawning grounds situated upstream a river course, the succession of dams, even if equipped with a fish ladder, delay reproduction or force the use of downstream spawning grounds that are not fully suitable for reproduction success (Acolas et al. 2006; Gao et al. 2014). Associated with hydraulic regime alteration, temperature may rise with the succession of dams, or with power plant implantation, and thus affecting the survival of cold-water related species, such as salmonids (Horne et al. 2014). Moreover, the importance of connectivity is not restricted to rivers, but also between estuarine and coastal habitats for species that actively use the estuarine environment (Ray 2005).

- Habitat degradation is often linked to the extraction of substrate in river and estuarine beds, which will remove suitable substrates either for spawning, or if they act as nursery and growth habitats (De Groot 2002; Gessner and Jaric 2014). Conversely, intensive agriculture and cattle grazing lead to soil erosion and increase the input of fine sediment into rivers which can alter the habitat's functionality (e.g., oxygen depletion, temperature modification), especially for the early life stages that live under gravel, as the salmonids (Sternecker et al. 2014).

Threats can also be directed on the species itself when the species is overexploited or extensively used for aquaculture purposes. Overfishing can occur in all environments (rivers, estuaries, oceans) and targeting both juveniles or adults, either for meat or egg (i.e., caviar) consumption. Overfishing leads to the extirpation of populations with high market value, as sturgeons (Rochard et al. 1990; Krykhtin and Svirskii 1997) or salmonids (Rand et al. 2012). For some species, the expansion of intensive aquaculture might promote the risk of gene introgression and competition between wild and domesticated fish, as well as disease transmission to wild populations (Naylor et al. 2005; Fisher et al. 2014). Alternatively, other sources of peril can affect anadromous fish population health, like pollutants (Johnson et al. 2013), invasive species (Vignon and Sasal 2010) and climate change (Lassalle and Rochard 2009). Unfortunately, most of these threats occur in tandem, either on the same life stage or along the migration journey (Edge and Gilhen 2001).

Endangered species and conservation projects

The conservation status of the 175 anadromous fish species is only known for 48% of them (extraction of the data with R interface Reol (Banbury and O'Meara 2014)), of which, 30.9% are listed as extinct or threatened (critically endangered, endangered or vulnerable) (Table 5.1). Two species are classified as extinct, one Salmoniforme and one Retropinniforme, and fourteen species are critically endangered, thirteen Acipenseriforme and one Salmoniforme. Two species are endangered, one Clupeiforme and one Osmeriforme. Eight species are considered vulnerable, four Salmoniforme, two Clupeiforme, one Acipenseriforme and one Salangiforme. These statuses are regularly updated, but some of them have evolved rapidly, such as the supposed extinct Salmoniforme *Coregonus oxyrinchus* that benefited from a successful reintroduction program in the Rhine (Borcherding et al. 2010).

In the study of 22 anadromous species of the North Atlantic Basin, Limburg and Waldmann (2009) highlighted that all species had suffered population extirpation since the 19th century. According to their study, and thanks to a 35 years time series analysis, the relative abundance dropped over 90%, except for striped bass, the northern population of Atlantic salmon and the Icelandic population of sea-run brown trout. For most of those 22 species, moderate to sharp declines occurred in the 1990's followed by low harvests or cessation fishing. For allis Shad, population extirpation in the southern part of its repartition range (Morocco) occurred in 1992 (Sabatié and Baglinière 2001), but it had already disappeared from large European rivers in the mid-20th century, such as in the Seine (Spillman 1961) and Rhine rivers (Degroot 1990). More recently a drastic decline of the population was observed in the Gironde basin, which lead to a fishing moratorium in 2008 (Rougier et al. 2012). In the Rhine River, the eight anadromous species once present are either extirpated or under threat (Degroot 2002; Freyhof and Schöter 2005).

Facing these threats and population declines, government and stakeholders have tackled the issue by different means. From habitat point of view, partial connectivity has been restored in a few rivers by selected dam removal which can lead to successful recolonization of anadromous species (Hogg et al. 2013), but dam destruction has high financial costs (Null et al. 2014). At a smaller scale, technological alternatives can be developed to limit juvenile mortality in turbines (Brown et al. 2012) and improve upstream fish passage (Muir and Williams 2012). When specific habitats have been

Table 5.1. 2012 IUCN status for anadromous species extracted with R interface REOL (Banbury and O'Meara 2014).

		Orders	Species
Known status	48.0 %		
Extinct or threatened among known status	30.9%		
Extinct (EX)	2	Salmoniformes Retropinniformes	*Coregonus oxyrinchus* (Linnaeus, 1758) *Prototroctes oxyrhynchus* (Günther, 1870)
Critically Endangered (CR)	14	Acipenseriformes	*Acipenser stellatus* (Pallas, 1771) *Acipenser sturio* (Linnaeus, 1758) *Acipenser naccarii* (Bonaparte, 1836) *Acipenser nudiventris* (Lovetsky, 1828) *Acipenser gueldenstaedtii* (Brandt and Ratzeburg, 1833) *Acipenser schrenckii* (Brandt, 1869) *Acipenser dabryanus* (Duméril, 1869) *Acipenser sinensis* (Gray, 1835) *Acipenser mikadoi* (Hilgendorf, 1892) *Acipenser persicus* (Borodin, 1897) *Huso dauricus* (Georgi, 1775) *Huso huso* (Linnaeus, 1758) *Scaphirhynchus suttkusi* (Williams and Clemmer, 1991)
		Salmoniformes	*Hucho perryi* (Brevoort, 1856)
Endangered (EN)	2	Clupeiformes Osmeriformes	*Alosa volgensis* (Berg, 1913) *Hypomesus transpacificus* (McAllister, 1963)
Vulnerable (VU)	8	Clupeiformes	*Alosa aestivalis* (Mitchill, 1814) *Alosa immaculata* (Bennett, 1835)
		Acipenseriformes Salmoniformes	*Acipenser brevirostrum* (Lesueur, 1818) *Acipenser oxyrinchus desotoi* (Vladykov, 1955) *Coregonus lavaretus* (Linnaeus, 1758) *Coregonus huntsmani* (Scott, 1987) *Salvelinus confluentus* (Suckley, 1859)
		Salangiformes	*Neosalanx reganius* (Wakiya and Takahashi, 1937)
Near Threatened (NT)	5	Acipenseriformes	*Acipenser medirostris* (Ayres, 1854) *Acipenser oxyrinchus oxyrinchus* (Mitchill, 1815)
		Cithariniformes Petromyzontiformes Retropinniformes	*Citharinus eburneensis* (Daget, 1962) *Caspiomyzon wagneri* (Kessler, 1870) *Prototroctes maraena* (Günther, 1864)
Least Concern (LC)	51	Acipenseriformes Ariiformes Bagriformes Claroteiformes Clupeiformes	Acipenser transmontanus (Richardson, 1836) *Arius madagascariensis* (Vaillant, 1894) *Mystus gulio* (Hamilton, 1822) *Clarotes laticeps* (Rüppell, 1829) *Alosa alosa* (Linnaeus, 1758) *Alosa fallax* (Lacepède, 1803) *Alosa kessleri* (Grimm, 1887) *Alosa mediocris* (Mitchill, 1814) *Alosa pseudoharengus* (Wilson, 1811) *Alosa sapidissima* (Wilson, 1811) *Alosa tanaica* (Grimm, 1901) *Anodontostoma thailandiae* (Wongratana, 1983)

Table 5.1. contd....

Table 5.1. contd....

	Orders	Species
Least Concern (LC)	51	*Clupeonella cultriventris* (Nordmann, 1840)
		Dorosoma cepedianum (Lesueur, 1818)
		Dorosoma petenense (Günther, 1867)
		Nematalosa galatheae (Nelson and Rothman, 1973)
		Nematalosa nasus (Bloch, 1795)
		Pellonula vorax (Günther, 1868)
	Cypriniformes	*Pelecus cultratus* (Linnaeus, 1758)
		Rutilus frisii (Nordmann, 1840)
		Vimba vimba (Linnaeus, 1758)
	Gadiformes	*Microgadus tomcod* (Walbaum, 1792)
	Gasterosteiformes	*Pungitius pungitius* (Linnaeus, 1758)
	Moroniformes	*Morone americana* (Gmelin, 1789)
		Morone saxatilis (Walbaum, 1792)
	Mugiliformes	*Rhinomugil corsula* (Hamilton, 1822)
	Ophichthiformes	*Pisodonophis boro* (Hamilton, 1822)
	Osmeriformes	*Hypomesus olidus* (Pallas, 1814)
		Osmerus eperlanus (Linnaeus, 1758)
		Osmerus mordax dentex (Steindachner and Kner, 1870)
		Spirinchus thaleichthys (Ayres, 1860)
		Thaleichthys pacificus (Richardson, 1836)
	Perciformes	*Perca fluviatilis* (Linnaeus, 1758)
	Petromyzontiformes	*Lampetra ayresii* (Günther, 1870)
		Lampetra fluviatilis (Linnaeus, 1758)
		Petromyzon marinus (Linnaeus, 1758)
		Lethenteron camtschaticum (Tilesius, 1811)
		Lethenteron reissneri (Dybowski, 1869)
	Salmoniformes	*Coregonus albula* (Linnaeus, 1758)
		Coregonus autumnalis (Pallas, 1776)
		Coregonus muksun (Pallas, 1814)
		Coregonus nasus (Pallas, 1776)
		Coregonus pallasii (Valenciennes, 1848)
		Coregonus peled (Gmelin, 1789)
		Coregonus pidschian (Gmelin, 1789)
		Coregonus sardinella (Valenciennes, 1848)
		Oncorhynchus nerka (Walbaum, 1792)
		Salmo labrax (Pallas, 1814)
		Salmo marmoratus (Cuvier, 1829)
		Stenodus leucichthys (Güldenstädt, 1772)
	Tetraodontiformes	*Takifugu obscurus* (Abe, 1949)
Lower Risk/ Least concern (LR/LC)	2	Salmoniformes	*Coregonus artedi* (Lesueur, 1818)
		Salmo salar (Linnaeus, 1758)
Data Deficient (DD)	7		...
Not evaluated (NE)	84		...
Total	175		

altered, such as spawning grounds, the reconstruction of habitats by means of gravel addition can be a solution (Barlaup et al. 2008), but at great expenses (Zeug et al. 2014). These restoration projects are usually accompanied by specific habitat protection measures, as well as by the implementation of new national policies on substrate extraction in rivers or estuaries. From a general point of view, anadromous fish can benefit from water quality amelioration in some rivers, as in the Seine (Belliard et al. 2009; Perrier et al. 2010) and Rhine rivers (Plum and Schulte-Wulwer-Leidig 2014).

For species targeted by fishing exploitation, a fishing ban can be implemented, as for European sturgeon in France in 1982, and then followed by an European ban in 1998. However, these measures are often considered to be implemented too late, despite the early alert calls dating back to the 1920's (Roule 1922). However in some cases, a fishing moratorium can thwart the development of water resources exploitation projects (Olney et al. 2008). When a species is at a low level of abundance, then several policies can be implemented upon fisheries to control or limit its impact, as fishing quotas, definition of fisheries season and/or net mesh size (Navodaru et al. 2001). However, identifying the stocks characteristics is still needed for many populations to ensure sustainable fisheries policies (McBride 2014).

In a situation of abundance decrease, stocking at a different age is a tool that has been widely used to sustain fisheries, maintain a population for conservation purposes, and introduce a species in a novel environment for conservation or fisheries purposes. This tool has been used extensively for salmonids with positive and negative results according to the sites (Prignon et al. 1999; Aprahamian et al. 2003; Ayllon et al. 2006; Perrier et al. 2013), and in a lower proportion for other endangered anadromous species, such as shads (Frank et al. 2011; Hasselman and Limburg 2012), sturgeons (St. Pierre 1999; Zhu et al. 2006), North Sea houting (Borcherding et al. 2006) and striped bass (Secor and Houde 1998), but with various rates of success (Ruzzante et al. 2004; Araki and Schmid 2010). For a few anadromous species, specific restoration plans with multiple action level plans (e.g., habitat protection, habitat restoration, stocking, monitoring) have been carried out thanks to the involvement of local populations, non-governmental organizations, advice of scientists and new governmental policies (Borcherding et al. 2010; Brown and St. Pierre 2001; De Groot 2002). For example, European sturgeon is classified as critically endangered by IUCN, and benefited from a European restoration plan (Rosenthal et al. 2007) declined in national action plans in France and Germany (Gessner et al. 2010; Ministère de l'écologie du développement durable des transports et du logement 2011) and relying on specific scientific programs (Acolas et al. 2011b).

Over the last two decades, publications about anadromous fish conservation and restoration have increased by three and five times, respectively (Fig. 5.5). The prediction for the actual decade (since 2010) is to double the publications focusing on conservation and to increase by 1.5 the works on restoration. These observations highlighted the increased interest of scientists and stakeholders, which financed research in the preservation and recovery of these anadromous species. However, the main conservation projects seem to happen in the Northern hemisphere, while scanty data is available from the Southern hemisphere.

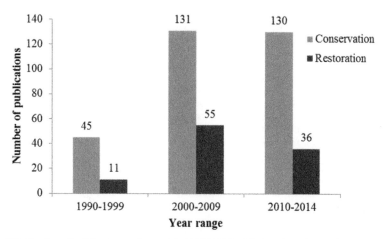

Figure 5.5. Number of publications referenced in WOS (Web of Science) dealing with conservation or restoration of anadromous fish since 1990.
Detailed queries:
Queries for conservation: conservation and anadrom* not restoration within each year range.
Queries for restoration: restoration and anadrom* not conservation within each year range.
Date of the queries: 23 October 2014.

References

Acolas ML, Veron V, Jourdan H, Begout ML, Sabatie MR and Bagliniere JL (2006) Upstream migration and reproductive patterns of a population of allis shad in a small river (L'Aulne, Brittany, France). ICES Journal of Marine Science 63: 476–484.

Acolas ML, Castelnaud G, Lepage M and Rochard E (2011a) Biological cycle and migrations. pp. 147–152. *In*: Williot P, Rochard E, Desse-Berset N, Kirschbaum F and Gessner J (eds.). Biology and Conservation of the Atlantic European Sturgeon *Acipenser sturio* L., 1758. Springer, Berlin Heidelberg.

Acolas ML, Gessner J and Rochard E (2011b) Population conservation requires improved understanding of *in-situ* life histories. pp. 585–592. *In*: Williot P, Rochard E, Desse-Berset, N Kirschbaum F and Gessner J (eds.). Biology and Conservation of the Atlantic European Sturgeon *Acipenser sturio* L., 1758. Springer Berlin Heidelberg.

Acolas ML, Labonne J, Baglinière J and Roussel J (2012) The role of body size versus growth on the decision to migrate: a case study with *Salmo trutta*. Naturwissenschaften 99: 11–21.

Allen P, McEnroe M, Forostyan T, Cole S, Nicholl M, Hodge B and Cech J (2011) Ontogeny of salinity tolerance and evidence for seawater-entry preparation in juvenile green sturgeon, *Acipenser medirostris*. Journal of Comparative Physiology B: Biochemical, Systemic, and Environmental Physiology 181: 1045–1062.

Almeida PR and Quintella BR (2013) Sea lamprey migration: a millenial journey. pp. 105–131. *In*: Ueda H and Tsukamoto K (eds.). Physiology and Ecology of Fish Migration. CRC Press, Boca Raton, FL, USA.

Amstutz U, Giger T, Champigneulle A, Day PJR and Largiader CR (2006) Distinct temporal patterns of transaldolase 1 gene expression in future migratory and sedentary brown trout (*Salmo trutta*). Aquaculture 260: 326–336.

Aprahamian MW, Bagliniere JL, Sabatié MR, Alexandrino P and Aprahamian CD (2002) Synopsis of biological data on *Alosa alosa* and *Alosa fallax* spp., 346. Warrington: Environment Agency.

Aprahamian MW, Martin Smith K, McGinnity P, McKelvey S and Taylor J (2003) Restocking of salmonids-opportunities and limitations. Fisheries Research 62: 211–227.

Araki H and Schmid C (2010) Is hatchery stocking a help or harm? Evidence, limitations and future directions in ecological and genetic surveys. Aquaculture 308: S2–S11.

Ayllon F, Moran P and Garcia-Vazquez E (2006) Maintenance of a small anadromous subpopulation of brown trout (*Salmo trutta* L.) by straying. Freshwater Biology 51: 351–358.

Baglinière JJ and Maisse G (1999) Biology and Ecology of the Brown and Sea Trout, Jointly published with Praxis Publishing, UK. Originally published by INRA, Paris, 1991, VII, 286pp.

Banbury BL and O'Meara BC (2014) Reol: R interface to the Encyclopedia of Life. Ecology and Evolution 4: 2577–2583.

Bangsgaard L, Cording R and Kjeldsen JH (2014) Efforts to enhance anadromous brown trout on Funen, Denmark. Journal of Coastal Conservation 18: 89–95.

Bardonnet A, Poncin P and Roussel JM (2006) Brown trout fry move inshore at night: a choice of water depth or velocity? Ecology of Freshwater Fish 15: 309–314.

Barlaup BT, Gabrielsen SE, Skoglund H and Wiers T (2008) Addition of spawning gravel—A means to restore spawning habitat of Atlantic salmon (*Salmo salar* L.) and anadromous and resident brown trout (*Salmo trutta* L.) in regulated rivers. River Research and Applications 24: 543–550.

Beamish FWH and Potter IC (1975) The biology of the anadromous Sea lamprey (*Petromyzon marinus*) in New Brunswick. J. Zool. 177: 57–72.

Bell KNI (2009) What comes down must go up: the migration cycle of juvenile-return anadromous taxa. pp. 321–341. *In*: Haro A, Smith KL, Rulifson RA, Moffitt CM, Klauda RJ, Dadswell MJ, Cunjak RA, Cooper JE, Beal KL and Avery TS (eds.). Challenges for Diadromous Fishes in a Dynamic Global Environment. Amer Fisheries Soc., Bethesda.

Belliard J, Marchal J, Ditche JM, Tales E, Sabatie R and Bagliniere JL (2009) Return of adult anadromous allis shad (*Alosa alosa* L.) in the river Seine, France: a sign of river recovery? River Research and Applications 25: 788–794.

Blair GR, Rogers DE and Quinn TP (1993) Variation in life history characteristics and morphology of sockeye salmon in the Kvichak River System, Bristol Bay, Alaska. Transactions—American Fisheries Society 122: 550–559.

Blinder TR, Cooke SJ and Hinch SG (2011) The biology of fish migration. pp. 1921–1927. *In*: Farell AP (ed.). Encyclopedia of Fish Physiology: From Genome to Environment. Academic Press, San Diego.

Bloom DD and Lovejoy NR (2014) The evolutionary origins of diadromy inferred from a time-calibrated phylogeny for Clupeiformes (herring and allies). Proceedings of the Royal Society B: Biological Sciences, 281.

Bohlin T, Dellefors C and Faremo U (1996) Date of smolt migration depends on body-size but not age in wild sea-run brown trout. Journal of Fish Biology 49: 157–164.

Bohlin T, Pettersson J and Degerman E (2001) Population density of migratory and resident brown trout (*Salmo trutta*) in relation to altitude: evidence for a migration cost. Journal of Animal Ecology 70: 112–121.

Boisneau C, Moatar F, Bodin M and Boisneau P (2008) Does global warming impact on migration patterns and recruitment of Allis shad (*Alosa alosa* L.) young of the year in the Loire River, France? Hydrobiologia 602: 179–186.

Borcherding J, Scharbert A and Urbatzka R (2006) Timing of downstream migration and food uptake of juvenile North Sea houting stocked in the Lower Rhine and the Lippe (Germany). Journal of Fish Biology 68: 1271–1286.

Borcherding J, Heynen M, Jäger-Kleinicke T, Winter HV and Eckmann R (2010) Re-establishment of the North Sea houting in the River Rhine. Fisheries Management and Ecology 17: 291–293.

Borcherding J, Breukelaar AW, Winter HV and König U (2014) Spawning migration and larval drift of anadromous North Sea houting (*Coregonus oxyrinchus*) in the River IJssel, the Netherlands. Ecology of Freshwater Fish 23: 161–170.

Brown JJ and St Pierre RA (2001) Restoration of American shad *Alosa sapidissima* populations in the Susquehanna and Delaware rivers, USA. Washington: Marine Technology Soc. 321–326.

Brown JJ, Limburg KE, Waldman JR, Stephenson K, Glenn EP, Juanes F and Jordaan A (2013) Fish and hydropower on the US Atlantic coast: failed fisheries policies from half-way technologies. Conservation Letters 6: 280–286.

Brown RS, Carlson TJ, Gingerich AJ, Stephenson JR, Pflugrath BD, Welch AE, Langeslay MJ, Ahmann ML, Johnson RL, Skalski JR, Seaburg AG and Townsend RL (2012) Quantifying mortal injury of juvenile Chinook salmon exposed to simulated hydro-turbine passage. Transactions of the American Fisheries Society 141: 147–157.

Bruford MW (2006) Review and perspectives on molecular genetic approaches to sea trout biology. pp. 248–256. *In*: Harris G and Milner N (eds.). Sea Trout: Biology, Conservation and Management. Proceedings of the First International Sea Trout Symposium, Cardiff, July 2004. Blackwell Publishing, Oxford, UK.

Byrne CJ, Poole R, Dillane A, Rogan G and Whelan KF (2004) Temporal and environmental influences on the variation in sea trout (*Salmo trutta* L.) smolt migration in the Burrishoole system in the west of Ireland from 1971 to 2000. Fisheries Research 66: 85–94.

Cassou-Leins JJ, Cassou-Leins F, Boisneau P and Baglinière JL (2000) La reproduction. pp. 73–92. *In*: Baglinière JL and Elie P (eds.). Les Aloses (*Alosa alosa* et *Alosa fallax* spp.): Ecobiologie et variabilité des Populations, Inra-Cemagref, Paris.

Chapman BB, Hulthen K, Blomqvist DR, Hansson LA, Nilsson JA, Brodersen J, Nilsson PA, Skov C and Bronmark C (2011) To boldly go: Individual differences in boldness influence migratory tendency in a cyprinid fish. Ecology Letters 14: 871–876.

Chapman BB, Skov C, Hulthen K, Brodersen J, Nilsson PA, Hansson LA and Bronmark C (2012) Partial migration in fishes: definitions, methodologies and taxonomic distribution. J. Fish Biol. 81: 479–499.

Cooke S, Crossin GT and Hinch SG (2011) Pacific salmon migration: completing the cycle. pp. 1945–1952. *In*: Farrell AP (ed.). Encyclopedia of Fish Physiology: From Genome to Environment, Volume 3. Academic Press, San Diego.

Crespi BJ and Teo R (2002) Comparative phylogenetic analysis of the evolution of semelparity and life history in salmonid fishes. Evolution 56: 1008–1020.

Cucherousset J, Ombredane D, Charles K, Marchand F and Baglinière JL (2005) A continuum of life history tactics in a brown trout (*Salmo trutta*) population. Can. J. Fish. Aquat. Sci. 62: 1600–1610.

Cury P (1994) Obstinate nature: an ecology of individuals. Thoughts on reproductive behavior and biodiversity. Can. J. Fish. Aquat. Sci. 51: 1664–1673.

Dadswell MJ (2006) A review of the status of Atlantic sturgeon in Canada, with comparisons to populations in the United States and Europe. Fisheries 31: 218–229.

De Groot SJ (2002) A review of the past and present status of anadromous fish species in the Netherlands: is restocking the Rhine feasible? Hydrobiologia 478: 205–218.

Degroot SJ (1990) The former allis and twaite shad fisheries of the lower Rhine. The Netherlands Journal of Applied Ichthyology-Zeitschrift Fur Angewandte Ichthyologie 6: 252–256.

Dionne PE, Zydlewski GB, Kinnison MT, Zydlewski J and Wippelhauser GS (2013) Reconsidering residency: Characterization and conservation implications of complex migratory patterns of shortnose sturgeon (*Acipenser brevirostrum*). Canadian Journal of Fisheries and Aquatic Sciences 70: 119–127.

Dittman AH and Quinn TP (1996) Homing in Pacific salmon: mechanisms and ecological basis. The Journal of Experimental Biology 199: 83–91.

Dittman AH, Quinn TP and Nevitt GA (1996) Timing of imprinting to natural and artificial odors by coho salmon (*Oncorhynchus kisutch*). Canadian Journal of Fisheries and Aquatic Sciences 53: 434–442.

Dodson JJ and Leggett WC (1974) Role of olfaction and vision in the behavior of American shad (*Alosa alosa*) homing to the Connecticur river from Long islang Sound. J. Fish. Res. Board Can. 31: 1607–1619.

Dodson JJ, Aubin-Horth N, Thériault V and Páez DJ (2013) The evolutionary ecology of alternative migratory tactics in salmonid fishes. Biological Reviews 88: 602–625.

Dunning DJ, Ross QE, McKown KA and Socrates JB (2009) Effect of striped bass larvae transported from the Hudson river on juvenile abundance in western Long Island Sound. Marine and Coastal Fisheries 1: 343–353.

Edge TA and Gilhen J (2001) Updated status report on the endangered Atlantic Whitefish, *Coregonus huntsmani*. Canadian Field-Naturalist 115: 635–652.

Eschmeyer WN and Fong JD (2014) Species by family/subfamily (http://research.calacademy.org/research/ichthyology/catalog/SpeciesByFamily.asp).

Fisher AC, Volpe JP and Fisher JT (2014) Occupancy dynamics of escaped farmed Atlantic salmon in Canadian Pacific coastal salmon streams: implications for sustained invasions. Biological Invasions 16: 2137–2146.

Fontaine M (1975) Physiological mechanisms in the migration of marine and amphihalin fish. Advances in Marine Biology 13: 241–355.

Forseth T, Naesje TF, Jonsson B and Harsaker K (1999) Juvenile migration in brown trout: a consequence of energetic state. Journal of Animal Ecology 68: 783–793.

Frank HJ, Mather ME, Smith JM, Muth RM and Finn JT (2011) Role of origin and release location in pre-spawning distribution and movements of anadromous alewife. Fisheries Management and Ecology 18: 12–24.

Freyhof J and Schoter C (2005) The houting *Coregonus oxyrinchus* (L.) (Salmoniformes: Coregonidae), a globally extinct species from the North Sea basin. Journal of Fish Biology 67: 713–729.

Gao X, Lin PC, Li MZ, Duan ZH and Liu HZ (2014) Effects of Water Temperature and Discharge on Natural Reproduction Time of the Chinese Sturgeon, *Acipenser sinensis*, in the Yangtze River, China and Impacts of the Impoundment of the Three Gorges Reservoir. Zoological Science 31: 274–278.

Gardner C, Coghlan SM, Jr. and Zydlewski J (2012) Distribution and abundance of anadromous sea lamprey spawners in a fragmented stream: current status and potential range expansion following barrier removal. Northeastern Naturalist 19: 99–110.

Gessner J and Jaric I (2014) A life-stage population model of the European sturgeon (*Acipenser sturio* L., 1758) in the Elbe River. Part II: assessment of the historic population decline. Journal of Applied Ichthyology 30: 267–271.

Gessner J, Tauteman M, Von Nordheim H and Borchers T (2010) Plan national d'actions pour la protection et la conservation de l'Esturgeon européen (*Acipenser sturio*) en Allemagne. Ministère fédéral de l'environnement de la protection de la nature et de la sureté nucléaire. 83p.

Glebe BD and Leggett WC (1981) Latitudinal differences in energy allocation and use during the freshwater migrations of American shad (*Alosa sapidissima*) and their life history consequences. Canadian Journal of Fisheries and Aquatic Sciences 38(7): 806–820.

Grobis MM, Pearish SP and Bell AM (2013) Avoidance or escape? Discriminating between two hypotheses for the function of schooling in three spine sticklebacks. Animal Behaviour 85: 187–194.

Gross MR (1987) Evolution of diadromy in fishes. American Fisheries Society Symposium 1: 14–25.

Gross MR, Coleman RM and McDowall RM (1988) Aquatic productivity and the evolution of diadromous fish migration. Science 239: 1291–1293.

Hall CJ, Jordaan A and Frisk MG (2011) The historic influence of dams on diadromous fish habitat with a focus on river herring and hydrologic longitudinal connectivity. Landscape Ecology 26: 95–107.

Hasler AD, Horrall RM, Wisby WJ and Braemer W (1958) Sun orientation and homing in fishes. Limnology and Oceanography 3: 355–361.

Hasselman DJ and Limburg KE (2012) Alosine restoration in the 21st century: challenging the status quo. Marine and Coastal Fisheries 4: 174–187.

Hasselman DJ, Ricard D and Bentzen P (2013) Genetic diversity and differentiation in a wide ranging anadromous fish, American shad (*Alosa sapidissima*), is correlated with latitude. Molecular Ecology 22: 1558–1573.

Heggenes J and Traaen T (1988) Downstream migration and critical water velocities in stream channels for fry of 4 salmonid species Journal of Fish Biology 32: 717–727.

Hendry AP and Stearns SC (2004) Evolution Illuminated, Salmon and their Relatives. University Press, Oxford. 510p.

Hendry AP, Bohlin T, Jonsson B and Berg OK (2004) To sea or not to sea. pp. 92–125. *In*: Hendry Andrew P and Stearns Stephen C (eds.). Evolution Illuminated. Salmon and their Relatives. Oxford University Press, New York.

Hinch SG and Rand PS (2000) Optimal swimming speeds and forward-assisted propulsion: energy-conserving behaviours of upriver-migrating adult salmon. Canadian Journal of Fisheries and Aquatic Sciences 57: 2470–2478.

Hoar WS (1988) The physiology of smolting salmonids. pp. 275–343. *In*: Hoar WS and Randall DJ (eds.). Fish Physiology, Vol. XI. Part B. Academic Press, San Diego, CA.

Hogg R, Coghlan SM and Zydlewski J (2013) Anadromous sea lampreys recolonize a maine coastal river tributary after dam removal. Transactions of the American Fisheries Society 142: 1381–1394.

Honda K, Arai T, Kobayashi S, Tsuda Y and Miyashita K (2012) Migratory patterns of exotic brown trout *Salmo trutta* in south-western Hokkaido, Japan, on the basis of otolith Sr:Ca ratios and acoustic telemetry. Journal of Fish Biology 80: 408–426.

Horne BD, Rutherford ES and Wehrly KE (2004) Simulating effects of hydro-dam alteration on thermal regime and wild steelhead recruitment in a stable-flow Lake Michigan tributary. River Research and Applications 20: 185–203.

Iafrate J and Oliveira K (2008) Factors affecting migration patterns of juvenile river herring in a coastal Massachusetts stream. Environmental Biology of Fishes 81: 101–110.

Jatteau P (1998) Etude bibliographique des principales caractéristiques de l'écologie des larves d'Acipenséridés. Bulletin Français de la Pêche et de la Pisciculture 350-51: 445–464.

Johnson L, Anulacion B, Arkoosh M, Olson OP, Sloan C, Sol SY, Spromberg J, Teel DJ, Yanagida G and Ylitalo G (2013) Persistent organic pollutants in juvenile Chinook salmon in the Columbia River Basin: implications for stock recovery. Transactions of the American Fisheries Society 142: 21–40.

Jokikokko E, Huhmarniemi A, Leskelä A and Vähä V (2012) Migration to the sea of river spawning whitefish (*Coregonus lavaretus* L.) fry in the northern baltic sea. Journal of Fish Biology 117–125.

Jonsson B and Jonsson N (1993) Partial migration—niche shift versus sexual-maturation in fishes. Reviews in Fish Biology and Fisheries 3: 348–365.

Jonsson B, Jonsson N, Brodtkorb E and Ingebrigtsen PJ (2001) Life-history traits of Brown Trout vary with the size of small streams. Functional Ecology 15: 310–317.

Junge C, Museth J, Hindar K, Kraabol M and Vollestad LA (2014) Assessing the consequences of habitat fragmentation for two migratory salmonid fishes. Aquatic Conservation-Marine and Freshwater Ecosystems 24: 297–311.

Kacem A, Baglinière JL and Meunier FJ (2013) Resorption of scales in Atlantic salmon (*Salmo salar*) during its anadromous migration: A quantitative study. Cybium 37: 199–206.

Keefer ML and Caudill CC (2014) Homing and straying by anadromous salmonids: a review of mechanisms and rates. Reviews in Fish Biology and Fisheries 24: 333–368.

Keefer ML, Taylor GA, Garletts DF, Helms CK, Gauthier GA, Pierce TM and Caudill CC (2012) Reservoir entrapment and dam passage mortality of juvenile Chinook salmon in the Middle Fork Willamette River. Ecology of Freshwater Fish. 21: 222–234.

Kerr LA and Secor DH (2010) Latent effects of early life history on partial migration for an estuarine-dependent fish. Environmental Biology of Fishes 89: 479–492.

Kerr LA and Secor DH (2012) Partial migration across populations of white perch (*Morone americana*): a flexible life history strategy in a variable estuarine environment. Estuaries and Coasts 35: 227–236.

Kerr LA, Secor DH and Piccoli PM (2009) Partial migration of fishes as exemplified by the estuarine-dependent white perch. Fisheries 34: 114–123.

Kinnison MT, Unwin MJ, Hendry AP and Quinn TP (2001) Migratory costs and the evolution of egg size and number in introduced and indigenous salmon populations. Evolution 55: 1656–1667.

Kinnison MT, Unwin MJ and Quinn TP (2003) Migratory costs and contemporary evolution of reproductive allocation in male chinook salmon. Journal of Evolutionary Biology 16: 1257–1269.

Kitano J, Ishikawa A, Kume M and Mori S (2012) Physiological and genetic basis for variation in migratory behavior in the three-spined stickleback, *Gasterosteus aculeatus*. Ichthyological Research 59: 293–303.

Klemetsen A, Amundsen PA, Dempson JB, Jonsson B, Jonsson N, O'Connell MF and Mortensen E (2003) Atlantic salmon *Salmo salar* L., brown trout *Salmo trutta* L. and Arctic charr *Salvelinus alpinus* (L.): a review of aspects of their life histories. Ecology of Freshwater Fish 12: 1–59.

Krykhtin ML and Svirskii VG (1997) Endemic sturgeons of the Amor River: Kaluga, *Huso dauricus*, and Amur sturgeon, *Acipenser schrenckii*. Environmental Biology of Fishes 48: 231–240.

Lassalle G and Rochard E (2009) Impact of twenty-first century climate change on diadromous fish spread over Europe, North Africa and the Middle East. Global Change Biology 15: 1072–1089.

Lassalle G, Trancart T, Lambert P and Rochard E (2008) Latitudinal variations in age and size at maturity among allis shad *Alosa alosa* populations. Journal of Fish Biology 73: 1799–1809.

Limburg KE, Hattala KA and Kahnle A (2003) American shad in its native range. *In*: Limburg KE and Waldmann JR (eds.). Biodiversity, Status and Conservation of the World's Shads. American Fisheries Symposium 35: 125–140.

Lochet A, Boutry S and Rochard E (2009) Estuarine phase during seaward migration for allis shad *Alosa alosa* and twaite shad *Alosa fallax* future spawners. Ecology of Freshwater Fish 18: 323–335.

Locke A, Hanson JM, Klassen GJ, Richardson SM and Aube CI (2003) The damming of the Petitcodiac River: Species, populations, and habitats lost. Northeastern Naturalist 10: 39–54.

Makiguchi Y, Nii H, Nakao K and Ueda H (2007) Upstream migration of adult chum and pink salmon in the Shibetsu River. Hydrobiologia 582: 43–54.

Maltais E, Daigle G, Colbeck G and Dodson JJ (2010) Spawning dynamics of American shad (*Alosa sapidissima*) in the St. Lawrence River, Canada-USA. Ecology of Freshwater Fish 19(4): 586–594.

Marco-Rius F, Caballero P, Moran P and de Leaniz CG (2013) Can migrants escape from density dependence? Ecology and Evolution 3: 2524–2534.

McBride RS (2014) Managing a marine stock portfolio: stock identification, structure, and management of 25 fishery species along the Atlantic coast of the United States. North American Journal of Fisheries Management 34: 710–734.

McCormick SD, Regish AM, Christensen AK and Bjornsson BT (2013) Differential regulation of sodium-potassium pump isoforms during smolt development and seawater exposure of Atlantic salmon. Journal of Experimental Biology 216: 1142–1151.

McDowall RM (1988) Diadromy in Fishes. Migration between Freshwater and Marine Environments. Timber Press, Croon Helm, London. 308p.

McDowall RM (1997) The evolution of diadromy in fishes (revisited) and its place in phylogenetic analysis. Reviews in Fish Biology and Fisheries 7: 443–462.

McDowall RM (2001) Anadromy and homing: two life-history traits with adaptive synergies in salmonid fishes? Fish and Fisheries 2: 78–85.

McDowall RM (2008) Why are so many boreal freshwater fishes anadromous? Confronting 'conventional wisdom'. Fish and Fisheries 9: 208–213.

Mennesson-Boisneau C, Aprahamian MW, Sabatié MR and Cassou-Leins JJ (2000) Remontée migratoire des adultes. pp. 55–72. *In*: Baglinière JL and Elie P (eds.). Les aloses (*Alosa alosa* et *Alosa fallax* spp.). Ecologie et variabilité des populations. Cemagref Editions/INRA Editions, Paris.

Miller AS, Sheehan TF, Renkawitz MD, Meister AL and Miller TJ (2012) Revisiting the marine migration of US Atlantic salmon using historical Carlin tag data. ICES Journal of Marine Science: Journal du Conseil 69: 1609–1615.

Ministère de l'écologie du développement durable des transports et du logement (2011) Plan national d'actions en faveur de l'esturgeon européen *Acipenser sturio* 2011–2015. 69p.

Moore A, Freake SM and Thomas IM (1990) Magnetic particle in the lateral line of the Atlantic salmon (*Salmo salar* L.). Philosophical Transactions of the Royal Society of London Series B-Biological Sciences 329: 11–15.

Moore A, Potter ECE, Milner NJ and Bamber S (1995) The migratory behavior of wild Atlantic salmon (*Salmo salar*) smolts in the estuary of the river Conwy, North Wales HE. Canadian Journal of Fisheries and Aquatic Sciences 52: 1923–1935.

Moore A, Privitera L and Riley WD (2013) The behaviour and physiology of migrating Atlantic salmon. pp. 28–55. *In*: Ueda H and Tsukamoto K (eds.). Physiology and Ecology of Fish Migration. CRC Press, Boca Raton, FL, USA.

Morasse S, Guderley H and Dodson JJ (2008) Paternal reproductive strategy influences metabolic capacities and muscle development of Atlantic salmon (*Salmo salar* L.) embryos. Physiological and Biochemical Zoology 81: 402–413.

Morita K, Yamamoto S and Hoshino N (2000) Extreme life history change of white-spotted char (*Salvelinus leucomaenis*) after damming. Canadian Journal of Fisheries and Aquatic Sciences 57: 1300–1306.

Muir WD and Williams JG (2012) Improving connectivity between freshwater and marine environments for salmon migrating through the lower Snake and Columbia River hydropower system. Ecological Engineering 48: 19–24.

Myers GS (1949) Usage of anadromous, catadromous and allied terms for migratory fishes. Copeia 89–97.

Navodaru I, Staras M and Cernisencu I (2001) The challenge of sustainable use of the Danube Delta fisheries, Romania. Fisheries Management and Ecology 8: 323–332.

Naylor R, Hindar K, Fleming IA, Goldburg R, Williams S, Volpe J, Whoriskey F, Eagle J, Kelso D and Mangel M (2005) Fugitive salmon: assessing the risks of escaped fish from net-pen aquaculture. Bioscience 55: 427–437.

Nesbø CL, Magnhagen C and Jakobsen KS (1998) Genetic differentiation among stationary and anadromous perch (*Perca fluviatilis*) in the Baltic Sea. Hereditas 129: 241–249.

Neuenfeld S, Hinrichsen HH, Nielsen A and Andersen KH (2007) Reconstructing migrations of individual cod (*Gadus morhua* L.) in the Baltic Sea by using electronic data storage tags. Fisheries Oceanography 16: 526–535.

Null SE, Medellin-Azuara J, Escriva-Bou A, Lent M and Lund JR (2014) Optimizing the dammed: Water supply losses and fish habitat gains from dam removal in California. Journal of Environmental Management 136: 121–131.

Ohms HA, Sloat MR, Reeves GH, Jordan CE and Dunham JB (2014) Influence of sex, migration distance, and latitude on life history expression in steelhead and rainbow trout (*Oncorhynchus mykiss*). Canadian Journal of Fisheries and Aquatic Sciences 71: 70–80.

Olney JE, Bilkovic DM, Hershner CH, Varnell LM, Wang H and Mann RL (2008) Six fish and 600,000 thirsty folks—A fishing moratorium on American shad thwarts a controversial municipal reservoir project in Virginia, USA. *In*: Nielsen J, Dodson JJ, Friedland K, Hamon TR, Musick J and Verspoor E (eds.). Reconciling Fisheries with Conservation, Vols. I and II, 1853–1863. American Fisheries Society, Bethesda.

Olsson IC and Greenberg LA (2004) Partial migration in a landlocked brown trout population. Journal of Fish Biology 65: 106–121.

Olsson IC, Greenberg LA, Bergman E and Wysujack K (2006) Environmentally induced migration: the importance of food. Ecology Letters 9: 645–651.

Otero J, L'Abee-Lund JH, Castro-Santos T, Leonardsson K, Storvik GO, Jonsson B, Dempson B, Russell IC, Jensen AJ, Bagliniere JL, Dionne M, Armstrong JD, Romakkaniemi A, Letcher BH, Kocik JF, Erkinaro J, Poole R, Rogan G, Lundqvist H, MacLean JC, Jokikokko E, Arnekleiv JV, Kennedy RJ, Niemela E, Caballero P, Music PA, Antonsson T, Gudjonsson S, Veselov AE, Lamberg A, Groom S, Taylor BH, Taberner M, Dillane M, Arnason F, Horton G, Hvidsten NA, Jonsson IR, Jonsson N, McKelvey S, Naesje TF, Skaala O, Smith GW, Saegrov H, Stenseth NC and Vollestad LA (2014) Basin-scale phenology and effects of climate variability on global timing of initial seaward migration of Atlantic salmon (*Salmo salar*). Global Change Biology 20: 61–75.

Peer AC and Miller TJ (2014) Climate change, migration phenology, and fisheries management interact with unanticipated consequences. North American Journal of Fisheries Management 34: 94–110.

Perrier C, Evanno G, Belliard J, Guyomard R and Bagliniere JL (2010) Natural recolonization of the Seine River by Atlantic salmon (*Salmo salar*) of multiple origins. Canadian Journal of Fisheries and Aquatic Sciences 67: 1–4.

Perrier C, Guyomard R, Bagliniere JL, Nikolic N and Evanno G (2013) Changes in the genetic structure of Atlantic salmon populations over four decades reveal substantial impacts of stocking and potential resiliency. Ecology and Evolution 3: 2334–2349.

Plum N and Schulte-Wulwer-Leidig A (2014) From a sewer into a living river: the Rhine between Sandoz and Salmon. Hydrobiologia 729: 95–106.

Prignon C, Micha JC, Rimbaud G and Philippart JC (1999) Rehabilitation efforts for Atlantic salmon in the Meuse basin area: Synthesis 1983–1998. Hydrobiologia 410: 69–77.

Pulido F (2007) The genetics and evolution of avian migration. BioScience 57: 165–174.

Putman NF, Lohmann KJ, Putman EM, Quinn TP, Klimley AP and Noakes DLG (2013) Evidence for geomagnetic imprinting as a homing mechanism in Pacific salmon. Current Biology 23: 312–316.

Quinn TP and Myers KW (2004) Anadromy and the marine migrations of Pacific salmon and trout: Rounsefell revisited. Reviews in Fish Biology and Fisheries 14: 421–442.

Quintella BR, Andrade NO and Almeida PR (2003) Distribution, larval stage duration and growth of the sea lamprey ammocoetes, *Petromyzon marinus* L., in a highly modified river basin. Ecology of Freshwater Fish 12: 286–293.

Rand PS, Goslin M, Gross MR, Irvine JR, Augerot X, McHugh PA and Bugaev VF (2012) Global assessment of extinction risk to populations of Sockeye salmon *Oncorhynchus nerka*. PLOS ONE 7.

Ray GC (2005) Connectivities of estuarine fishes to the coastal realm. Estuarine Coastal and Shelf Science 64: 18–32.

Reebs SG (2002) Plasticity of diel and circadian activity rhythms in fishes. Reviews in Fish Biology and Fisheries 12: 349–371.

Reis-Santos P, McCormick SD and Wilson JM (2008) Ionoregulatory changes during metamorphosis and salinity exposure of juvenile sea lamprey (*Petromyzon marinus* L.). Journal of Experimental Biology 211: 978–988.

Riede K (2004) Global register of migratory species—from global to regional scales. Final Report of the R&D-Projekt 808 05 081. Federal Agency for Nature Conservation, Bonn, Germany. 329p.

Rochard E, Castelnaud G and Lepage M (1990) Sturgeons (Pisces: Acipenseridae); threats and prospects. Journal of Fish Biology 37: 123–132.

Rochard E, Lepage M, Dumont P, Tremblay S and Gazeau C (2001) Downstream migration of juvenile European sturgeon *Acipenser sturio* L. in the Gironde Estuary. Estuaries 24: 108–115.

Rolls RJ, Stewart-Koster B, Ellison T, Faggotter S and Roberts DT (2014) Multiple factors determine the effect of anthropogenic barriers to connectivity on riverine fish. Biodiversity and Conservation 23: 2201–2220.

Romer JD, Leblanc CA, Clements S, Ferguson JA, Kent ML, Noakes D and Schreck CB (2013) Survival and behavior of juvenile steelhead trout (*Oncorhynchus mykiss*) in two estuaries in Oregon, USA. Environmental Biology of Fishes 96: 849–863.

Rosenthal H, Bronzi P, Gessner J, Moreau D, Rochard E and Lasen C (2007) Draft action plan for the conservation and restoration of the European sturgeon (*Acipenser sturio*). 47. Strasbourg: Council of Europe, Convention on the conservation of European wildlife and natural habitats.

Rougier T, Lambert P, Drouineau H, Girardin M, Castelnaud G, Carry L, Aprahamian M, Rivot E and Rochard E (2012) Collapse of allis shad, *Alosa alosa*, in the Gironde system (southwest France): environmental change, fishing mortality, or Allee effect? ICES Journal of Marine Science 69: 1802–1811.

Roule L (1922) Etude sur l'esturgeon du Golfe de Gascogne et du bassin girondin. Office Scientifique et Technique Des Pêches Maritimes 20: 12.

Rounsefell GA (1958) Anadromy in North American Salmonidae. Fishery Bulletin 131: 171–185.

Royce WF (1968) Models of oceanic migrations of pacific salmon and comments on guidance mechanisms. Fishery Bulletin 66: 441–462.

Rundio DE, Williams TH, Pearse DE and Lindley ST (2012) Male-biased sex ratio of nonanadromous *Oncorhynchus mykiss* in a partially migratory population in California. Ecology of Freshwater Fish 21: 293–299.

Ruzzante DE, Hansen MM, Meldrup D and Ebert KM (2004) Stocking impact and migration pattern in an anadromous brown trout (*Salmo trutta*) complex: where have all the stocked spawning sea trout gone? Molecular Ecology 13: 1433–1445.

Sabatie R and Bagliniere JL (2001) Some ecobiological traits in Morrocan shads; a cultural and socio-economic value interest which has disappeared. Bulletin Francais de la Peche et de la Pisciculture 903–917.

Sackville M, Wilson JM, Farrell AP and Brauner CJ (2012) Water balance trumps ion balance for early marine survival of juvenile pink salmon (*Oncorhynchus gorbuscha*). Journal of Comparative Physiology B: Biochemical, Systemic, and Environmental Physiology 182: 781–792.

Sahashi G and Morita K (2013) Migration costs drive convergence of threshold traits for migratory tactics. Proceedings of the Royal Society B-Biological Sciences 280: 20132539.

Sand O and Karlsen HE (2000) Detection of infrasound and linear acceleration in fishes. Philosophical Transactions of the Royal Society of London Series B-Biological Sciences 355: 1295–1298.

Secor DH and Houde ED (1998) Use of larval stocking in restoration of Chesapeake Bay striped bass. Ices Journal of Marine Science 55: 228–239.

Silva S, Servia MJ, Vieira-Lanero R and Cobo F (2013) Downstream migration and hematophagous feeding of newly metamorphosed sea lampreys (*Petromyzon marinus* Linnaeus, 1758). Hydrobiologia 700: 277–286.

Sommer T, Mejia FH, Nobriga ML, Feyrer F and Grimaldo L (2011) The spawning migration of delta smelt in the upper san Francisco estuary. San Francisco Estuary and Watershed Science 9: 1–16.

Sorensen PW, Vrieze LA and Fine JM (2003) A multi-component migratory pheromone in the sea lamprey. Fish Physiology and Biochemistry 28: 253–257.

Spillmann CJ (1961) Faune de France 65: Poissons d'Eau Douce. Lechevalier: Paris.

Sternecker K, Denic M and Geist J (2014) Timing matters: species-specific interactions between spawning time, substrate quality, and recruitment success in three salmonid species. Ecology and Evolution 4: 2749–2758.

St Pierre RA (1999) Restoration of Atlantic sturgeon in the northeastern USA with special emphasis on culture and restocking. Journal of Applied Ichthyology-Zeitschrift Fur Angewandte Ichthyologie 15: 180–182.

Taverny C, Lepage M, Piefort S, Dumont P and Rochard E (2002) Habitat selection by juvenile European sturgeon *Acipenser sturio* in the Gironde estuary (France). Journal of Applied Ichthyology 18: 536–541.

Taylor EB (1991) A review of local adaptation in Salmonidae, with particular reference to Pacific and Atlantic salmon. Aquaculture 98: 185–207.

Tentelier C, Larrieu M, Aymes JC and Labonne J (2011) Male antagonistic behaviour after spawning suggests paternal care in brown trout, *Salmo trutta*. Ecology of Freshwater Fish 20: 580–587.

Thomson KA, Ingraham WJ, Healey MC, Leblond PH, Groot C and Healey CG (1994) Computer-simulations of the influence of ocean currents on Fraser-river sockeye salmon (*Oncorhynchus nerka*) return times. Canadian Journal of Fisheries and Aquatic Sciences 51: 441–449.

Thorpe JE, Mangel M, Metcalfe NB and Huntingford FA (1998) Modelling the proximate basis of salmonid life-history variation, with application to Atlantic salmon, *Salmo salar* L. Evolutionary Ecology 12: 581–599.

Tsuda Y, Kawabe R, Tanaka H, Mitsunaga Y, Hiraishi T, Yamamoto K and Nashimoto K (2006) Monitoring the spawning behaviour of chum salmon with an acceleration data logger. Ecology of Freshwater Fish 15: 264–274.

Ueda H (2012) Physiological mechanisms of imprinting and homing migration in Pacific salmon *Oncorhynchus* spp. J. Fish Biol. 81: 543–558.

Vecsei P, Sucui R and Peterson D (2002) Threatened fishes of the world: *Huso huso* (Linnaeus, 1758) (Acipenseridae). Environmental Biology of Fishes 65(3): 363–365.

Vignon M and Sasal P (2010) Fish introduction and parasites in marine ecosystems: a need for information. Environmental Biology of Fishes 87: 1–8.

Waldman JR, Grunwald C, Stabile J and Wirgin I (2002) Impacts of life history and biogeography on the genetic stock structure of Atlantic sturgeon *Acipenser oxyrinchus oxyrinchus*, Gulf sturgeon *A-oxyrinchus desotoi*, and shortnose sturgeon *A-brevirostrum*. Journal of Applied Ichthyology 18: 509–518.

Waldman J, Grunwald C and Wirgin I (2008) Sea lamprey *Petromyzon marinus*: an exception to the rule of homing in anadromous fishes. Biology Letters 4: 659–662.

Walther BD, Thorrold SR and Olney JE (2008) Geochemical signatures in otoliths record natal origins of American shad. Transactions of the American Fisheries Society 137: 57–69.

Waters JM, Epifanio JM, Gunter T and Brown BL (2000) Homing behaviour facilitates subtle genetic differentiation among river populations of *Alosa sapidissima*: microsatellites and mtDNA. Journal of Fish Biology 56: 622–636.

Weitkamp LA, Goulette G, Hawkes J, O'Malley M and Lipsky C (2014) Juvenile salmon in estuaries: comparisons between North American Atlantic and Pacific salmon populations. Reviews in Fish Biology and Fisheries 24(3): 713–736.

Williams JG, Smith SG and Muir WD (2001) Survival estimates for downstream migrant yearling juvenile salmonids through the Snake and Columbia rivers hydropower system, 1966–1980 and 1993–1999. North American Journal of Fisheries Management 21: 310–317.

Yako LA, Mather ME and Juanes F (2002) Mechanisms for migration of anadromous herring: an ecological basis for effective conservation. Ecological Applications 12: 521–534.

Yi YJ, Wang ZY and Yang ZF (2010) Two-dimensional habitat modeling of Chinese sturgeon spawning sites. Ecological Modelling 221: 864–875.

Zeug SC, Sellheim K, Watry C, Rook B, Hannon J, Zimmerman J, Cox D and Merz J (2014) Gravel augmentation increases spawning utilisation by anadromous salmonids: a case study from California, USA. River Research and Applications 30: 707–718.

Zhou JZ, Zhao Y, Song LX, Bi S and Zhang HJ (2014) Assessing the effect of the Three Gorges reservoir impoundment on spawning habitat suitability of Chinese sturgeon (*Acipenser sinensis*) in Yangtze River, China. Ecological Informatics 20: 33–46.

Zhu Y, Wei Q, Yang D, Wang K, Chen X, Liu J and Li L (2006) Large-scale cultivation of fingerlings of the Chinese Sturgeon *Acipenser sinensis* for re-stocking: a description of current technology. Journal of Applied Ichthyology 22: 238–243.

Zlokovitz ER, Secor DH and Piccoli PM (2003) Patterns of migration in Hudson River striped bass as determined by otolith microchemistry. Fisheries Research 63: 245–259.

CHAPTER 6

Life Histories of Catadromous Fishes

Michael J. Miller

Introduction

Diadromous fishes differ from other types of fishes because they migrate between freshwater or estuarine habitats and the marine environment at specific times in their lives. There are only about 250 species of diadromous fishes compared to the likely more than 30,000 or more species of purely freshwater, estuarine or marine fishes (McDowall 1988, 1997; Nelson 2006), so they are a unique subset of the world's fishes. The categories of diadromous fishes have been defined based on which life history stages use the marine and freshwater habitats and for what purpose (Myers 1949; McDowall 1988, 1997). Catadromy is a diadromous life history with the fish using the marine environment for reproduction and larval growth and using freshwater for juvenile growth. Or more specifically, McDowall (1997) provided a detailed definition of catadromy as: "Diadromous fishes in which most feeding and growth are in fresh water prior to migration of fully grown, adult fish to sea to reproduce; there is either no subsequent feeding at sea, or any feeding is accompanied by little somatic growth; the principal feeding and growing biome (freshwater) differs from the reproductive biome (the sea)." This definition however seems to have expanded somewhat to include estuaries as a separate biome, which would then include fish that migrate from freshwater to estuarine habitats to reproduce, or from estuarine to marine habitats to reproduce, with the juvenile growth biome being either freshwater or estuarine habitats. Further complicating this issue is that many seemingly catadromous species have plasticity in their life history patterns especially in terms of how much they may use the pure freshwater environment (Tsukamoto and Arai 2001; Daverat et al. 2006, 2012; Walther et al. 2011).

Department of Marine Science and Resources, College of Bioresource Sciences, Nihon University, 1866 Kameino, Fujisawa-shi, Kanagawa, 252-0880, Japan.
Email: michael.miller@nihon-u.ac.jp

In general, however, fishes that can be considered to be catadromous must show some sort of migration to the brackish or marine environment to spawn, and their larvae then feed and grow in higher salinity habitats before entering estuaries or freshwater as juveniles for growth until maturity. The distance each species migrates into the ocean to reproduce varies widely though, from thousands of kilometers in the case of anguillid eels (Aoyama 2009), to just near the transition to saline waters within estuaries or in nearshore waters in the case of other catadromous fishes (McDowall 1988). As will be apparent later in this chapter, and is seen from previous overviews of catadromous or freshwater fishes and their migrations (McDowall 1987, 1988, 1997; Lucas and Baras 2001), there is little evidence of any catadromous fishes having spawning areas far out in the ocean like anguillid eels; so the catadromous anguillid life history seems to be unique, with most other species being only marginally catadromous in comparison.

The kinds of catadromous fishes other than eels seem to mostly consist of a range of mullets, kuhliids, and flatfishes, or one or a few species of percyichthyid, centropomid, cottid, scorpaenid, galaxiid or other species (McDowall 1987, 1988). These various fishes and eels have a wide range of body forms and maximum body sizes (Fig. 6.1). McDowall (1987) could find evidence of 41 catadromous fish species in the world, and this included 15 species of anguillid eels and about 12 mullets. Feutry et al. (2013)

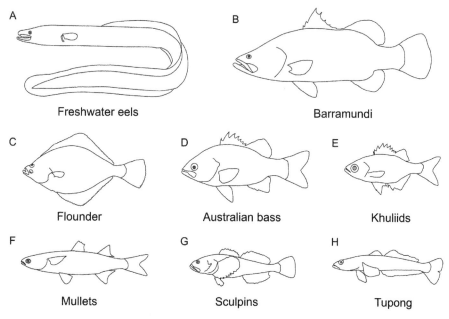

Figure 6.1. The body forms of different types of catadromous fishes showing (A) the giant mottled eel, *Anguilla marmorata*, the largest tropical anguillid species that is widely distributed across the Indo-Pacific, (B) the barramundi *Lates calcarifer*, an important sport fish and aquaculture species found from northern Australia and Southeast Asia to East Africa, (C) the European flounder *Platichthys flesus*, (D) an Australian bass *Macquaria novemaculeata*, a sport fish from eastern Australia, (E) flagtails of the Genus *Kuhlia*, which are distributed in the tropical Pacific and Indian ocean region, (F) mullets of the Family Mugilidae, which are distributed worldwide, (G) the fourspine sculpin *Cottus kazika* from southern Japan, and (H) the tupong *Pseudaphritis urvillii* from southeastern Australia.

listed five species of kuhliids as being catadromous compared to the two species known by McDowall (1987), and a new anguillid eel species has been discovered (Watanabe et al. 2009); but even after several decades of research, few additional species of fishes seem to have been found to have catadromous life histories.

The species with catadromous life histories are distributed at tropical to temperate latitudes, with the greatest number being present at tropical to subtropical latitudes due the larger numbers of anguillids there (Fig. 6.2, McDowall 1997). This latitudinal distribution is in contrast with anadromous species, which are most abundant at higher latitudes in the Northern Hemisphere (McDowall 1997).

Catadromy may have typically evolved in marine fishes that developed the ability to enter freshwater for feeding and juvenile growth (McDowall 1997). This has been considered to be the case for anguillid eels (Tsukamoto et al. 2002), which have the most distinct catadromous life histories. Phylogenetic analysis of the Anguilliformes, which includes all freshwater and marine eels, has shown that the family Anguillidae was derived from a lineage comprised of mesopelagic families of marine eels that

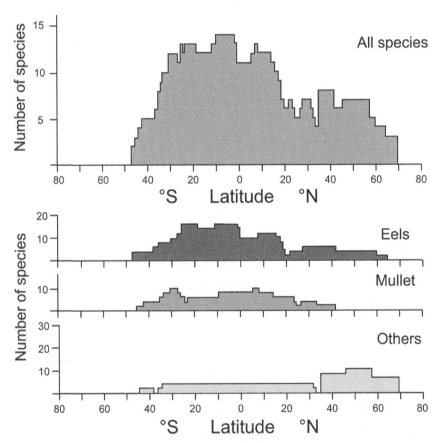

Figure 6.2. The number of species of all catadromous fishes at each latitude or separated into different taxonomic groups. Modified from McDowall (1997).

live in the open ocean (Inoue et al. 2010). The ancestral anguillid eel species appears to have originated in the tropical Indo-Pacific where its juveniles entered freshwater presumably because of there being a vacant niche for eels with greater feeding success and likely fewer predators (Tsukamoto et al. 2002, 2009; Inoue et al. 2010). For catadromous fishes not from the deep-ocean, their evolutionary patterns may not be so clear as for anguillid eels, but many appear to be derived from marine ancestors. Catadromous kuhliid species for example appear to have given rise to purely marine species (Feutry et al. 2013), although the genus itself appears to be derived from a marine ancestor (Yagishita et al. 2009). The Centropomidae (snooks and barramundi) appears to be derived from marine fish groups even though it includes some freshwater species (Li et al. 2011).

This chapter briefly overviews catadromous fish life histories and their migrations or movements. It places more emphasis on anguillids, which make the most distinct catadromous migrations and have been studied in more detail. Because there have been reviews or books published about most aspects of anguillid eel life history (e.g., McCleave 1993, 2003; Aida et al. 2003; Feunteun et al. 2003; Tesch 2003; Jellyman 2003; van Ginneken and Maes 2005; Jessop et al. 2008, 2010; Aoyama 2009; McCleave and Edeline 2009; Miller et al. 2009a,b; Tsukamoto 2009; van den Thillart et al. 2009; Bonhommeau et al. 2010; Watanabe and Miller 2012), this chapter does not exhaustively cover the literature about anguillid life history and places emphasis on more recent studies. The other types of fishes with catadromous life histories have also been described previously (McDowall 1988; Lucas and Baras 2001) and are outlined here to give an overall view of the present state of knowledge about these interesting fishes that use freshwater or estuarine habitats, but migrate to spawn in higher salinity environments.

Anguillid life histories

Freshwater eels have been the classic example of catadromous fishes ever since Schmidt (1922) showed the world what interesting life histories they have because of the long migrations they make in the ocean. Their offshore spawning areas and the random return of their larvae to their growth habitats result in temperate anguillids consisting of single panmictic populations (Wirth and Bernachez 2003; Dannewitz et al. 2005; Han et al. 2010). Some tropical anguillid eels with widespread distributions consist of multiple spawning populations or metapopulations (Ishikawa et al. 2004; Minegishi et al. 2008; Gagnaire et al. 2011). Other tropical anguillids have local spawning areas and may have single populations or have more than one spawning area (Miller et al. 2009a). These population structures appear to be formed by the establishment of specific migration loops between the growth habitats and spawning areas, which depend on the geography of landmasses and patterns of ocean currents that transport the larvae (Tsukamoto et al. 2002; Ishikawa et al. 2004; Aoyama 2009).

In addition to interest in their long migrations and population structures, in recent years attention has been focused on their 'facultative catadromy' in which not all individuals of each species actually enter freshwater for juvenile growth (Tsukamoto et al. 1998, Tsukamoto and Arai 2001, Daverat et al. 2006). However, all anguillid eel species have many individuals that migrate into freshwater and move far inland, wherever they have not been blocked by dams, so their catadromy cannot be doubted. Regardless of whether or not all the juveniles of the species migrate upstream into freshwater from the estuary, all the adults migrate far offshore to spawn, and all their larvae, called leptocephali live exclusively in the ocean environment (Tesch 2003; Aoyama 2009). Their life histories are characterized by four distinct stages that consist of larvae (non-feeding preleptocephali and feeding leptocephali) that metamorphose into glass eels (recruitment stage), and yellow eels (juvenile growth stage), which metamorphose into silver eels (adult stage) (Fig. 6.3).

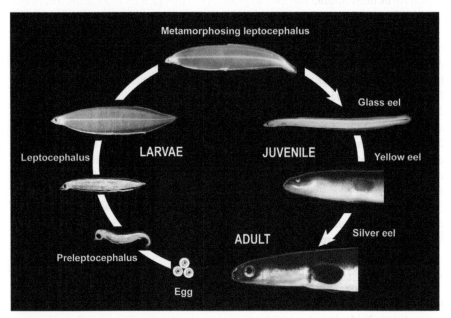

Figure 6.3. The life cycle of anguillid eels showing each life history stage. The adult silver eels spawn offshore in the ocean, the leptocephali live only in the ocean, and the glass eels and yellow eels are found in freshwater and estuarine habitats.

Silver eels and their spawning areas

The life cycles of anguillid eels both end and begin with the actions of the reproductive-stage silver eels that migrate out of freshwater and estuarine habitats into the ocean, where they swim to their offshore spawning areas for reproduction. After having grown for many years as yellow eels and reaching a general age, size, or amount of energy has been accumulated (Vollestad 1992), something triggers the maturation process and the silver eels start gonadal maturation, their eyes enlarge, and their body and fin coloration changes (Tesch 2003; Aoyama and Miller 2003; Durif et al. 2005; Okamura et al. 2007; Tsukamoto 2009). These types of changes appear to be associated with facilitating the long migration through the ocean and for eventual reproduction, but changes in their osmoregulatory systems also occur (Sasai et al. 1998). The silver eel migrations of northern temperate anguillid eels occur during the fall season (Haro 2003; Tesch 2003), whereas the migration season of tropical eels is less well understood and possibly more linked to climatic cycles of rainfall (Wouthuyzen et al. 2009). Silver eels in temperate regions tend to migrate downstream during periods of stormy weather or rainfall when water discharge increases in rivers, or during phases of the lunar cycle when there is little moonlight (Haro 2003; Durif et al. 2008; Bruijs and Durif 2009).

The spawning areas of temperate anguillid eels are located along the low latitude edges of subtropical gyres, which have westward currents to transport their larvae to their recruitment areas (Miller et al. 2009b; Aoyama 2009). This requires the adult silver eels of these species to migrate thousands of kilometers from the continental margins to reach these spawning areas. Little is known about these oceanic migrations (Tsukamoto 2009), except that the silver eels show distinct patterns of diel vertical migration from shallow depths (a few hundred meters) during the night, to much greater depths (600–800 m) during the day, while they are migrating towards their spawning areas (Jellyman and Tsukamoto 2005, 2010; Aarestrup et al. 2009; Manabe et al. 2011; Schabetsberger et al. 2013). Although silver eels are not often considered to be great swimmers, recent research has shown that they are probably highly efficient swimmers, capable of swimming continuously for many months without feeding (van den Thillart et al. 2009). There is little light at the migration depths used by silver eels, and temperate species must all cross through a variety of currents or water masses to reach their spawning areas, so there are few clues about how they navigate back to their spawning areas (Tsukamoto 2009). This suggests that their geomagnetic sense (Nishi et al. 2004; Nishi and Kawamura 2005; Durif et al. 2013) may play an important role in their migration back to the spawning area where they were born, before other senses are used to determine where exactly they will spawn within the spawning area.

The locations where the Atlantic eels (European eel *Anguilla anguilla*; American eel *A. rostrata*) and the Japanese eel *Anguilla japonica* spawn in these areas appear to be determined by landmarks, such as temperature or salinity fronts or a seamount ridge (Kleckner and McCleave 1988; Munk et al. 2010; Tsukamoto et al. 2011; Aoyama et al. 2014). In contrast, tropical anguillid species can have shorter migration distances to more local spawning areas (Aoyama et al. 2003, 2007; Miller et al. 2009a; Wouthuyzen et al. 2009), or they can also migrate offshore like eels from the northern population of the giant mottled eel *Anguilla marmorata* (Miller et al. 2002; Kuroki et al. 2009). The locations of the spawning areas of many tropical anguillid species are not known yet though, because few of their leptocephali have been collected (Miller 2003; Kuroki et al. 2008).

Collections of all the oceanic life history stages of the Japanese eel suggest that spawning by anguillid eels occurs in the upper few hundred meters of the ocean. Adult eels of both *A. japonica* and *A. marmorata* appeared to be caught in the 250–300 m depth layer by large midwater trawls (Chow et al. 2009; Tsukamoto et al. 2011; Kurogi et al. 2011). Eggs and preleptocephali were also all collected in the surface layer, with both being found in highest abundance in the layers near the top of the thermocline centered at about 160 m (Tsukamoto et al. 2011; Kurogi et al. 2011; Aoyama et al. 2014). Small anguillid leptocephali in the Sargasso Sea spawning area of the Atlantic eels were also present at similar depths (Castonguay and McCleave 1987), which along with the similarities in the diel vertical migration behaviors of silver eels suggest the possibility of similar spawning ecology for other anguillid eels. The Japanese eel is also known to only spawn during new moon periods (Tsukamoto et al. 2003, 2011), but it is not known if other anguillids also have this characteristic. All anguillid eels are thought to be semelparous and die after their spawning season (Tesch 2003; Tsukamoto et al. 2011).

The leptocephalus stage and larval migrations

After anguillid eggs are fertilized they may develop for about 1.5–2 days before hatching, depending on temperature (Ahn et al. 2012). Japanese eel eggs (embryos) have been collected recently at a size of about 1.6 mm in diameter (Tsukamoto et al. 2011; Yoshinaga et al. 2012; Aoyama et al. 2014). After the eggs hatch, all anguillid eels have very similar leptocephalus larvae (Miller and Tsukamoto 2004) (Fig. 6.3). The leptocephali are unusual in comparison to other types of fish larvae because they have a long larval duration, a large size increase, and they have an unusual diet. The larval duration of anguillid eels seems to range from about three months to almost a year or likely more (Marui et al. 2001; Kuroki et al. 2006, 2014). Larvae seem to feed exclusively on various types of particulate organic matter, such as marine snow and discarded appendicularian houses and not on zooplankton (Otake et al. 1993; Mochioka and Iwamizu 1996; Miller et al. 2013). This type of diet has resulted in them having a very low trophic position within the marine food web (Miller et al. 2013). Anguillid and other leptocephali all live in the upper few hundred meters of the ocean (Castonguay and McCleave 1987; Miller 2009). The growth rates and maximum larval sizes of temperate and tropical anguillid leptocephali differ with tropical species being shorter and with higher growth rates (Kuroki et al. 2006, 2014). Accordingly, the tropical species of leptocephali have shorter larval durations than the temperate species (Marui et al. 2001; Kuroki et al. 2014). As they feed and grow, anguillid leptocephali are transported from their oceanic spawning areas to the continental margins, or near islands, where they metamorphose into the glass eel stage, which then enter estuaries and freshwater habitats (Tesch 2003).

It is presently unknown if the leptocephali of anguillid eels engage in active horizontal migration as larvae, or if they are just passively transported by ocean currents as discussed previously (McCleave et al. 1998; Miller 2009). Leptocephali swim during their vertical migrations between different depths (Castonguay and McCleave 1987), and laboratory swimming studies on conger eel *Conger oceanicus* leptocephali indicate they are capable swimmers for short periods of time (Wuenschel and Able 2008). The spawning areas of temperate anguillids are all located in westward flowing currents that enter western boundary currents, which can transport their leptocephali towards their recruitment areas (Miller et al. 2009b). However, at the latter stages of their migrations, as leptocephali or glass eels, it is likely that some species must use active swimming to reach coastal areas. For example, American eel leptocephali originate on the east side of the Gulf Stream, but must eventually move west or later north to cross the powerful current (Kleckner and McCleave 1982). There seems to be no continuously occurring mechanism to allow large numbers of larvae to cross the Gulf Stream without swimming though. European eel larvae that also enter the Gulf

Stream (Kleckner and McCleave 1982) do not cross over towards the coastline of North America (Kleckner and McCleave 1985), but continue on to the east towards Europe, so this also implies active migration by American eel larvae to cross the current. Both species then metamorphose into glass eels, which may also need to actively swim to reach the coastline. However, this does not imply that anguillid leptocephali show active migratory swimming throughout their larval stage, which they likely do not, since that could interfere with feeding and growth.

Near the end of their larval period there may be some active migration, as anguillid leptocephali approach the areas near the continental shelf or islands. At this stage, something triggers the metamorphosis process, in which their laterally compressed leaf-like and totally transparent bodies transform into the rounded body shape of eels (Otake 2003; Tesch 2003) (Fig. 6.3). As glass eels reach estuaries and begin their movements upstream, they become increasingly pigmented, and the various stages of pigmentation have been classified (see Fukuda et al. 2013). The glass eels must also develop the ability to osmoregulate in freshwater rather than seawater when they migrate upstream (Wilson et al. 2004; Sasai et al. 2007).

The recruitment and growth stages of anguillid eels

Glass eels are the recruitment stage of anguillid eels that enter estuaries and then subsequently migrate upstream into freshwater (Tesh 2003). In large estuaries, with strong tidal flows that include upstream flow of water into the lower river reaches, glass eels can use selective tidal stream transport (STST) to move upstream (McCleave and Kleckner 1982). This behavior involves the glass eels resting at the bottom during ebb tides, and then entering the water column during flood tides so the current will transport them upstream. The timing of their entering the water column may be regulated at least in part by a biological rhythm entrained to the tidal/lunar cycle (Wippelhauser and McCleave 1988). If not blocked by dams, or other obstructions, anguillid eels will often keep moving upstream as they grow larger (Feunteun et al. 2003).

Estuaries are productive feeding habitats for anguillid eels though, so some glass eels may remain in the brackish environment without moving upstream into freshwater if there is suitable habitat available for them. Others may move back downstream into estuarine bays after initially settling in the lower reaches of rivers. Otolith microchemistry studies using Sr:Ca ratios have consistently found individuals that likely remained in the estuary, or high salinity habitats, during their entire juvenile growth phase without residing in freshwater, while other eels were freshwater residents their whole lives or switched from one type of habitat to another (e.g., Tzeng et al. 1997; Tsukamoto et al. 1998; Tsukamoto and Arai 2001; Jessop et al. 2002, 2008; Daverat et al. 2006; Kaifu et al. 2010). What determines if an eel will choose to migrate upstream or remain in the estuary is difficult to determine, but it may be related to factors such as the energy reserves or physiological state of each individual, or behavioral interactions resulting from the density of eels in the estuary already (Edeline 2007; McCleave and Edeline 2009). In the present-day time period of lower eel abundances and at higher latitudes with low productivity in freshwater, the growth rates of eels has been found to be faster in the brackish water habitats (Jessop et al. 2004; Morrison and Secor 2003; Daverat and Tomas 2006;

Carins et al. 2009; Jessop 2010; Daverat et al. 2012a; Kaifu et al. 2013a). Possibly as a result of this, the majority of eels that switch between habitats are those that move back downstream from freshwater into the estuary (e.g., Jessop et al. 2002; Daverat et al. 2006; Kaifu et al. 2010). At the northern extreme of the species range of the American eel though, yellow eels in brackish waters may move into freshwater during winter to avoid colder waters in the estuary (Thibault et al. 2007). In less extreme environments, why eels switch habitats by moving back downstream is unclear, but evidence from a recent otolith microchemistry study on the Japanese eel suggested that eels that switched back to the estuary were those that appeared to have entered freshwater later than those that migrated directly into freshwater with no delay of first residing in the estuary (Yokouchi et al. 2012).

Eels can make a variety of small-scale movements within estuaries that are not recorded very well by otolith microchemistry, probably because there may be as much as a two month time-lag for ambient water chemistry to be clearly reflected in their otoliths (Yokouchi et al. 2011). This means that eels can move between the freshwater/saltwater boundaries for short periods of time, which might not be clearly recorded in their otoliths. These types of movements have been seen in tracking studies of eels in estuaries and can be substantial, especially in large river systems with strong tidal currents that allows them to use STST (Parker 1995; Parker and McCleave 1997; Hedger et al. 2010). It should be emphasized that these types of behaviors or long-term habitat shifts should be referred to as movements not migrations, because they do not fit the generally accepted definition of a migration (a regular movement at a certain life history stage by the majority of the population for a particular purpose).

Regardless of whether yellow eels are living in freshwater or estuarine habitats, all species of anguillid eels appear to basically be opportunistic predators on aquatic animals and they likely grow at various rates related to food availability and temperature. Anguillid eels feed on things like aquatic insects, crustaceans, gastropods annelids, and small fishes depending on the habitat and season (Dörner et al. 2009; Kaifu et al. 2012 and references within). The proportion of fishes in their diet often increases with the age of the eel. The growth rates of eels also seem to vary considerably among freshwater and brackish habitats, with latitude, and between sexes, which may be related to eel densities, water temperatures, or low food availability (see Jellyman 1995; Morrison and Secor 2003; Oliveria and McCleave 2002; Yokouchi et al. 2008; Carins et al. 2009; Jessop 2010 and references within).

What determines the start of the silvering process of yellow eels is not clear, but it is associated with the initiation of their downstream and oceanic migration (Durif et al. 2005; van Ginneken et al. 2007). It has been hypothesized that male yellow eels use a time-minimizing strategy of reaching maturity sooner to migrate at a younger age than females, due to the low cost of producing sperm compared to producing eggs (Helfman et al. 1987). Females would use a size maximizing strategy to produce a large number of eggs, with both strategies being designed to maximize reproductive fitness. Because sex determination in anguillid eels appears to be environmentally determined by the density of eels and their growth rates (Krueger and Oliveira 1999; Davey and Jellyman 2005), the different strategies likely evolved in response to selective pressures resulting from large numbers of eels recruiting to the same habitats each year.

Other catadromous fish life histories

Compared to the anguillid eels there seem to be no other catadromous fishes that make such impressive migrations (McDowall 1988; Lucas and Baras 2001), but the variety of other catadromous fishes are also interesting for their ability to move across the freshwater/seawater boundaries. These other fishes migrate from their freshwater or estuarine growth habitats to higher salinity estuarine or marine habitats to reproduce. The common features of these life histories are the movement of the adults to higher salinity environments for spawning, and that their larvae then feed in brackish or marine habitats before entering estuarine or freshwater habitats as larvae or juveniles. For many species it is not yet clear how much variability there is among individuals, or exactly what the adults do after each spawning. These fishes also have seemingly overall greater variation in their life history patterns than anguillid eels (McDowall 1988), likely due to the presence of some different reproductive ecologies as mentioned below.

Characteristics of other catadromous fishes

There are a variety of species with life histories that are considered to be catadromous in eastern Australia, which include the Australian bass *Macquaria novemaculeata* (Fig. 6.1D) and the estuary perch *Macquaria colonorum*. These species migrate from freshwater or the upper reaches of estuaries to spawn in the estuary and seem to be the only known catadromous percichthyids (Walsh et al. 2011). The Australian bass uses freshwater to a much greater degree than the estuary perch, and migrates out of rivers during periods of strong flow to spawn in the estuary, but Australian bass may delay reproduction during low-flow drought periods (Harris 1986). The reproduction of estuary perch may also be linked to similar environmental conditions and they showed poor recruitment in some years (Walsh et al. 2010). The eggs of the Australian bass would not hatch in freshwater and have the best survival in higher salinities (van der Wal 1985). The larvae are present in estuaries and may be transported further into the estuary by tidal currents (Trnski et al. 2005). The estuary perch can reach ages of 40 years old (Walsh et al. 2010), so they may have the chance to reproduce several times in their lives (Harris 1986; Walsh et al. 2011). Like anguillid eels though, the Australian bass may not be an obligatory catadromous species based on a tracking study of adult fish in a river and estuary (Walsh et al. 2012).

A quite different fish that appears to be catadromous is the tupong *Pseudaphritis urvillii* (Fig. 6.1H). This species is the only member of the Pseudaphritidae (closely related to the Antarctic icefishes and temperate thornfishes) and is found in freshwater and estuarine regions of Southeastern Australia (Crook et al. 2010). The mature tupong females migrate downstream to spawn in the estuary or marine environment (Crook et al. 2010). However, the males may be estuarine/marine residents that are not diadromous, but further research is needed on this species. Otolith microchemistry studies on other catadromous or diadromous fishes in Southeastern Australia also show diversity in habitat-use patterns, not always conforming to the traditional expectations (Miles et al. 2009).

The barramundi *Lates calcarifer* (Fig. 6.1B) is an important commercial fisheries species, sport fish, and aquaculture species found in northern Australia and in other parts of the Indo-West Pacific, and is also considered to have a catadromous life history (Moore 1982; Russell and Garrett 1988; Athauda et al. 2012). Movements of adult barramundi downstream for spawning, and of juveniles into freshwater or between adjacent river systems have been documented in Australia and Papua New Guinea using tagging studies (Moore and Reynold 1982; Davis 1986; Russell and Garrett 1988). Spawning occurs in brackish water in or near river mouths, and the juveniles use various habitats such as tidal creeks and swamps as nursery areas (Davis 1985; Russell and Garrett 1985; Blaber et al. 2008). The larvae are collected within and outside of estuaries (Moore 1982; Russell and Garrett 1985). Telemetry has been used recently to examine their movements in freshwater during two different seasons (Heupel et al. 2011), and their growth rates in different river flow regimes have been studied (Robins et al. 2006). Otolith microchemistry has been used in studies on this species to examine individual habitat-use patterns, which show that not all individuals exhibit the same types of movements between freshwater and brackish habitats (see McCulloch et al. 2005; Milton and Chenery 2005; Walther et al. 2011). The life history of barramundi is unusual though, because it is a protandrous hermaphrodite that changes sequentially from males to females, spawning for three–four years as males before becoming females (Moore 1979; Blaber et al. 2008; Athauda et al. 2012).

A closely related and moderately catadromous species, that is also a sport fish, is the common snook *Centropomus undecimalis*. It is distributed from Florida to Brazil and is also a protandrous hermaphrodite (Taylor et al. 2000). This species lives in brackish and freshwater areas and migrates slightly offshore from its primary estuarine habitats for spawning (Adams et al. 2009; Trotter et al. 2012). The larvae initially feed in nearshore waters before moving into vegetated shorelines for early juvenile growth (Peters et al. 1998). Some fish move into freshwater, but the reason for those movements are not yet well understood (Blewett et al. 2009).

A group of smaller-sized tropical fishes are the 12 species of flagtails of the family Kuhliidae, genus *Kuhlia* (Fig. 6.1E). These fishes inhabit tropical islands in the Indo-Pacific and include some catadromous species that migrate from rivers to spawn in saline habitats (Randall and Randall 2001; Feutry et al. 2013). They are probably typically found in many of the same tropical island rivers as the amphidromous gobies that use the freshwater habitats not only for both juvenile and adult growth, but also for spawning (Keith 2003; Closs and Warburton 2016). The kuhliids however, seemingly must move to higher salinity regions near the river mouths to spawn (Hogan and Nicholson 1987; Oka and Tachihara 2001). This has been confirmed by otolith Sr:Ca studies on a few species that show the larval period occurs in brackish or marine waters before the fish enter freshwater for growth, with some variability in habitat use among individuals (Benson and Fitzsimons 2002; Oka and Tachihara 2008; Feutry et al. 2011, 2012a). *Kuhlia xenura* in Hawaii appears to use freshwater, estuarine, rocky shore, and tide pools for juvenile growth, but was most abundant in the estuary (McRae et al. 2011). The larval growth phase in saline waters appears to be at least a month or more in some species (Feutry et al. 2012b).

A few species other than anguillid eels also appear to have some form of at least semi-catadromous life histories in more temperate waters based on their movements towards brackish water for spawning. On the Atlantic coast of Europe it appears that the European flounder *Platichthys flesus* (Fig. 6.1C) might use estuaries as spawning areas, as well as growth habitats. Larvae can also be found in freshwater during early life stage (Morais et al. 2011; Daverat et al. 2012b). The larvae of the European flounder may be able to use STST to move upstream in river estuaries (Bos 1999; Jager 1999), and in the laboratory the flounder larvae showed a preference for freshwater in a choice experiment (Bos and Thiel 2006). This suggests that larvae enter the estuaries and then try to move upstream towards freshwater, with the smaller juveniles likely showing a preference for lower salinities and larger ones preferring higher salinities within estuaries (Primo et al. 2013; Souza et al. 2013). Some other flounder species also enter freshwater to various degrees in Australia, New Zealand and elsewhere, with reproduction occurring in the marine environment (McDowall 1988). Otolith microchemistry studies have shown that flounders use estuarine habitats, but a wide range of freshwater and marine habitat-use patterns exist and some spawning by flounders may occur in brackish water and also in freshwater (Daverat et al. 2011 2012b; Morais et al. 2011).

The Atlantic coast of Europe also has a species of mullet, the thinlip mullet *Liza ramada*, that use estuaries for growth and may enter freshwater before they spawn in marine waters (Daverat et al. 2011). Other species of catadromous mullets, the flathead grey mullet *Mugil cephalus*, and white mullet *M. curema*, have been studied using otolith microchemistry, showing that these species used estuaries after being born in higher salinities, but few individuals resided very long in freshwater (Chang et al. 2004; Wang et al. 2010; Ibáñez et al. 2012). Mullet species clearly enter estuarine habitats as new recruits and use them as nursery areas and for later growth habitats (e.g., DeSiva 1980; Chub et al. 1981; Torricelli et al. 1982; Lebreton et al. 2011). If the adults only use offshore marine habitats as spawning areas, then they would meet the basic criteria for being considered catadromous, rather than simply being estuarine-dependent fishes that use marine waters for both feeding and reproduction. Mullets appear to have a diversity of life histories around the world relative to how they use freshwater, or if they use it. However, some species might have an obligate freshwater juvenile growth stage (McDowall 1988; Lucas and Baras 2001).

A somewhat similar situation may exist for tarpon, such as *Megalops cyprinoides*, which has been found to enter freshwater in places such as Papua New Guinea (Coates 1987; McDowall 1988); but they probably also use the marine environment for feeding and not just reproduction and larval development. Tarpon leptocephali have a shorter larval duration and do not grow as large as most anguilliform leptocephali, but they have a comparatively fast growth rate (Crabtree et al. 1992; Zerbi et al. 2001). Otolith microchemistry showed evidence that tarpon are a facultative freshwater user in Taiwan, but some fish did not enter freshwater much or moved back and forth (Shen et al. 2009).

Two species of catadromous sculpins are present in southern Japan and nearby regions that spawn in the lower reaches of estuaries in benthic nests, with their larvae remaining close to river mouths and moving upstream into the estuaries. The fourspine sculpin *Cottus kazika* (Fig. 6.1G) uses freshwater habitats for juvenile growth (Takeshita et al. 2005) and spawns in nests in the estuary (Takeshita et al. 1999), with the larvae and juveniles being present in estuaries (Harada et al. 1999; Kinoshita et al. 1999). The roughskin sculpin *Trachidermus fasciatus* spawns in nests in the lower estuary (Onikura et al. 2002), and the larvae and early juveniles are mainly present in the brackish section of the estuary, though there is a tendency for these stages to move upstream towards the freshwater part of the estuary (Islam et al. 2007). The juveniles then move upstream into freshwater (Onikura et al. 1999). Comparative observations of the larval distribution behavior of *C. kazika* and *T. fasciatus* made in the laboratory suggest that there may be differences in the dispersal and settlement characteristics of these species (Takeshita et al. 2004). There may be a few other species or coastal forms of sculpins that also move from freshwater to estuaries to spawn, such as some populations of prickly sculpin *Cottus asper* on the west coast of North America (McDowall 1988).

McDowall (1988) also lists several species of fishes that appear to be catadromous, but most of these have not been studied much yet. The endemic Papuan black snapper *Lutjanus goldiei* is an important sport fish living in freshwater in Papua New Guinea and can reach one meter in length, but apparently reproduces in the marine environment. It is a member of the Lutjanidae, which are almost all marine fishes, but little appears to be published about this species. Similarly, an endemic lutjanid in Fiji, the Orange-spotted therapon *Mesopristes kneri*, appears to live in freshwater and migrate to the estuary to reproduce (McDowall 1988). The Lutjanidae colonized freshwater in Australia, where it diversified into various freshwater species according to trophic specializations (Davis et al. 2012), but it is unclear if any of the Australian species are catadromous or not. The Australian bullrout *Notesthes robusta* is a scorpaniform fish with venomous spines that is found in freshwater, which may move to the estuary to reproduce, but this does not seem to be well documented (McDowall 1988). Although there are various species of amphidromous galaxiids (Closs and Warburton 2016), perhaps only one species is marginally catadromous, because mature adults of the common galaxias *Galaxias maculatus* migrate downstream to the estuary to spawn and die, but may also sometimes spawn in freshwater or have landlocked populations (McDowall 1987, 1988; Barbee et al. 2011).

Advantages of catadromous life histories

From this overview of the diversity of types of fishes that have some type of catadromous life history and previous evaluations of diadromy (McDowall 1987, 1988, 1993), it is apparent that these species groups have derived their diadromy independently, and this type of life history is not a phylogenetic trait restricted to particular lineages of fishes, or fishes with a particular body form (Fig. 6.1). How and why diadromy has evolved is an interesting subject that has been examined by authors such as Gross (1987), McDowall (1993, 1997, 2002), Parenti (2008), Dodson et al. (2009), Tsukamoto et al. (2009) and Feutry et al. (2013). It has been hypothesized that diadromy evolved in response to differences in productivity between the freshwater and marine environments, which vary latitudinally (higher productivity in freshwater in the tropics and higher productivity in the ocean at high latitudes) and that it would evolve when the collective benefit of switching habitats outweighed the cost of migration between habitats (Gross 1987; Gross et al. 1988). Differences in productivity between environments do not seem to be the only factors affecting the distribution of the occurrence of diadromy though (McDowall 2008). There are various differences in competition and predation among fishes and other animals between freshwater and marine habitats that also can likely support the establishment of diadromy (McCleave and Edeline 2009).

Diadromy may be beneficial though, because using different environments for particular parts of the life history allows catadromous species to benefit from each habitat's advantages and avoid their particular disadvantages. For example, eels can use freshwater and estuarine habitats for juvenile feeding and growth, but their reproduction and larval growth occurs far away and is not linked to the ecology of their juvenile growth habitats. This provides them several advantages over purely freshwater fishes and over those living only in the marine environment. Marine habitats would have had many other eels present that could be competitors, because anguillid eels are one of the most recently derived groups of anguilliform eels, whereas freshwater environments likely lacked any eels (Inoue et al. 2010). Anguillid eels may, in a sense, have been 'escaping' the marine environment to obtain the benefits of less competition for food, and to experience lower predation pressure (Tsukamoto et al. 2009). Once anguillid eels had established their use of freshwater as juvenile growth habitats, they would have begun to compete with other non-eel predatory fishes in those habitats for food and to have predator-prey interactions with them. However, one major advantage of having a catadromous life history, compared to pure freshwater fishes, would be that the eels would have the potential to prey upon the eggs and young stages of the non-diadromous freshwater species. But because the eels migrate offshore into the ocean to reproduce, those freshwater species could not prey upon the eggs and youngest stages of eels. Similarly, by leaving the marine environment for their juvenile growth, anguillids are probably exposed to fewer predators and do not have to compete with marine eels for food (Kaifu et al. 2013b).

Other catadromous fishes that developed life histories in which they migrate out of freshwater or estuaries to reproduce, would also likely obtain benefits in competitive interactions with freshwater fishes and also the benefits of using marine habitats for larval development and growth. Leaving freshwater to reproduce in brackish or marine habitats, likely results in greater food resources being available for larval feeding and growth. Rivers may have lower quality food resources for fish larvae, or at least can be more unpredictable due to flooding or drought, compared to the marine environment. In parts of Australia, for example, some freshwater fishes must time their reproduction with periods of high water that provide nursery habitats for their larvae and juveniles (Roberts et al. 2008). These types of unpredictable environmental factors would favor catadromous species to migrate to the estuary to spawn. After reproduction, the nearshore brackish waters and freshwater rivers and lakes, where juveniles grow, would likely have fewer large predators to prey on the juvenile and adult stages of catadromous fishes than the coastal marine environment, so returning to freshwater may be advantageous. As discussed for amphidromous fishes (McDowall 2001, 2010), a catadromous life history of using freshwater or brackish water for juvenile growth, but having larval development and growth occurring in marine habitats, would enable larval dispersal through the ocean and colonization of new and unstable freshwater systems, such as newly formed islands. Spawning offshore by anguillids in a relatively stable environment compared to freshwater provides the chance for higher larval survival and for much wider dispersal. Even spawning in coastal waters by other catadromous fishes will allow some larvae to be transported to nearby river systems, or if transported offshore, much wider dispersal is possible. This is likely an advantage over a purely freshwater life history in areas with frequent environmental changes over evolutionary time.

Concluding remarks

Catadromous fishes include some interesting and economically important fishes around the world, as well as some unusual and poorly known species. The use of purely freshwater habitats unrelated to estuaries by most of these species, including anguillid eels, appears to be a conditional strategy, as overviewed by McCleave and Edeline (2009), in which individuals use freshwater when it is advantageous for them to do so. Thus, these species may be facultatively catadromous in their use of freshwater (e.g., Daverat et al. 2012b). Knowledge is steadily accumulating for many of these species, but we are far from a clear understanding of many aspects of their life histories. It is important to note though, that what we see in the present day must be considered in the context of the extensive anthropogenic impacts that have occurred in both the aquatic and terrestrial realms of the Earth through habitat loss and degradation, changes in community structure due to deforestation, development, agriculture, and overfishing (Vitousek et al. 1997; Lotze et al. 2006). For example, in highly populated parts of the world, or in heavily channelized and polluted areas, it would seem that there might be little advantage for some species to consistently enter freshwaters for long periods of time during their catadromous life histories. In many places, fishes are blocked by dams that prevent them from migrating upstream or safely returning back downstream, and populations of many diadromous fishes have declined dramatically (Haro et al. 2000; Limburg and Waldman 2009), so the present-day populations of catadromous fishes may be showing us their ability to adapt to changing environments more than they are showing us how much they may have used freshwater environments before their world was changed by humans.

Acknowledgements

The contributions of the late Robert McDowall to help us understand diadromy are hard to underestimate, and his insights and reflections on these fishes will long be remembered. Pedro Morais and Françoise Daverat gave comments on the chapter. Katsumi Tsukamoto and Tsuguo Otake provided support while writing the chapter.

References

Aarestrup K, Okland F, Hansen MM, Righton D, Gargan P, Castonguay M, Bernatchez L, Howey P, Sparholt H, Pedersen MI and McKinley RS (2009) Oceanic spawning migration of the European eel (*Anguilla anguilla*). Science 325: 1660.
Adams A, Wolfe RK, Barkowski N and Overcash D (2009) Fidelity to spawning grounds by a catadromous fish, *Centropomusun decimalis*. Marine Ecology Progress Series 389: 213–222.

Ahn H, Yamada Y, Okamura A, Horie N, Mikawa N, Tanaka S and Tsukamoto K (2012) Effect of water temperature on embryonic development and hatching time of the Japanese eel *Anguilla japonica*. Aquaculture 330-333: 100–105.

Aida K, Tsukamoto K and Yamauchi K (eds.) (2003) Eel Biology. Springer Verlag, Tokyo.

Aoyama J (2009) Life history and evolution of migration in catadromous eels (Genus *Anguilla*). Aqua-BioScience Monographs 2: 1–42.

Aoyama J and Miller MJ (2003) The silver eel. pp. 107–117. *In*: Aida K, Tsukamoto K and Yamauchi K (eds.). Eel Biology. Springer, Tokyo.

Aoyama J, Wouthuyzen S, Miller MJ, Inagaki T and Tsukamoto K (2003) Short-distance spawning migration of tropical freshwater eels. The Biological Bulletin 204: 104–108.

Aoyama J, Wouthuyzen S, Miller MJ, Minegishi Y, Minagawa G, Kuroki M, Suharti SR, Kawakami T, Sumardiharga KO and Tsukamoto K (2007) Distribution of leptocephali of the freshwater eels, genus *Anguilla*, in the waters off west Sumatra in the Indian Ocean. Environmental Biology of Fishes 80: 445–452.

Aoyama J, Watanabe S, Miller MJ, Mochioka N, Otake T, Yoshinaga K and Tsukamoto K (2014) Spawning Sites of the Japanese Eel in Relation to Oceanographic Structure and the West Mariana Ridge. PLOS ONE 9: e88759.

Athauda S, Anderson T and de Nys R (2012) Effect of rearing water temperature on protandrous sex inversion in cultured Asian Seabass (*Lates calcarifer*). General and Comparative Endocrinology 175: 416–423.

Barbee NC, Hale R, Morrongiello J, Hicks A, Semmens D, Downes BJ and Swearer SE (2011) Large-scale variation in life history traits of the widespread diadromous fish, *Galaxias maculatus*, reflects geographic differences in local environmental conditions. Marine & Freshwater Research 62: 790–800.

Benson LK and Fitzsimons JM (2002) Life history of the Hawaiian fish *Kuhlia sandvicensis* as inferred from daily growth rings of otoliths. Environmental Biology of Fishes 65: 131–137.

Blaber SJM, Milton DA and Salini JP (2008) The biology of barramundi (*Lates calcarifer*) in the Fly River System. *In*: Bolton B (ed.). Developments in Earth and Environmental Sciences, Elsevier 9: 411–426.

Blewett DA, Stevens PW, Champeau TR and Taylor RG (2009) Use of rivers by common snook *Centropomusun decimalis* in southwest Florida: a first step in addressing the overwintering paradigm. Florida Scientist 72: 310–324.

Bonhommeau S, Castonguay M, Rivot E, Sabatié R and LePape O (2010) The duration of migration of Atlantic *Anguilla* larvae. Fish and Fisheries 11: 289–306.

Bos AR (1999) Tidal transport of flounder larvae (*Pleuronectes flesus*) in the Elbe River. Germany. Archive of Fishery and Marine Research 47: 47–60.

Bos AR and Thiel R (2006) Influence of salinity on the migration of postlarval and juvenile flounder (*Pleuronectes flesus* L.) in a gradient experiment. Journal of Fish Biology 68: 1411–1420.

Bruijs MCM and Durif CMF (2009) Silver eel migration and behaviour. *In*: van den Thillart G, Dufour S and Rankin JC (eds.). Spawning Migration of the European Eel. Fish and Fisheries Series. Springer 30: 65–95.

Cairns DK, Secor DA, Morrison WE and Hallett JA (2009) Salinity-linked growth in anguillid eels and the paradox of temperate-zone catadromy. Journal of Fish Biology 74: 2094–2114.

Castonguay M and McCleave JD (1987) Vertical distributions, diel and ontogenetic vertical migrations and net avoidance of leptocephali of *Anguilla* and other common species in the Sargasso Sea. Journal of Plankton Research 9: 195–214.

Chang CW, Iizuka Y and Tzeng WN (2004) Migratory environmental history of the grey mullet *Mugil cephalus* as revealed by otolith Sr:Ca ratios. Marine Ecology Progress Series 269: 277–288.

Chubb CF, Potter IC, Grant CJ, Lenanton RCJ and Wallace J (1981) Age structure, growth rates and movements of sea mullet, *Mugil cephalus* L., and yellow-eyed mullet, *Aldrichetta forsteri* (Valenciennes), in the Swan-Avon River system, Western Australia. Australina Journal of Marine & Freshwater Research 32: 605–628.

Chow S, Kurogi H, Mochioka N, Kaji S, Okazaki M and Tsukamoto K (2009) Discovery of mature freshwater eels in the open ocean. Fisheris Science 75: 257–259.

Closs GP and Warburton M (2016). Life histories of amphidromous fishes. pp. 102–122. *In*: Morais P and Daverat F (eds.). An Introduction to Fish Migration. CRC Press, Boca Raton, FL, USA (this book).

Coates D (1987) Observations on the biology of tarpon, *Megalops cyprinoides* (Broussonet) (Pisces: Megalopidae), in the Sepik River, northern Papua New Guinea. Australian Journal of Marine & Freshwater Research 38: 529–35.

Crabtree RE, Cyr EC, Bishop RE, Falkenstein LM and Dean JM (1992) Age and growth of tarpon *Megalops atlanticus*, larvae in the eastern Gulf of Mexico, with notes on relative abundance and probable spawning areas. Environmental Biology of Fishes 35: 361–370.

Crook DA, Koster WM, Macdonald JI, Nicol SJ, Belcher CA, Dawson DR, O'Mahony DJ, Lovett D, Walker A and Bannam L (2010) Catadromous migrations by female tupong (*Pseudaphritis urvillii*) in coastal streams in Victoria, Australia. Marine & Freshwater Research 61: 474–483.

Dannewitz J, Maes GE, Johansson L, Wickström H, Volckaert FAM and Järvi T (2005) Panmixia in the European eel: a matter of time... Proceedings of the Royal Society of London B 272: 1129–1137.

Daverat F and Tomás J (2006) Tactics and demographic attributes in the European eel *Anguilla anguilla* in the Gironde watershed, SW France. Marine Ecology Progress Series 307: 247–257.

Daverat F, Limburg KE, Thibault I, Shiao J-C, Dodson JJ, Caron F, Tzeng W-N, Iizuka Y and Wickström H (2006) Phenotypic plasticity of habitat use by three temperate eel species, *Anguilla anguilla*, *A. japonica* and *A. rostrata*. Marine Ecology Progress Series 308: 231–241.

Daverat F, Martin J, Fablet R and Pécheyran C (2011) Colonisation tactics of three temperate catadromous species, eel *Anguilla anguilla*, mullet *Liza ramada* and flounder *Platichthys flesus*, revealed by Bayesian multielemental otolith microchemistry approach. Ecology of Freshwater Fish 20: 42–51.

Daverat F, Beaulaton L, Poole R, Lambert P, Wickström H, Andersson J, Aprahamian M, Hizem B, Elie P, Yalçın-Özdilek S and Gumus A (2012a) One century of eel growth: changes and implications. Ecology of Freshwater Fish 21: 325–336.

Daverat F, Morais P, Dias E, Babaluk J, Martin J, Eon M, Fablet R, Pécheyran C and Antunes C (2012b) Plasticity of European flounder life history patterns discloses alternatives to catadromy. Marine Ecology Progress Series 465: 267–280.

Davey AJH and Jellyman DJ (2005) Sex determination in freshwater eels and management options for manipulation of sex. Reviews in Fish Biology and Fisheries 15: 37–52.

Davis AM, Unmack PJ, Pusey JB, Johnson JB and Pearson RG (2012) Marine–freshwater transitions are associated with the evolution of dietary diversification in terapontid grunters (Teleostei: Terapontidae). Journal of Evolutionary Biology 25: 1163–1179.

Davis TLO (1985) Seasonal changes in gonad maturity, and abundance of larvae and early juveniles of barramundi, *Latescalcarifer* (Bloch), in Van Diemen Gulf and the Gulf of Carpentaria. Australian Journal of Marine & Freshwater Research 36: 177–190.

Davis TLO (1986) Migration patterns in barramundi, *Latescalcarifer* (Bloch), in Van Diemen Gulf, Australia, with estimates of fishing mortality in specific areas. Fisheries Research 4: 243–258.

De Silva SS (1980) Biology of juvenile grey mullet: a short review. Aquaculture 19: 21–36.

Dodson JJ, Laroche J and Frédéric L (2009) Contrasting evolutionary pathways of anadromy in euteleostean fishes. *In*: Haro AJ, Smith KL, Rulifson RA, Moffitt CM, Klauda RJ, Dadswell MJ, Cunjak RA, Cooper JE, Beal KL and Avery TS (eds.). Challenges for Diadromous Fishes in a Dynamic Global Environment. Symp 69. American Fisheries Society, Bethesda, Maryland.

Dörner H, Skov C, Berg S, Schulze T, Beare DJ and Velde VdG (2009) Piscivory and trophic position of *Anguilla anguilla* in two lakes: importance of macrozoobenthos density. Journal of Fish Biology 74: 2115–2131.

Durif C, Dufour S and Elie P (2005) The silvering process of *Anguilla anguilla*: a new classification from the yellow resident to the silver migrating stage. Journal of Fish Biology 66: 1025–1043.

Durif CMF, Travade F, Rives J, Elie P and Gosset C (2008) Relationship between locomotor activity, environmental factors, and timing of the spawning migration in the European eel, *Anguilla anguilla*. Aquatic Living Resources 21: 163–170.

Durif CMF, Browman HI, Phillips JB, Skiftesvik AB, Vøllestad LA and Stockhausen HH (2013) Magnetic compass orientation in the European eel. PLOS ONE 8: e59212. doi:10.1371/journal.pone.0059212.

Edeline E (2007) Adaptive phenotypic plasticity of eel diadromy. Marine Ecology Progress Series 341: 229–232.

Feunteun E, Laffaille P, Robinet T, Briand C, Baisez A, Oliver JM and Acou A (2003) A review of upstream migration and movements in inland waters by anguillid eels: toward a general theory. pp. 191–213. *In*: Aida K, Tsukamoto K and Yamauchi K (eds.). Eel Biology. Springer-Verlag, Tokyo.

Feutry P, Keith P, Pécheyran C, Claverie F and Robinet T (2011) Evidence of diadromy in the French Polynesian *Kuhliamalo* (Teleostei: Percoidei) inferred from otolith microchemistry analysis. Ecology of Freshwater Fish 20: 636–645.

Feutry P, Tabouret H, Maeda K, Pécheyran C and Keith P (2012a) Diadromous life cycle and behavioural plasticity in freshwater and estuarine Kuhliidae species (Teleostei) revealed by otolith microchemistry. Aquatic Biology 15: 195–204.

Feutry P, Valade P, Ovenden JR, Lopez PJ and Keith P (2012b) Pelagic larval duration of two diadromous species of Kuhliidae (Teleostei: Percoidei) from Indo-Pacific insular systems. Marine & Freshwater Research 63: 397–402.

Feutry P, Castelin M, Ovenden JR, Dettaï A, Robinet T, Cruaud C and Keith P (2013) Evolution of diadromy in fish: insights from a tropical genus (*Kuhlia* Species). American Naturalist 181: 52–63.

Fukuda N, Miller MJ, Aoyama J, Shinoda A, Suzuki Y and Tsukamoto K (2013) Evaluation of the pigmentation stages and body proportions from the glass eel to yellow eel stage in *Anguilla japonica*. Fisheries Science 79: 425–438.

Gagnaire PA, Minegishi Y, Zenboudji S, Valade P, Aoyama J and Berrebi P (2011) Within-population structure highlighted by differential introgression across semipermeable barriers to gene flow in *Anguilla marmorata*. Evolution 65: 3413–3427.

van Ginneken VJT and Maes GE (2005) The European eel (*Anguilla anguilla*, Linnaeus), its lifecycle, evolution and reproduction: a literature review. Reviews in Fish Biology and Fisheries 15: 367–398.

van Ginneken V, Durif C, Balm SP, Boot R, Verstegen MWA, Antonissen E and van den Thillart G (2007) Silvering of European eel (*Anguilla anguilla* L.): seasonal changes of morphological and metabolic parameters. Animal Biology 57: 63–77.

Gross MR (1987) The evolution of diadromy in fishes. American Fisheries Society Symposium 1: 14–25.

Gross MR, Coleman RM and McDowall RM (1988) Aquatic productivity and the evolution of diadromous fish migration. Science 239: 1291–1293.

Han YS, Hung CL, Liao YF and Tzeng WN (2010) Population genetic structure of the Japanese eel *Anguilla japonica*: evidence for panmixia in spatial and temporal scales. Marine Ecology Progress Series 401: 221–232.

Harada S, Kinoshita I, Omi H and Tanaka M (1999) Distribution and migration of a catadromous sculpin, *Cottus kazika*, larvae and juveniles in the Yura estuary and neighboring waters, facing Wakasa Bay. Japanese Journal of Ichthyology 46: 91–99 (In Japanese with an English abstract).

Harris JH (1986) Reproduction of the Australian bass, *Macquarian ovemaculeata* (Perciformes: Percichthyidae) in the Sydney Basin. Australian Journal of Marine & Freshwater Research 37: 209–235.

Haro A, Richkus W, Whalen K, Hoar A, Busch W-D, Lary S, Brush T and Dixon D (2000) Population decline of the American eel: implications for research and management. Fisheries 25: 7–16.

Hedger RD, Dodson JJ, Hatin D, Caron F and Fournier D (2010) River and estuary movements of yellow-stage American eels *Anguilla rostrata*, using a hydrophone array. Journal of Fish Biology 76: 1294–1311.

Helfman GS, Facey DE, Hales S, Jr. and Bozeman, EL, Jr. (1987) Reproductive ecology of the American eel. pp. 42–56. *In*: Dadswell MJ, Klauda RJ, Moffitt CM, Saunders RL, Rulifson RA and Cooper JE (eds.). Common Strategies of Anadromous and Catadromous Fishes. American Fisheries Society Symposium 1, American Fisheries Society, Bethesda, Maryland.

Heupel MR, Knip DM, de Lestang P, Allsop QA and Grace BS (2011) Short-term movement of barramundi in a seasonally closed freshwater habitat. Aquatic Biology 12: 147–155.

Hogan AE and Nicholson JC (1987) Sperm motility of sooty grunter *Hephaestus fuliginosus* (Macleay), and jungle perch, *Kuhlia rupestris* (Lacépède), in different salinities. Australian Journal Marine & Freshwater Research 38: 523–528.

Ibáñez AL, CW Chang, Hsu CC, Wang CH, Iizuka Y and Tzeng WN (2012) Diversity of migratory environmental history of the mullets *Mugil cephalus* and *M. curema* in Mexican coastal waters as indicated by otolith Sr:Ca ratios. Ciencias Marinas 38: 73–87.

Inoue JG, Miya M, Miller MJ, Sado T, Hanel R, López JA, Hatooka K, Aoyama J, Minegishi Y, Nishida M and Tsukamoto K (2010) Deep-ocean origin of the freshwater eels. Biology Letters 6: 363–366.

Ishikawa S, Tsukamoto K and Nishida M (2004) Genetic evidence for multiple geographic populations of the giant mottled eel *Anguilla marmorata* in the Pacific and Indian oceans. Ichthyological Research 51: 343–353.

Islam M, Hibino M and Tanaka M (2007) Distribution and diet of the roughskin sculpin, *Trachidermus fasciatus*, larvae and juveniles in the Chikugo River estuary, Ariake Bay, Japan. Ichthyological Research 54: 160–167.

Jager Z (1999) Selective tidal stream transport of flounder larvae (*Platichthys flesus* L.) in the Dollard (Ems estuary). Estuarine Coastal and Shelf Science 49: 347–362.

Jager Z (2001) Transport and retention of flounder larvae (*Platichthys flesus* L.) in the Dollard nursery (Ems estuary). Journal of Sea Research 45: 153–171.

Jellyman DJ (1995) Longevity of longfinned eels *Anguilla dieffenbachii* in a New Zealand high country lake. Ecology of Freshwater Fish 4: 106–112.

Jellyman DJ (2003) The distribution and biology of the South Pacific species of *Anguilla*. pp. 275–292. *In*: Aida K, Tsukamoto K and Yamauchi K (eds.). Eel Biology. Springer, Tokyo.

Jellyman D and Tsukamoto K (2005) Swimming depths of offshore migrating longfin eels *Anguilla dieffenbachii*. Marine Ecology Progress Series 286: 261–267.

Jellyman D and Tsukamoto K (2010) Vertical migrations may control maturation in migrating female *Anguilla dieffenbachii*. Marine Ecology Progress Series 404: 241–247.

Jessop BM (2010) Geographic effects on American eel (*Anguilla rostrata*) life history characteristics and strategies. Canadian Journal of Fisheries and Aquatic Sciences 67: 326–346.

Jessop BM, Shiao JC, Iizuka Y and Tzeng WN (2002) Migratory behaviour and habitat use by American eels *Anguilla rostrata* as revealed by otolith microchemistry. Marine Ecology Progress Series 233: 217–229.

Jessop BM, Shiao JC, Iizuka Y and Tzeng WN (2004) Variation in the annual growth, by sex and migration history, of silver American eels *Anguilla rostrata*. Marine Ecology Progress Series 272: 231–244.

Jessop BM, Cairns, DK, Thibault I and Tzeng WN (2008) Life history of American eel *Anguilla rostrata*: new insights from otolith microchemistry. Aquatic Biology 1: 205–216.

Kaifu K, Tamura M, Aoyama J and Tsukamoto K (2010) Dispersal of yellow phase Japanese eels *Anguilla japonica* after recruitment in the Kojima Bay-Asahi River system, Japan. Environmental Biology of Fishes 88: 273–282.

Kaifu K, Miyazaki S, Aoyama J, Kimura S and Tsukamoto K (2012) Diet of Japanese eels *Anguilla japonica* in the Kojima Bay-Asahi River system, Japan. Environmental Biology of Fishes 96: 439–446.

Kaifu K, Miller MJ, Yada T, Aoyama J, Washitani I and Tsukamoto K (2013a) Growth differences of Japanese eels *Anguilla japonica* between fresh and brackish water habitats in relation to annual food consumption in the Kojima Bay-Asahi River system, Japan. Ecology of Freshwater Fish 22: 127–136.

Kaifu K, Miller MJ, Aoyama J, Washitani I and Tsukamoto K (2013b) Evidence of niche segregation between freshwater eels and conger eels in Kojima Bay Japan. Fisheries Science 79: 593–603.

Keith P (2003) Biology and ecology of amphidromous Gobiidae of the Indo-Pacific and the Caribbean regions. Journal of Fish Biology 63: 831–847.

Kleckner RC and McCleave JD (1982) Entry of migrating American eel leptocephali into the Gulf Stream system. Helgoländer wiss Meeresunters 35: 329–339.

Kleckner RC and McCleave JD (1985) Spatial and temporal distribution of American eel larvae in relation to North Atlantic Ocean current systems. Dana 4: 67–92.

Kleckner RC and McCleave JD (1988) The northern limit of spawning by Atlantic eels (*Anguilla* spp.) in the Sargasso Sea in relation to thermal fronts and surface water masses. Journal of Marine Research 46: 647–667.

Kinoshita I, Azuma K, Fujita S, Takahashi I, Niimi K and Harada S (1999) Early life history of a catadromous sculpin in western Japan. Environmental Biology of Fishes 54: 135–149.

Kurogi H, Okazaki M, Mochioka N, Jinbo T, Hashimoto H, Takahashi M, Tawa A, Aoyama J, Shinoda A, Tsukamoto K, Tanaka H, Gen K, Kazeto Y and Chow S (2011) First capture of post-spawning female of the Japanese eel *Anguilla japonica* at the southern West Mariana Ridge. Fisheries Science 77: 199–205.

Kuroki M, Aoyama J, Miller MJ, Wouthuyzen S, Arai T and Tsukamoto K (2006) Contrasting patterns of growth and migration of tropical anguillid leptocephali in the western Pacific and Indonesian Seas. Marine Ecology Progress Series 309: 233–246.

Kuroki M, Aoyama J, Miller MJ, Watanabe S, Shinoda A, Jellyman DJ, Feunteun E and Tsukamoto K (2008) Distribution and early life history characteristics of anguillid leptocephali in the western South Pacific. Marine & Freshwater Research 59: 1035–1047.

Kuroki M, Aoyama J, Miller MJ, Yoshinaga T, Shinoda S, Hagihara K and Tsukamoto K (2009) Sympatric spawning of *Anguilla marmorata* and *Anguilla japonica* in the western North Pacific Ocean. Journal of Fish Biology 74: 1853–1865.

Kuroki M, Miller MJ and Tsukamoto K (2014) Diversity of early life history traits in freshwater eels and the evolution of their oceanic migrations. Canadian Journal of Zoology 92: 749–770.

Krueger WH and Oliveira K (1999) Evidence for environmental sex determination in the American eel. Environmental Biology of Fishes 55: 381–389.

Lebreton B, Richard P, Parlier EP, Guillou G and Blanchard GF (2011) Trophic ecology of mullets during their spring migration in a European saltmarsh: a stable isotope study. Estuarine, Coastal and Shelf Science 91: 502–510.

Li C, Ricardo BR, Smith WL and Ortí G (2011) Monophyly and interrelationships of snook and barramundi (Centropomidaesensu Greenwood) and five new markers for fish phylogenetics. Molecular Phylogenetics and Evolution 60: 463–71.

Limburg KE and Waldman JR (2009) Dramatic declines in North Atlantic diadromous fishes. Bioscience 59: 955–965.

Lotze HK, Lenihan HS, Bourque BJ, Bradbury RH, Cooke RG, Kay MC, Kidwell SM, Kirby MX, Peterson CH and Jackson JBC (2006) Depletion, degradation, and recovery potential of estuaries and coastal seas. Science 312: 1806–1809.

Lucas MC and Baras E (2001) Migration of Freshwater Fishes. Blackwell Science Ltd., London.

Manabe R, Aoyama J, Watanabe K, Kawai M, Miller MJ and Tsukamoto K (2011) First observations of the oceanic migration of the Japanese eel using pop-up archival transmitting tags. Marine Ecology Progress Series 437: 229–240.

Marui M, Arai T, Miller MJ, Jellyman DJ and Tsukamoto K (2001) Comparison of the early life history between New Zealand temperate eels and Pacific tropical eels revealed by otolith microstructure and microchemistry. Marine Ecology Progress Series 213: 273–284.

McCleave JD (1993) Physical and behavioral controls on the oceanic distribution and migration of leptocephali. Journal of Fish Biology 43(Suppl A.): 243–273.

McCleave JD (2003) Spawning areas of Atlantic eels. pp. 141–156. *In*: Aida K, Tsukamoto K and Yamauchi K (eds.). Eel Biology. Springer-Verlag, Tokyo.

McCleave JD and Kleckner RC (1982) Selective tidal stream transport in the estuarine migration of glass eels of the American eel (*Anguilla rostrata*). Journal du Conseil International Pour l'Exploration de la Mer 40: 262–271.

McCleave JD and Edeline E (2009) Diadromy as a conditional strategy: patterns and drivers of eel movements in continental habitats. pp. 97–119. *In*: Haro AJ, Avery T, Beal K, Cooper J, Cunjak R, Dadswell M, Klauda R, Moffitt C, Rulifson R and Smith K (eds.). Challenges for Diadromous Fishes in a Dynamic Global Environment. American Fisheries Society Symposium, Vol. 69, Bethesda.

McCleave JD, Brickley PJ, O'Brien KM, Kistner-Morris DA, Wong MW, Gallagher M and Watson SM (1998) Do leptocephali of the European eel swim to reach continental waters? Status of the question. Journal of the Marine Biological Association of the United Kingdom 78: 285–306.

McCulloch M, Cappo M, Aumend J and Müller W (2005) Tracing the life history of individual barramundi using laser ablation MC-ICP-MS Sr-isotopic and Sr/Ba ratios in otoliths. Marine & Freshwater Research 56: 637–644.

McDowall RM (1987) The occurrence and distribution of diadromy among fishes. pp. 1–13. *In*: Dadswell MJ, Klauda RJ, Moffitt CM, Saunders RL, Rulifson RA and Cooper JE (eds.). Common Strategies of Anadromous and Catadromous Fishes. American Fisheries Society Symposium 1, American Fisheries Society, Bethesda, Maryland.

McDowall RM (1988) Diadromy in Fishes: Migrations between Freshwater and Marine Environments. Croom Helm, London. 308pp.

McDowall RM (1993) A recent marine ancestry for diadromous fishes? Sometimes yes, but mostly no. Environmental Biology of Fishes 37: 329–335.

McDowall RM (1997) The evolution of diadromy in fishes (revisited) and its place in phylogenetic analysis. Reviews in Fish Biology and Fisheries 7: 443–462.

McDowall RM (2001) Diadromy, diversity and divergence: implications for speciation processes in fishes. Fish and Fisheries Series 2: 278–285.

McDowall RM (2002) The origin of the salmonid fishes: marine, freshwater… or neither? Reviews in Fish Biology and Fisheries 11: 171–179.

McDowall RM (2008) Why are so many boreal freshwater fishes anadromous? Confronting 'conventional wisdom'. Fish and Fisheries 9: 208–213.

McDowall RM (2010) Why be amphidromous: Expatrial dispersal and the place of source and sink population dynamics? Reviews in Fish Biology and Fisheries 20: 87–100.

McRae MG, McRae LB and Fitzsimons JM (2011) Habitats used by juvenile flagtails (*Kuhlia* spp.; Perciformes: Kuhliidae) on the island of Hawai'i. Pacific Science 65: 441–450.

Miles NG, West RJ and Norman MD (2009) Does otolith chemistry indicate diadromous lifecycles for five Australian riverine fishes? Marine & Freshwater Research 60: 904–911.

Miller MJ (2003) The worldwide distribution of anguillid leptocephali. pp. 157–168. *In*: Aida K, Tsukamoto K and Yamauchi K (eds.). Eel Biology. Springer-Verlag, Tokyo.

Miller MJ (2009) Ecology of anguilliform leptocephali: remarkable transparent fish larvae of the ocean surface layer. Aqua-BioScience Monographs 2: 1–94.

Miller MJ and Tsukamoto K (2004) An introduction to leptocephali: Biology and identification. Ocean Research Institute, University of Tokyo. pp. 96.

Miller MJ, Aoyama J and Tsukamoto K (2009a) New perspectives on the early life history of tropical anguillid eels: Implications for management. pp. 71–84. *In*: Casselman J and Cairns D (eds.). Eels at the Edge. American Fisheries Society, Symposium 58, Bethesda, Maryland.

Miller MJ, Kimura S, Friedland KD, Knights B, Kim H, Jellyman DJ and Tsukamoto K (2009b) Review of ocean-atmospheric factors in the Atlantic and Pacific oceans influencing spawning and recruitment of anguillid eels. pp. 231–249. *In*: Haro AJ et al. (eds.). Challenges for Diadromous Fishes in a Dynamic Global Environment. American Fisheries Society, Symposium 69, Bethesda, Maryland.

Miller MJ, Chikaraishi Y, Ogawa NO, Yamada Y, Tsukamoto K and Ohkouchi N (2013) A low trophic position of Japanese eel larvae indicates feeding on marine snow. Biology Letters 9: 20120826.

Milton DA and Chenery SR (2005) Movement patterns of barramundi *Lates calcarifer*, inferred from [87]Sr/[86]S and Sr/Ca ratios in otoliths, indicate non-participation in spawning. Marine Ecology Progress Series 301: 279–291.

Minegishi Y, Aoyama J and Tsukamoto K (2008) Multiple population structure of the giant mottled eel, *Anguilla marmorata*. Molecular Ecology 17: 3109–3122.

Mochioka N and Iwamizu M (1996) Diet of anguillid larvae: leptocephali feed selectively on larvacean houses and fecal pellets. Marine Biology 125: 447–452.

Moore R (1979) Natural sex inversion in the giant perch (*Lates calcarifer*). Australian Journal of Marine & Freshwater Research 30: 803–813.

Moore R (1982) Spawning and early life history of barramundi, *Lates calcarifer* (Bloch), in Papua New Guinea. Australian Journal of Marine & Freshwater Research 33: 647–61.

Moore R and Reynold LF (1982) Migration patterns of barramundi, *Lates calcarifer* (Bloch), in Papua New Guinea. Australian Journal of Marine & Freshwater Research 33: 671–682.

Morais P, Dias E, Babaluk J and Antunes C (2011) The migration patterns of the European flounder *Platichthys flesus* (Linnaeus, 1758) (Pleuronectidae, Pisces) at the southern limit of its distribution range: ecological implications and fishery management. Journal of Sea Research 65: 235–246.

Morrison WE and Secor DH (2003) Demographic attributes of yellow-phase American eels (*Anguilla rostara*) in the Hudson River estuary. Canadian Journal of Fisheries and Aquatic Sciences 60: 1487–1501.

Munk P, Hansen MM, Maes GE, Nielsen TG, Castonguay M, Riemann L, Sparholt H, Als TD, Aarestrup K, Andersen NG and Bachler M (2010) Oceanic fronts in the Sargasso Sea control the early life and drift of Atlantic eels. Proceedings of the Royal Society of London B 277: 3593–3599.

Myers GS (1949) Usage of anadromous, catadromous and allied terms for migratory fishes. Copeia 1949: 89–97.

Nelson JS (2006) Fishes of the World. 4th Edition. John Wiley & Sons, Inc. Hoboken, New Jersey.

Nishi T and Kawamura G (2005) *Anguilla japonica* is already magnetosensitive at the glass eel phase. Journal of Fish Biology 67: 1213–1224.

Nishi T, Kawamura G and Matsumoto K (2004) Magnetic sense in the Japanese eel *Anguilla japonica* as determined by conditioning and electrocardiography. The Journal of Experimental Biology 207: 2965–2970.

Oka S-i and Tachihara K (2001) Estimation of spawning sites in the spotted flagtail, *Kuhlia marginata*, based on sperm motility. Ichthyological Research 48: 425–427.

Oka S-i and Tachihara K (2008) Migratory history of the spotted flagtail, *Kuhlia marginata*. Environmental Biology of Fishes 81: 321–327.

Okamura A, Yamada Y, Yokouchi K, Horie N, Mikawa N, Utoh T, Tanaka S and Tsukamoto K (2007) A silvering index for the Japanese eel *Anguilla japonica*. Environmental Biology of Fishes 80: 77–89.

Oliveria K and McCleave JD (2002) Sexually different growth histories of the American eel in four rivers in Maine. Transactions of the American Fisheries Society 131: 203–211.

Onikura N, Takeshita N, Matsui S and Kimura S (1999) Growth and migration of the rough skin sculpin, *Trachidermus fasciatus*, in the Kashima River, Kyushu Island, Japan. Japanese Journal of Ichthyology 46: 31–37 (In Japanese with an English abstract).

Onikura N, Takeshita N, Matsui S and Kimura S (2002) Spawning grounds and nests of *Trachidermus fasciatus* (Cottidae) in the Kashima and Shiota estuaries system facing Ariake Bay, Japan. Ichthyological Research 49: 198–201.

Otake T (2003) Metamorphosis. pp. 61–74. *In*: Aida K, Tsukamoto K and Yamauchi K (eds.). Eel Biology. Springer-Verlag, Tokyo.

Otake T, Nogami K and Maruyama K (1993) Dissolved and particulate organic matter as possible food sources for eel leptocephali. Marine Ecology Progress Series 92: 27–34.

Parenti LR (2008) Life history patterns and biogeography: an interpretation of diadromy in fishes. Annals of the Missouri Botanical Garden 95: 232–257.

Parker SJ (1995) Homing ability and home range of yellow-phase American eels in a tidally dominated estuary. Journal of the Marine Biological Association of the United Kingdom 75: 127–140.

Parker SJ and McCleave JD (1997) Selective tidal stream transport by American eels during homing movements and estuarine migration. Journal of the Marine Biological Association of the United Kingdom 77: 881–889.

Peters KM, Matheson RE, Jr. and Taylor RG (1998) Reproduction and early life history of common snook, *Centropomus undecimalis* (Bloch), in Florida. Bulletin of Marine Science 62: 509–529.

Primo AL, Azeiteiro UM, Marques SC, Martinho F, Baptista J and Pardal MA (2013) Colonization and nursery habitat use patterns of larval and juvenile flatfish species in a small temperate estuary. Journal of Sea Research 76: 126–134.

Randall JE and HA Randall (2001) Review of the fishes of the genus *Kuhlia* (Perciformes: Kuhliidae) of the Central Pacific. Pacific Science 55: 227–256.

Roberts DT, Duivenvoorden LJ and Stuart IG (2008) Factors influencing recruitment patterns of golden perch (*Macquaria ambigua oriens*) within a hydrologically variable and regulated Australian tropical river system. Ecology of Freshwater Fish 17: 577–589.

Robins J, Mayer D, Staunton-Smith J, Halliday I, Sawynok B and Sellin M (2006) Variable growth rates of the tropical estuarine fish barramundi *Lates calcarifer* (Bloch) under different freshwater flow conditions. Journal of Fish Biology 69: 379–391.

Russell DJ and Garrett RN (1985) Early life history of barramundi, *Lates calcarifer* (Bloch), in North-eastern Queensland. Australian Journal of Marine & Freshwater Research 36: 191–201.

Russell DJ and Garrett RN (1988) Movements of juvenile barramundi, *Lates calcarifer* (Bloch), in North-eastern Queensland. Australian Journal of Marine & Freshwater Research 39: 117–123.

Sasai S, Kaneko T, Hasegawa S and Tsukamoto K (1998) Morphological alteration in two types of gill chloride cells in Japanese eels (*Anguilla japonica*) during catadromous migration. Canadian Journal of Zoology 76: 1480–1487.

Sasai S, Katoh F, Kaneko T and Tsukamoto K (2007) Ontogenic change of gill chloride cells in leptocephalus and glass eel stages of the Japanese eel *Anguilla japonica*. Marine Biology 150: 487–496.

Schabetsberger R, Økland F, Aarestrup K, Kalfatak D, Sichrowsky U, Tambets M, Dall'Olmo G, Kaiser R and Miller PI (2013) Oceanic migration behaviour of tropical Pacific eels from Vanuatu. Marine Ecology Progress Series 75: 177–190.

Schmidt J (1922) The breeding places of the eel. Philosophical Transactions of the Royal Society 211: 179–208.

Shen KN, Chang CW, Iizuka Y and Tzeng WN (2009) Facultative habitat selection in Pacific tarpon *Megalops cyprinoides* as revealed by otolith Sr:Ca ratios. Marine Ecology Progress Series 387: 255–263.

Souza AT, Dias E, Nogueira A, Campos J, Marques JC and Martins I (2013) Population ecology and habitat preferences of juvenile flounder *Platichthys flesus* (Actinopterygii: Pleuronectidae) in a temperate estuary. Journal of Sea Research 79: 60–69.

Takeshita N, Onikura N, Nagata S, Matsui S and Kimura S (1999) A note on the reproductive ecology of the catadromous fourspine sculpin, *Cottuskazika* (Scorpaeniformes: Cottidae). Ichthyological Research 46: 309–313.

Takeshita N, Onikura N, Matsui S and Kimura S (2004) Comparison of early life history in two catadromous sculpin, *Trachidermus fasciatus* and *Cottus kazika*. Journal of National Fisheries University 52: 83–92.

Takeshita N, Ikeda I, Onikura N, Nishikawa M, Nagata S, Matsui S and Kimura S (2005) Growth of the fourspine sculpin *Cottus kazika* in the Gonokawa River, Japan, and effects of water temperature on growth. Fisheries Science 71: 784–790.

Taylor RG, Whittington JA, Grier HJ and Crabtree RE (2000) Age, growth, maturation, and protandric sex reversal in common snook, *Centropomus undecimalis*, from the east and west coasts of south Florida. Fishery Bulletin 98: 612–624.

Tesch FW (2003) The Eel Biology and Management of Anguillid Eels. Blackwell Publishing, London.

Thibault I, Dodson JJ, Caron F, Tzeng WN, Iizuka Y and Shiao JC (2007) Facultative catadromy in American eels: testing the conditional strategy hypothesis. Marine Ecology Progress Series 344: 219–229.

Torricelli P, Tongiorgi P and Almansi P (1982) Migration of grey mullet fry into the Arno River: seasonal appearance, daily activity and feeding rhythms. Fisheries Research 1: 219–234.

Trnski T, Hay AC and Fielder DS (2005) Larval development of estuary perch (*Macquaria colonorum*) and Australian bass (*M. novemaculeata*) (Perciformes: Percichthyidae), and comments on their life history. Fishery Bulletin 103: 183–194.

Trotter AA, Blewett DA, Taylor RG and Stevens PW (2012) Migrations of common snook from a tidal river with implications for skipped spawning. Transactions of the American Fisheries Society 141: 1016–1025.

Tsukamoto K (2009) Oceanic migration and spawning of anguillid eels. Journal of Fish Biology 74: 1833–1852.

Tsukamoto K and Arai T (2001) Facultative catadromy of the eel *Anguilla japonica* between freshwater and seawater habitats. Marine Ecology Progress Series 220: 265–276.

Tsukamoto K, Nakai I and Tesch WV (1998) Do all freshwater eels migrate? Nature 396: 635–636.

Tsukamoto K, Aoyama J and Miller MJ (2002) Migration, speciation and the evolution of diadromy in anguillid eels. Canadian Journal of Fisheries and Aquatic Sciences 59: 1989–1998.

Tsukamoto K, Otake T, Mochioka N, Lee TW, Fricke H, Inagaki T, Aoyama J, Ishikawa S, Kimura S, Miller MJ, Hasumoto H, Oya M and Suzuki Y (2003) Seamounts, new moon and eel spawning: the search for the spawning site of the Japanese eel. Environmental Biology of Fishes 66: 221–229.

Tsukamoto K, Miller MJ, Kotake A, Aoyama J and Uchida K (2009) The origin of fish migration: the random escapement hypothesis. pp. 45–61. *In*: Haro A, Avery T, Beal K, Cooper J, Cunjak R, Dadswell M, Klauda R, Moffitt C, Rulifson R and Smith K (eds.). Challenges for Diadromous Fishes in a Dynamic Global Environment. American Fisheries Society Bethesda, MD.

Tsukamoto K, Chow S, Otake T, Kurogi H, Mochioka N, Miller MJ, Aoyama J, Kimura S, Watanabe S, Yoshinaga T, Shinoda A, Kuroki M, Oya M, Watanabe T, Hata K, Ijiri S, Kazeto Y, Nomura K and Tanaka H (2011) Oceanic spawning ecology of freshwater eels in the western North Pacific. Nature Communications 2: 179.

Tzeng WN, Severin KP and Wickström H (1997) Use of otolith microchemistry to investigate the environmental history of European eel *Anguilla anguilla*. Marine Ecology Progress Series 149: 73–81.

van den Thillart G, Dufour S and Rankin JC (eds.) (2009) Spawning Migration of the European Eel. Fish and Fisheries Series 30. Springer Science+Business Media, Berlin. 477p.

van der Wal E (1995) Effects of temperature and salinity on the hatch rate and survival of Australian bass (*Macquaria novemaculeata*) eggs and yolk-sac larvae. Aquaculture 47: 239–244.

Vitousek PM, Mooney HA, Lubchenco J and Melillo JM (1997) Human domination of Earth's ecosystems. Science 277: 494–499.

Vøllestad LA (1992) Geographic variation in age and length at metamorphosis of maturing European eel—environmental effects and phenotypic plasticity. Journal of Animal Ecology 61: 41–48.

Walsh CT, Reinfelds IV, Gray CA, West RJ, van der Meulen DE and Williams LFG (2010) Growth, episodic recruitment and age truncation in populations of a catadromous percichthyid, *Macquaria colonorum*. Marine & Freshwater Research 61: 397–407.

Walsh CT, Gray CA, West RJ and Williams LFG (2011) Reproductive biology and spawning strategy of the catadromous percichthyid, *Macquaria colonorum*. Environmental Biology of Fishes 91: 471–486.

Walsh CT, Reinfelds IV, Gray CA, West RJ, van der Meulen DE and Craig JR (2012) Seasonal residency and movement patterns of two co-occurring catadromous percichthyids within a south-eastern Australian river. Ecology of Freshwater Fish 21: 145–159.

Walther BD, Dempster T, Letnic M and McCulloch MT (2011) Movements of diadromous fish in large unregulated tropical rivers inferred from geochemical tracers. PLOS ONE 6: e18351.

Wang CH, Hsu CC, Chang CW, You CF and Tzeng WN (2010) The migratory environmental history of freshwater resident flathead mullet *Mugil cephalus* L. in the Tanshui River, northern Taiwan. Zoological Studies 49: 504–514.

Watanabe S and Miller MJ (2012) Species, geographic distribution, habitat and conservation of freshwater eels. pp. 1–43. *In*: Sachiko N and Fujimoto M (eds.). Eels: Physiology, Habitat and Conservation. Nova Science Publishers, New York.

Watanabe S, Aoyama J and Tsukamoto K (2009) A new species of freshwater eel *Anguilla luzonensis* (Teleostei: Anguillidae) from Luzon Island of the Philippines. Fisheries Science 75: 387–392.

Wilson JM, Antunes JC, Bouça D and Coimbra J (2004) Osmoregulatory plasticity of the glass eel of *Anguilla anguilla*: freshwater entry and changes in branchial ion-transport protein expression. Canadian Journal of Fisheries and Aquatic Sciences 61: 432–442.

Wippelhauser GS and McCleave JD (1988) Rhythmic activity of migrating juvenile American eels (*Anguilla rostrata* LeSueur). Journal of the Marine Biological Association of the United Kingdom 68: 81–91.

Wirth T and Bernachez L (2003) Decline of North Atlantic eels: a fatal synergy? Proceedings of the Royal Society of London B 270: 681–688.

Wouthuyzen S, Aoyama J, Sugeha YH, Miller MJ, Kuroki M, Minegishi Y, Suharti S and Tsukamoto K (2009) Seasonality of spawning by tropical anguillid eels around Sulawesi Island, Indonesia. Naturwissenschaften 96: 153–158.

Wuenschel MJ and Able KW (2008) Swimming ability of eels (*Anguilla rostrata, Conger oceanicus*) at estuarinein gress: contrasting patterns of cross-shelf transport? Marine Biology 154: 775–786.

Yagishita N, Miya M, Yamanoue Y, Shirai SM, Nakayama K, Suzuki N, Satoh TP, Mabuchi K, Nishida M and Nakabo T (2009) Mitogenomic evaluation of the unique facial nerve pattern as a phylogenetic marker within the perciform fishes (Teleostei: Percomorpha). Molecular Phylogenetics and Evolution 53: 258–266.

Yoshinaga T, Miller MJ, Yokouchi K, Otake T, Kimura S, Aoyama J, Watanabe S, Shinoda A, Oya M, Miyazaki S, Zenimoto K, Sudo R, Takahashi T, Ahn H, Manabe R, Hagihara S, Morioka H, Itakura H, Machida M, Ban K, Shiozaki M, Ai B and Tsukamoto K (2011) Genetic identification and morphology of naturally spawned eggs of the Japanese eel *Anguilla japonica* collected in the western North Pacific. Fisheries Science 77: 983–992.

Yokouchi K, Aoyama J, Oka HP and Tsukamoto K (2008) Variation in the demographic characteristics of yellow phase Japanese eels in different habitats of the Hamana Lake system, Japan. Ecology of Freshwater Fish 17: 639–652.

Yokouchi K, Fukuda N, Shirai K, Aoyama J, Daverat F and Tsukamoto K (2011) Time lag of the response on the otolith strontium/calcium ratios of the Japanese eel, *Anguilla japonica* to changes in strontium/calcium ratios of ambient water. Environmental Biology of Fishes 92: 469–478.

Yokouchi K, Fukuda N, Miller MJ, Aoyama J, Daverat F and Tsukamoto K (2012) Influences of early habitat use on the migratory plasticity and demography of Japanese eels in central Japan. Estuarine, Coastal and Shelf Science 107: 132–140.

Zerbi A, Aliaume C and Joyeux J-C (2001) Growth of juvenile tarpon in Puerto Rican estuaries. ICES Journal of Marine Science 58: 87–95.

CHAPTER 7

Life Histories of Amphidromous Fishes

Gerard P. Closs* and Manna Warburton

Definition, recognition and history of amphidromy

The term 'amphidromy' was first proposed by Myers (1949) to describe diadromous life histories where migrations between fresh and sea water are not directly associated with reproduction. Even though the term was first proposed in 1949, it took several decades for it to be widely accepted, with only very limited, if any, use of the term prior to 1970 (McDowall 1992). The term 'amphidromous' only begins to appear with any regularity in the titles and abstracts of published papers in the 1970s (Web of Knowledge; search conducted on 22 August 2012) as Japanese ichthyologists began employing the term to describe the life histories of various, mostly gobioid fishes, of the Japanese archipelago. Clearly, based on the number of publications (see Goto 1990; Iguchi and Mizuno 1990), the term was familiar to many Japanese researchers by 1990, however much of this literature was published in Japanese and hence relatively inaccessible to English speaking researchers.

Regular use of the term in the English-language scientific literature only began in the late 1980s with McDowall's cross-taxon review of the distribution of amphidromy (McDowall 1988) and Kinzie's use of the term to describe the life cycles of the Hawaiian stream fish fauna (Kinzie 1988). However, the term is now in regular use with a keyword search (Web of Knowledge; search conducted on 22 August 2012) using 'amphidromous' yielding 32 papers for 2011 alone, out of a total of 401 papers with a publication record starting in 1976. In comparison, the terms anadromous and catadromous have their origins in the fisheries literature over 200 years ago (McDowall 1992). Keyword searches using anadromous or catadromous (Web of Knowledge; conducted on 22 August 2012) produce publication records beginning in 1930 and 1937 respectively. Over 8,500 published papers use the term anadromous, but only 387 papers are identified as using the term catadromous.

Department of Zoology, University of Otago, PO Box 56, Dunedin, New Zealand.
* Corresponding author

The slow acceptance of the term 'amphidromous' was most likely due to the somewhat vague original definition proposed by Myers (1949), plus a general lack of exposure of North American and European fisheries researchers to fish that might be considered to be amphidromous (McDowall 2007). Myers' (1949) paper clarified the use of the terms anadromous and catadromous, defining them as life histories exhibited by fish that spent most of their time in either fresh or sea water respectively, and migrated to the alternate environment to breed. Using Myers' (1949) definition, both anadromous and catadromous involve a clearly defined spawning migration between marine and freshwater habitats. In contrast, amphidromy was defined as, "Diadromous fishes whose migration from fresh water to the sea, or vice-versa, is not for the purpose of breeding, but occurs regularly at some other definite stage of the life-cycle" (Fig. 7.1a). As McDowall (1988) comments, this definition is something of a "catch-all for left-over groups" not classified as either anadromous or catadromous. However, Myers (1949) also used the goby genus *Sicydium* to exemplify his concept of an amphidromous life history, describing them as fish that "live and spawn in swift water, but the larvae float downstream to salt water where they remain for a time before migrating back upstream as small fry", thus providing, perhaps unintentionally, a framework from which a more refined definition of amphidromy could be developed. Of the various authors that have used the term amphidromy, it is McDowall more than any other who progressively refined (perhaps, arguably redefined) the concept of amphidromy (McDowall 1988, 1992, 1998, 2007, 2010a) as a life history exclusively characterized by freshwater spawning, an immediate post-hatch larval migration to the sea, a return juvenile migration to freshwater, followed by growth to maturity over a period of months to years in freshwater (Fig. 7.1b,c).

McDowall (2007) considered the strong cultural differences in the use of the term 'amphidromous' reflected the likelihood of fisheries biologists' exposure to fishes that might be viewed as amphidromous. Certainly, amphidromous fish form a diverse and abundant component of the Japanese ichthyofauna, and it is Japanese ichthyologists who first made regular use of the term (see Goto 1990; Iguchi and Mizuno 1990), although much of their early work was published in Japanese language journals (e.g., Goto 1981). As mentioned previously, the first researchers to regularly use the term in the English-language scientific literature were McDowall (1988) and Kinzie (1988) working in New Zealand and Hawaii respectively, and hence, regions with a diverse amphidromous ichthyofauna. Whilst the diadromous migrations of many of the Australian and New Zealand species were widely appreciated and understood (e.g., Humphries 1989; Rowe et al. 1992), they were not described as amphidromous. Indeed, even McDowall only used the term tentatively at two points in his text "New Zealand Freshwater Fishes", published in 1990 (McDowall 1990). However, since 2000, the use of the term has become far more widespread, with the publication of papers describing various aspects of the biology of species that are explicitly described as being amphidromous from Australia (Miles et al. 2009; Rolls 2011; Ebner et al. 2010), Caribbean (Flevet et al. 2001; Debrot 2003a; Keith 2003; Cook et al. 2009, 2010), China (Nip 2010), Madagascar (Loiselle 2005; Keith et al. 2011a), Mexico and Central America (Lyons 2005), North America (Nordlie 2012) and West Africa (Keith et al. 2011b). The term is also increasingly being used to describe the life

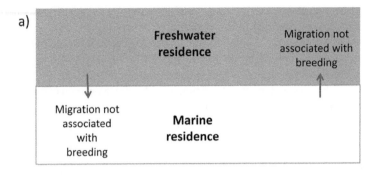

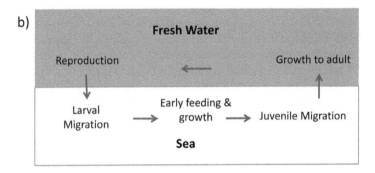

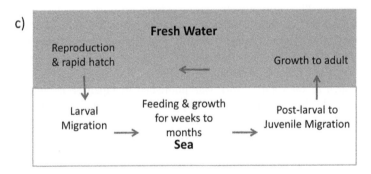

Figure 7.1. Increasing levels of detail and precision in published descriptions of amphidromy by (a) Myers 1949; (b) McDowall 1987; and (c) McDowall 2010a.

history of some atyid crustacean shrimps and gastropod molluscs in the genus *Neritina* (March et al. 1998; Flevet et al. 2001; Debrot 2003b; Cook 2004; McDowall 2004; Page et al. 2005; Kano 2009; Crandall et al. 2010; Gorbach et al. 2012). However, the amphidromous fish literature is still dominated by papers relating to fish from the Indo-Pacific region (e.g., Iida et al. 2010; Maeda and Tachibara 2010; Hoese and Allen 2011; Keith et al. 2011b; Maeda et al. 2011; Tamada 2011; Feutry et al. 2012a,b; Lord et al. 2012; Walter et al. 2012).

Taxonomic and biogeographic distribution of amphidromous fishes

The full extent of the diversity of amphidromous fishes has yet to be fully realized. When Myers (1949) first proposed the term 'amphidromous', he commented that it was "for a small known (but undoubtedly really large) number of species". McDowall (1988) represents the first attempt to collate a list of all known diadromous species, listing a total of 109 anadromous, 53 catadromous and 52 amphidromous species. A feature of McDowall's (1988) list of amphidromous fish is that it is primarily dominated by comparatively small species in the families Aplochitonidae, Clupeidae, Cottidae, Galaxiidae, Gobiidae, Eleotridae, Pinguipedidae (formerly Mugiloidae) and Prototroctidae, most of which have their centres of diversity remote from North America and Europe. Amphidromous gobies, particularly in the genera *Lentipes*, *Sicyopterus*, *Sicydium*, *Sicyopus*, *Stiphodon*, *Awaous*, *Stenogobius*, *Schismatogobius* and *Rhinogobius*, dominate the diversity of amphidromous fish species, comprising approximately 170 species (Keith 2003). Given the small and often cryptic nature of amphidromous fishes, the taxonomy of many groups is poorly resolved, and since McDowall (1988), the number of species considered to be amphidromous has steadily risen with 273 species listed by Riede (2004). Descriptions of new species of amphidromous Gobiidae, particularly amongst the Sicydiinae, continue to be published regularly (Watson et al. 2001, 2002, 2005, 2007; Chen and Tan 2005; Keith et al. 2002, 2004a,b, 2005a, 2007a,b, 2009, 2010, 2011a; Keith and Marquet 2007; Hoese and Allen 2011; Maeda et al. 2011; Suzuki et al. 2011). The taxonomy of *Rhinogobius* is also poorly understood, with various undescribed species, often with restricted distributions and poorly known life histories referred to in the literature (e.g., Ito et al. 2010; Tamada 2011; Tsunagawa and Arai 2011). The use of otolith microchemical techniques has also confirmed amphidromous life-histories in families and species for which data was previously lacking or unclear, including ninespine stickleback (Gasterosteidae, *Pungitius pungitius*) (Arai and Goto 2008), freshwater mullet (Mugilidae, *Myxus petardi*) (Miles et al. 2009) and the southern smelts (Retropinnidae) (Crook et al. 2008). Amphidromy has also recently been reported from the freshwater pipefish *Microphis leiaspis* (Ishihara and Tachihara 2008) and Indo-Pacific flagtails (Feutry et al. 2012a,b), adding two additional families (Syngnathidae and Kuhliidae respectively) to the list collated by McDowall (1988).

Amphidromous fish have a broad global distribution, although they are particularly prevalent on islands across the Indo-Pacific and Caribbean, and in the coastal streams of southern Australia and South America, and New Zealand, with taxonomic composition varying greatly between regions (McDowall 2007; Keith 2011). Amphidromous Gobiidae, particularly in the subfamily Sicydiinae, are widespread across the Indo-Pacific and Caribbean, often dominating the ichthyofauna of small islands and coastal streams on larger land masses throughout these regions (Keith 2003; Lyons 2005; Thuesen et al. 2011). Although they are most typically associated with small oceanic islands (Keith 2003), they can also be found on larger landmasses including the small coastal streams of Central America, Mexico, West Africa, Madagascar, New Caledonia, Japan (Lyons 2005; Keith et al. 2011a, 2011b), and have recently been found in coastal streams in the wet tropics of northeast Australia (Thuesen et al. 2011) and mainland China (Nip 2010). Around the northern and north-eastern Pacific coastlines, several amphidromous sculpins (cottids) occur (Patten 1971; Goto 1990), the number of which is likely to increase as the taxonomic and life-history knowledge of this group expands (Goto and Arai 2006; Tsukagoshi et al. 2011). Around the coastlines of New Zealand, and southern Australia and South America, various species of amphidromous galaxiid occur, with the greatest diversity being found in southeast Australia and New Zealand (Allen et al. 2002; McDowall 2010b). Amphidromous eleotrids also occur throughout the Indo-Pacific region, although understanding of the life history of many species is limited (Miles 2009; Nordlie 2012).

Larval dispersal has played a key role in the biogeographic distribution of amphidromous fishes (McDowall 2004; Lyons 2005; Keith et al. 2011b). Amphidromous fishes include some of the most widely distributed of any species; *Sicyopterus lagocephalus* has a distribution extending from the western Indian Ocean across to the western and eastern Pacific Ocean, including Japan and northeastern Australia (Lord et al. 2012), and *Galaxias maculatus* has a circumpolar distribution including New Zealand and southern Australia and South America (Barbee et al. 2011). Larval dispersal is the only plausible mechanism that could have facilitated such a distribution (McDowall 2004; Keith et al. 2011b). Genetic studies on *S. lagocephalus* suggest high levels of gene flow across much of the species' extensive range, while results for *G. maculatus* suggest extensive intra-continental gene flow, but only limited inter-continental exchange (Waters et al. 2000). The biogeography of the Sicydiinae gobies is also best explained by successive dispersal, colonization and divergence events (Keith et al. 2011b). The expansive distribution of some species contrasts with the very limited distribution of others species (Keith et al. 2011b; Lord et al. 2012), with some indication that the duration of the larval phase influences the extent of dispersal and hence distribution (Lord et al. 2010). However, the duration of the larval phase bears little or no relationship with the extent of distribution amongst some groups of species (Taillebois et al. 2012).

At finer spatial scales, the distribution and highest diversity of amphidromous fish species is most commonly associated with oceanic islands that have a high geographic relief with steep, fast flowing streams (Keith 2003, 2011; McDowall 2004, 2007), and more recently with steep small coastal streams on continental land masses (Lyons 2005; Iguchi 2007; Thuesen et al. 2011). The dominance of amphidromous fish on isolated islands has been attributed to their ability to disperse, and colonize either recently formed or highly disturbed habitats (McDowall 2007; Keith 2011; Thueusen et al. 2011). However, their absence from low gradient, and often larger river systems, may be related to the slow and limited transport of newly hatched larvae to the sea prior to a pelagic rearing phase (Lyons 2005; Iguchi 2007; Closs et al. 2013). Amphidromous fish typically exhibit high levels of individual fecundity, producing relatively small eggs and often hatching at a very early stage of development (Goto 1990; Keith 2003; McDowall 2009). However, due to the small size of the larvae, death from starvation occurs relatively rapidly following hatching, necessitating rapid transfer from their place of hatching in a fluvial environment to the pelagic feeding and rearing environment (Moriyama et al. 1998; Iguchi and Mizuno 1999; Tamada and Iwata 2005). Increasing female and egg size in *Rhinogobius* sp. CB with distance upstream has been proposed as a mechanism enabling increased larval resistance to starvation, and hence survival whilst drifting downstream to the sea (Tamada 2009). Alternatively, downstream migration by adults to spawn close to the sea, thus reducing the distance larvae must travel to their pelagic rearing habitat, may be a mechanism that enables some amphidromous species to occupy low gradient river systems (McDowall 2009).

Key features and characteristics of amphidromy

The diadromous migratory syndrome first defined as amphidromy by Myers (1949), and later refined by McDowall (McDowall 1997, 2007), is exemplified by the Sicydiine gobies (e.g., *Sicyopterus*, *Sicyopus*, *Smilosicyopus* and *Stiphodon*) of the Indo-Pacific and Caribbean, which are often found inhabiting tropical freshwater streams on rocky islands, often of geologically recent volcanic origin (Keith 2003; McDowall 2007). Freshwaters in these regions are characterized by small disturbance prone streams and rivers, which are often punctuated by spectacular waterfalls and dramatic elevation changes (Poff and Ward 1989; Keith 2003). Amongst the taxonomically diverse Sicydiine gobies, amphidromy occurs in the following fashion. Adults occupying freshwater habitats in these systems are often herbivorous (although this is not a characteristic of most amphidromous fish), thus offering access to an abundant food resource and avoiding reliance on macroinvertebrates, a prey item which is often scarce in these depauperate systems (Benstead et al. 2009). The adults may spawn in place, or make coordinated downstream movements associated with spawning (Keith 2003; Goto 1986; Iguchi et al. 2005; Iguchi and Mizuno 1999). Adults will pair off and spawn on the undersides of rocks and crevices using a variety of strategies for mate selection including body size, nest location (Ito and Yanagisawa 2003), and swimming ability (Takahashi and Kohda 2004). Fecundity is high, and large numbers of small eggs are produced. Small egg size is a shared characteristic of most amphidromous fishes (McDowall 2009a), and often contrasts with larger egg sizes found in closely related non-migratory conspecifics (Goto 1990; Elgar 1990; Closs et al. 2013). Amongst the Sicydiine gobies there is some limited parental care, with nests sometimes being guarded by the male up to the point where the eggs hatch and larvae are released. Following an abbreviated development in the egg, newly hatched larvae emerge and mostly rely on downstream currents to sweep them out of the adult habitat and into the proximate pelagic habitat (Keith 2003; McDowall 2008; Yamasaki et al. 2011).

Amongst the Sicydiine gobies, the pelagic rearing habitat is the near shore marine environment, but for other amphidromous fishes the pelagic habitat could be a downstream freshwater lake or other large water body providing the pelagic conditions necessary for larval development (Closs et al. 2003; David et al. 2004; Closs et al. 2013). There is recent evidence supporting an abbreviated migration in some species, whereby larvae are retained within the influence of the freshwater plume of the natal stream (Sorensen and Hobson 2005) or the nearshore coastal environment (Hicks 2012; Kondo et al. 2013). Details of the larval movements for most species are unknown, but for at least some, the larvae will undergo broader movements into adjacent longshore currents where they may be carried for long distances, sometimes between continental land masses (Waters et al. 2000; Lord et al. 2010). During the pelagic rearing phase, the larvae feed actively on co-occurring zooplankton and make the transition from larvae to juvenile (Kondo et al. 2013). A common morphological trait of amphidromous fish larvae during this period is a lack of body pigmentation, making the larvae translucent (Kondo et al. 2013). This translucent appearance is associated with occupation of the pelagic habitat and is assumed to be for predator avoidance. When the larvae achieve a certain body size or competency, they then undergo their second migration, cued by some zeitgeber (cues of temperature, salinity, light, developmental stage or combination thereof) to transition from the larval marine habitat into the adult freshwater habitat (Boehlert and Mundy 1988; Keith 2003). There is some evidence that the odour of adult conspecifics may act as a cue, guiding juveniles back into suitable streams on their return from marine habitats (Baker and Montgomery 2001; Baker 2003; Baker and Hicks 2003; Hale et al. 2008, 2009a), although the importance of these cues in natural situations has been questioned (Hale et al. 2009b).

Body size at the time of the migration back to freshwater is species-specific, or even population-specific in the case of *Galaxias maculatus* (Barbee et al. 2011). Some goby species migrate to freshwater at little more than 13 mm (Keith 2003), whilst galaxiid species may transition to freshwater at lengths greater than 50 mm (McDowall 2000). This leg of the migration is marked by multiple physiological transformations, such as changes in body and fin morphology, pigmentation, behaviour and diet (Keith et al. 2008; Kondo et al. 2013). In the case of Sicydiine gobies, the change in feeding is particularly marked with juveniles making a transition from a pelagic planktivorous habit to a benthic algal grazing habit (Keith 2003). Transitions of this nature are often reflected in the micro-structure of the fish's otoliths or ear bones (Campana 1999; Hale et al. 2008, 2009a), through a visible 'check mark' that will signify the transition from the pelagic to the benthic habitat (Lord et al. 2010; Shafer 2000). Post-larval fishes often make this transition within the estuarine interface, and are miniature versions of the adults less than a month after entering the non-tidal freshwater habitat (Keith et al. 2008). Once in the non-tidal freshwater habitat, they will continue their migration into the adult habitat to complete the life cycle, often overcoming extreme vertical drop structures, aided by the ventral fin sucker in the case of the Sicydiine gobies (Keith 2003), or strong pectoral and pelvic fins in the case of some galaxiids (David et al. 2009; David and Hamer 2012).

Downstream migration of larvae

The transition of newly hatched larvae to their pelagic rearing environment, be it sea or lake, appears to be a particularly vulnerable stage in the amphidromous life cycle (McDowall 2009). The eggs and larvae of many amphidromous fish are very small, usually less than 1.0 mm in diameter (McDowall 2009a), with newly hatched and poorly resourced larvae needing to gain access to small planktonic prey within a few days if they are not to starve (Moriyama et al. 1998; Iguchi and Mizuno 1999; Iguchi 2007). Various mechanisms appear to facilitate the rapid downstream transport of larvae to a pelagic habitat, including the downstream migrations by adults prior to spawning (Iguchi et al. 2005; Goto 1990; McDowall 2008). In many species, spawning takes place within the adult habitat (Fitzsimons et al. 2007; Allibone and Caskey 2000), hence newly hatched larvae must rely on passive transport on river currents into the pelagic nursery habitat (Goto 1990; Luton et al. 2005; McDowall

2009). In various gobiids and galaxiids, spawning and or hatching is timed to coincide with flood events (O'Connor and Koehn 1998; Charteris et al. 2003; Fitzsimons et al. 2007; McDowall 2009), perhaps increasing the likelihood of rapid transit to the sea. In some species of galaxiids, eggs are deposited so that they remain stranded amongst bankside vegetation as the floodwaters recede, with hatching triggered only when the eggs are resubmerged by floodwaters (O'Connor and Koehn 1998; Charteris et al. 2003). The larvae of many amphidromous fish are poorly developed, and only have limited capacity to maintain their position in the water column and drift downstream (Goto 1990; McDowall 2008; Ellien et al. 2011). McDowall (2008) argued that this pattern of early hatch, occurring as early as within 24–48 hours of fertilization in some species, ensures that migrating larvae have sufficient yolk to survive the downstream passage. As mentioned previously, slow flowing river reaches may act as barriers to the passively transported larvae, thus restricting most amphidromous species to steep, fast flowing coastal streams (Lyons 2005; Iguchi 2007; Thuesen et al. 2011).

Downstream migration of adults

Downstream migration of adults to spawning habitats close to the marine interface are known for a number of amphidromous species in multiple families including *Galaxias maculatus* (Galaxiidae) (McDowall 1990), *Awaous guamensis* (Gobiidae) (Ha and Kinzie 1996), *Cheimarrichthys fosteri* (Scrimgeour and Eldon 1989) and *Plecoglossus altivelis* (Osmeridae) (Iguchi et al. 1998). This downstream migration prior to spawning may be a mechanism to reduce the distance newly hatched larvae must drift to their pelagic rearing habitat (McDowall 2008). In several species, it appears that the spawning sites, whilst still in freshwater, are located as close as is possible to the marine larval rearing habitat. For example, *Awaous guamensis* spawns in the first riffle upstream of the estuary (Ha and Kinzie 1996), and *Galaxias maculatus* deposit their eggs amongst vegetation on spring high tides, leaving the developing eggs stranded above the tideline until the next spring high tide (McDowall 1990; McDowall and Charteris 2006). Such strategies are likely to leave larvae well placed for a rapid transfer to the marine pelagic habitat (McDowall 2008). The life cycle of these downstream migrating amphidromous species, including *Galaxias maculatus*, has been described as being 'marginally catadromous' (McDowall 1990), and the similarity between amphidromous and catadromous life histories has been noted by other authors (Fitzsimons et al. 2002; Kondo et al. 2013). Downstream migration of adults may allow some amphidromous species, such as *Galaxias maculatus* and *Prototroctes maraena*, to overcome the barriers to downstream larval migration that exist in low gradient habitats (see Lyons 2005; Iguchi 2007). Physiological constraints may restrict egg, sperm and early larval survival to freshwater in some groups, including galaxiids and *Plecoglossus altivelis* (Hicks et al. 2010; Iguchi and Takeshima 2011; Wylie 2011), thus preventing a full transition to a catadromous life history.

Facultative amphidromy and land-locked populations

In several amphidromous genera (*Gobiomorphus, Galaxias, Cottus, Rhinogobius, Retropinna* amongst others), the phenomena of land-locking is a common and repeated phenomena (Goto et al. 2002; Closs et al. 2003; David et al. 2004; Crook et al. 2008; Tsunagawa and Arai 2008; Michel et al. 2008; Chapman et al. 2009). Amphidromous land-locking is the process whereby a species completes its life cycle entirely within freshwater, with the pelagic larval phase occurring in a lentic rather than marine environment (McDowall 1988). This has been observed even in systems where natural lakes are connected to the sea by river channels, sometimes resulting in mixed amphidromous and non-migratory populations (Closs et al. 2003; Hicks 2012). It has also been observed to occur in newly created lake habitats produced by human impoundment (McDowall and Allibone 1994; Kawakami and Tachihara 2011), suggesting that at least for some species, the marine phase of the life-cycle is facultative. There appears to be considerable variation in the propensity of species and particular groups to form land-locked populations; for example, within the galaxiids, *Galaxias truttaceus* and *Galaxias brevipinnis*, readily form land-locked populations (Humphries 1989; Hicks 2012), whereas none are known for *Galaxias postvectus*. Similarly, whilst various *Rhinogobius* species readily form land-locked populations, there are no known examples of land-locked Sicydiine gobies (Keith 2003) suggesting that there may be physiological constraints dictating whether a species has the ability to form non-diadromous populations. Exposure to seawater appears to be a critical factor in triggering the final stages of larval development in some Sicydiine gobies (Ellien et al. 2011). The propensity for some species or taxonomic groups to form landlocked populations may promote speciation into non-diadromous forms, as in the Galaxiidae, where amphidromous ancestors have given rise to a suite of closely related non-amphidromous species (Waters and Wallis 2001). However, land-locking does not appear to be a prerequisite for speciation, as there are examples of amphidromous gobioids radiating into non-amphidromous variants without any evidence for the prior formation of landlocked populations (Iguchi 2007; Kondo et al. 2013).

What is the ecological function of amphidromy?

McDowall (2007) discussed eight possible reasons for the existence of amphidromy including: (i) enhanced dispersal to vacant habitats; (ii) enhanced recolonization of previously extirpated habitats; (iii) predator avoidance; (iv) adult habitat suitability; (v) adult adaptation to steep vertical drops in island catchments; (vi) maintenance of high levels of fecundity; (vii) beneficial source/sink population dynamics; and (viii) broad dispersal ability through a planktonic life stage. McDowall (2007, 2010) argued that, among these reasons, dispersal to vacant habitat was the most likely reason for amphidromy, based on the widespread distribution of many species and the dominance of amphidromous fish on isolated islands, many of which are geologically recent, isolated and volcanic in origin. The observation of fish communities dominated by amphidromous fish on geologically recent volcanic islands, in catchments defaunated by volcanic eruptions or ice sheets, or small tropical ephemeral streams, have all been presented as evidence supporting amphidromy as a mechanism for dispersal and colonization of vacant habitat (McDowall 2007, 2010b; Leathwick et al. 2008; Iida et al. 2009; Keith et al. 2011b; Thuesen et al. 2011). Communities in such recently formed or frequently disturbed streams are often depauperate, with few competitors or predators, thus there are potentially significant rewards for any individuals that manage to colonize them (Keith 2003; McDowall 2007). However, whilst amphidromous fish may possess traits that enable them to colonize and exploit habitats that have been recently created or significantly disturbed (e.g., volcanism, drought or an ice age), it is difficult to conceive of a mechanism that would allow species to evolve a trait that is specifically adapted to take advantage of such rare and unpredictable events (Lytle 2001; Closs et al. 2013). Hence, whilst the biogeographic distribution of amphidromous groups of fish can be partly explained by dispersal and colonization of disturbed and isolated habitats (McDowall 2007, 2010b; Leathwick et al. 2008; Iida et al. 2009; Keith et al. 2011b; Thuesen et al. 2011), a dispersive function for amphidromy does not provide a coherent explanation as to why amphidromous fish undertake such arduous and risky migrations to complete their life cycle, generation after generation and often in habitats not subjected to any obvious recent disturbance, nor how such pelagic larvae evolved in the first instance (Closs et al. 2013). There is also accumulating evidence that rather than dispersing, the larvae of many amphidromous fish actively avoid dispersal by rearing in freshwater river plumes (Sorensen and Hobson 2005; Hicks 2012; Kawakami and Tachihara 2011; Kondo et al. 2013), or in coastal freshwater and estuarine lagoons (Closs et al. 2003; David et al. 2004; Ohara et al. 2009; Hicks 2012). Furthermore, the restricted distribution and endemicity of many insular amphidromous species (Ikeda et al. 2003; Keith 2003; Murphy and Cowan 2007; Lord et al. 2010; Keith et al. paper) suggests widespread dispersal is not a feature of the life cycle of many species.

The repeated observation that non-migratory species of *Cottus*, *Rhinogobius* and *Galaxias*, which evolved from amphidromous species, have relatively larger eggs and are less fecund than their amphidromous sister-taxa (McDowall 1970, 2009, 2010a; Goto 1990; Katoh and Nishida 1994; Goto et al. 2002; Goto and Arai 2003; Maruyama et al. 2003; Yokoyama and Goto 2005) suggests a possible egg size/ fecundity life-history trade-off (see Smith and Fretwell 1974) may be influencing the reproductive ecology of these fish (Closs et al. 2013). The newly hatched larvae of the non-migratory conspecifics are relatively larger and more developmentally advanced, and are largely benthic in habit, compared to the newly hatched larvae of amphidromous species (Goto 1990; Mcdowall 2007). There are likely to be several reasons as to why larger, more developmentally advanced larvae are likely to be advantageous in freshwater, and particularly in stream environments (Closs et al. 2013). Such larvae have a greater capacity to resist currents in streams (McDowall 2007; Jellyman and McIntosh 2010), larger gape size and hence access to a wider range of prey (McDowall 2007), greater tolerance to starvation (Moriyama et al. 1998; Iguchi and Mizuno 1999; Maruyama et al. 2003; Tamada and Iwata 2005) and reduced susceptibility to spinal malformations caused by trematode infections (Kelly et al. 2010; Poulin et al. 2012). However, whilst the production of large eggs may be required for the survival of larvae in difficult freshwater environments (Einum and Fleming 1999, 2000), species that evolve wholly freshwater life cycles tend to be significantly less fecund than their amphidromous sister-species (McDowall 1970, 2007, 2010a; Patten 1971; Humphries 1989; Goto 1990).

A striking feature of the distribution of amphidromous species and their closely related non-migratory conspecifics is that they are largely allopatric, with non-migratory species mostly upstream of the amphidromous species (Goto 1990; Leathwick et al. 2008). Whilst the downstream migration of larvae (Goto 1990; Moriyama et al. 1998; Iguchi and Mizuno 1999; Maruyama et al. 2003; Tamada and Iwata 2005; Ellien et al. 2011) and the upstream migration of juveniles (McDowall 1998) place obvious constraints on the distribution of amphidromous species, it is not obvious why the distribution of non-migratory species is restricted to habitats upstream of the amphidromous species. However, the non-migratory species are likely to be at a significant competitive disadvantage in habitats close to the pelagic rearing habitat of their significantly more fecund amphidromous relatives. These patterns of distribution, fecundity and larval ecology suggest that amphidromy may be a life-history strategy that enables small freshwater fish (mostly fluvial) to maintain a high level of fecundity by producing small eggs and pelagic larvae, at least in habitats relatively close to suitable marine or lentic pelagic habitats (Closs et al. 2013).

Conservation of amphidromous species

Disruption of migration corridors

Amphidromous fish face similar challenges and vulnerabilities as other diadromous fish species due to their migration between freshwater and marine biomes (Keith 2003; Nordlie 2012; Ramirez et al. 2012; Walter et al. 2012). Of particular importance is the risk of disruption of long migratory corridors due to multiple impacts, including poorly constructed road culverts and dams (Joy and Death 2001; David et al. 2009; David and Hamer 2012), water abstraction, pollution, and instream habitat degradation including deforestation, particularly of the lower reaches of rivers (Luton et al. 2005; McEwan and Joy 2009; Gorbach et al. 2012; Walter et al. 2012; Ramirez et al. 2012; Nordlie 2012). Recent work suggests that remediation of passage through poorly designed culverts may be achieved by the strategic placement of climbing ropes (David et al. 2009; David and Hamer 2012), however there is little or no information on how larger barriers to migration may be overcome. In some regions, over-collecting by aquarists may also pose a threat (Ebner et al. 2011). Whilst multiple populations connected by larval dispersal may confer a degree of resilience on amphidromous populations that are subjected to the occasional disturbance (Thuesen et al. 2011), there is concern that the loss of critical habitat for either adults, juveniles or for spawning may also create population sinks, whereby juveniles are attracted in from the sea but are unable to complete their life cycle in degraded habitats (Hickford and Schiel 2011). The effects of such diffuse impacts on multiple inter-connected populations distributed across a broad landscape are largely unknown (Hickford and Schiel 2011). Whilst some amphidromous species are widespread (Keith 2003; McDowall 2007; Lord et al. 2012), the distribution of many amphidromous fish is comparatively restricted, often to a single island or limited number of catchments (Ishihara and Tachihara 2008; Maeda et al. 2011; Lord et al. 2012), thus making them vulnerable to land-use intensification and roading developments that may degrade migration corridors and habitat across their entire range (e.g., MacKenzie 2008).

Fisheries

Juveniles of various gobies and galaxiids, returning to freshwater after the marine pelagic phase of their life cycle, form the basis of artisanal, recreational and semi-commercial fisheries in various regions, including New Zealand (McDowall 1990), Reunion island (Hoareau et al. 2007; Valade et al. 2009; Teichert et al. 2012), Caribbean, Colombia (Castellanos-Galindo et al. 2011) and the Philippines (Myers 1949). In New Zealand, whitebait fisheries are based on the return migration of the post-larval stage of five species of *Galaxias* (*G. maculatus, G. argenteus, G. brevipinnis, G. fasciatus* and *G. postvectis*) (McDowall 1990), although the fishery is generally dominated by *G. maculatus* (Rowe et al. 1992). Historically, the fishery has undergone a significant decline, with limited understanding of the relationships between adult and juvenile populations (Hickford and Schiel 2011). In 1955, the annual whitebait catch on the West Coast, New Zealand peaked at 322,000 kg (McDowall and Eldon 1980), and given that the weight of each juvenile fish is approximately 0.45 g, the total catch represents a truly staggering number of individuals being removed from the population. However, the impact of fisheries-induced mortality is largely unknown, and declines in the abundance of galaxiid whitebait have mostly been attributed to habitat degradation associated with intensification of agriculture and draining of coastal wetlands (Hickford and Schiel 2011). In Colombia, on the tropical eastern Pacific coastline, the post larvae of *Sicydium salvini* are collected in a goby-fry fishery that represents an important seasonal food source for coastal populations (Castellanos-Galindo et al. 2011). As much as 1.37 ton per month may be harvested by a single village of 5000 inhabitants, representing the removal of approximately 20 million fish from the population (Castellanos-Galindo et al. 2011). Globally, goby-fry fisheries are largely unmanaged with little or no data on historic or current yields (Bell 1999). Even in New Zealand, known globally for its proactive management of marine fisheries, the amphidromous whitebait fishery is largely unregulated with no accurate collection of catch data. Some management of the fishery comes in the form of seasonal and daylight limits on fishing activities. These limits are designed to allow passage of some fraction of the catchable returning larval population, even during the active fishing season, although at present there is no published data on the value of these conservation measures.

Techniques for meeting research challenges

A major challenge in understanding the extent and dynamics of amphidromy is that the critical migratory stages primarily occur in the larval and post-larval life history stages (McDowall 2007). Due to the small size of these migratory life history stages, the more traditional methods used to tag and track the movements of larger fish cannot be used to track the migration of amphidromous fish (McDowall et al. 1975; Walther and Limburg 2012), and hence amphidromous migrations remain poorly described and understood (Keith 2003; McDowall 2007). Otolith microstructure analysis of daily increments and annuli (Campana 1999) can provide some indication of the age of larval and post-larval transitions between marine and freshwater habitats

(McDowall et al. 1994; Hale 2008, 2009a; Lord et al. 2010; Shafer 2000), however such methods cannot provide information on where the larvae have been.

Chemical analyses of various hard and soft body tissues are providing valuable insights into the ecology of amphidromous fish, potentially providing information on both movements and their timing (Walther and Limburg 2012). For example, stable isotope analysis ($\delta^{13}C$, $\delta^{15}N$) of post-larvae of amphidromous gobies entering freshwater streams in Hawaii suggests the larvae reared within the freshwater plume of streams entering the inshore environment, most likely without dispersing widely (Sorensen and Hobsen 2005). Such information is at odds with assumptions and evidence, gathered mostly through genetic techniques, of broad mixing of amphidromous larvae in the pelagic offshore environment (Waters et al. 2000; Keith et al. 2005b; Schmidt et al. 2010; Lord et al. 2012).

Otolith microchemistry is a rapidly developing field of research whereby trace element composition and isotopic ratios within the layers of the otolith are analyzed, allowing researchers to reconstruct life histories (Elsdon et al. 2008; Walther and Limburg 2012), and potentially describe fine scale movement within habitats across the lifetime of a single animal (Elsdon and Gillanders 2005). Otolith microchemical analyses are providing significant insights into the larval and post-larval life-history phases in many amphidromous species, including confirmation of amphidromy and duration of larval phase in species for which detailed life history information is lacking (Goto and Arai 2003, 2006; Iguchi et al. 2005; Crook et al. 2008; Tsunagawa and Arai 2008; Miles et al. 2009; Lord et al. 2010; Feutry et al. 2012), evidence of facultative amphidromy in a variety of species (Closs et al. 2003; David et al. 2004; Hicks et al. 2005; Arai and Goto 2008; Crook et al. 2008; Ohara et al. 2009; Hicks 2012), and indications of regional philopatry in species of *Galaxias* and *Gobiomorphus* (Hicks 2012).

Specialized field sampling techniques

Because of the broad range of habitats and large-scale migrations that define amphidromous life histories, specialized sampling techniques may be required to answer research questions. As larval life history is perhaps the biggest missing piece in the puzzle of amphidromous life cycles, specialized sampling gear focused on detecting and capturing larval fishes may provide additional insights. Specialized plankton and drift nets commonly employed in plankton and freshwater macroinvertebrate sampling may find use when focusing on amphidromous fishes (Luton et al. 2005; Bell 2007). Setting drift nets to capture downstream movements of larval fishes during the first few hours after dark may provide insights into community composition and reproductive timing (Luton et al. 2005). Snorkel surveys have also proved to be effective at detecting these otherwise difficult to sample fish species (Thuesen et al. 2011). Night spotlighting using a powerful lamp or torch has also been found to be an excellent alternative to more traditional methods to both locate and catch fish in hand nets, resulting in greater detection of species and providing better estimates of density, particularly in low water velocity conditions (Hickey and Closs 2006; David et al. 2007; Hansen and Closs 2005).

Conclusions

Recognition of amphidromy as a distinct form of diadromous migration has been slow (McDowall 2010). However based on the increasingly widespread and frequent use of the term, amphidromy now has a well-recognized place in the lexicon of fisheries biology. Myers' (1949) original definition of amphidromy was rather vague, simply referring to it as migration not associated with breeding between the sea and freshwaters. However, Myers' use of *Sicydium* to illustrate what he meant by amphidromy clearly suggests that his views were not that dissimilar to the conceptual refinement of amphidromy as later proposed and developed by McDowall (1997, 1998, 2007, 2010). It is the late Bob McDowall who deserves much of the credit for taking Myers' (1949) original definition and refining the concept (McDowall 2007, 2010) into a form that is readily recognized and clearly applies to a wide variety of mostly small freshwater fish that often dominate fish communities in coastal streams in many regions of the world. The availability of rapidly developing analytical techniques, particularly in the area of otolith microchemistry (Walther et al. 2012) and genetics (Lord et al. 2012), is also driving research in amphidromy, enabling analyses of population-level patterns of movement, recruitment and source-sink dynamics (Closs et al. 2003; Crook et al. 2008; Ohara et al. 2009; Lord et al. 2010, 2012; Hicks 2012). These techniques will enable researchers to increasingly address critical knowledge gaps in the ecology of amphidromous fishes, particularly in our understanding of the extent of movement and exchange of larvae between adult populations, and the ecological requirements of larvae in marine and lentic pelagic environments.

References

Allen GR, Midgley SH and Allen M (2002) Freshwater Fishes of Australia. Western Australian Museum, Perth.

Allibone RM and Caskey D (2000) Timing and habitat of koaro (*Galaxias brevipinnis*) spawning in streams draining Mt. Taranaki, New Zealand. New Zealand Journal of Marine and Freshwater Research 34: 593–595.

Arai T and Goto A (2008) Diverse migratory histories in a brackish water type of the ninespine stickleback, *Pungitius pungitius*. Environmental Biology of Fishes 83: 349–353.

Baker CF (2003) Effect of adult pheromones on the avoidance of suspended sediment by migratory banded kokopu juveniles. Journal of Fish Biology 62: 386–394.

Baker CF and Montgomery JC (2001) Species-specific attraction of migratory banded kokopu juveniles to adult pheromones. Journal of Fish Biology 58: 1221–1229.

Baker CF and Hicks BJ (2003) Attraction of migratory inanga (*Galaxias maculatus*) and koaro (*Galaxias brevipinnis*) juveniles to adult galaxiid odours. New Zealand Journal of Marine and Freshwater Research 37: 291–299.

Barbee NC, Hale R, Morrongiello J, Hicks A, Semmens D, Downes BJ and Swearer SE (2011) Large-scale variation in life history traits of the widespread diadromous fish, *Galaxias maculatus*, reflects geographic differences in local environmental conditions. Marine & Freshwater Research 62: 790–800.

Bell KNI (1999) An overview of goby-fry fisheries. Naga 22: 30–36.

Bell KNI (2007) Opportunities in stream drift: methods, goby larval types, temporal cycles, *in situ* mortality estimation, and conservation implications. Bishop Museum Bulletin in Cultural & Environmental Studies 3: 35–61.

Benstead JP, March JG, Pringle CM, Ewel KC and Short JW (2009) Biodiversity and ecosystem function in species-poor communities: Community structure and leaf litter breakdown in a Pacific island stream. Journal of the North American Benthological Society 28: 454–465.

Boehlert GW and Mundy BC (1988) Roles of behavioral and physical factors in larval and juvenile fish recruitment to estuarine nursery areas. American Fisheries Society Symposium 3: 1–67.

Campana SE (1999) Chemistry and composition of fish otoliths: pathways, mechanisms and applications. Marine Ecology Progress Series 188: 263–297.

Castellanos-Galindo GA, Sanchez GC, Beltran-Leon BS and Zapata L (2011) A goby-fry fishery in the northern Colombian Pacific Ocean. Cybium 35: 391–395.

Chapman A, Morgan DL and Gill HS (2009) Description of the larval development of *Galaxias maculatus* in landlocked lentic and lotic systems in Western Australia. New Zealand Journal of Marine and Freshwater Research 43: 563–569.

Charteris SC, Allibone RM and Death RG (2003) Spawning site selection, egg development, and larval drift of *Galaxias postvectis* and *G. fasciatus* in a New Zealand stream. New Zealand Journal of Marine and Freshwater Research 37: 493–505.

Chen IS and Tan HH (2005) A new species of freshwater goby (Teleostei: Gobiidae: *Stiphodon*) from PulauTioman, Pahang, Peninsular Malaysia. Raffles Bulletin of Zoology 53: 237–242.

Closs GP, Smith M, Barry B and Markwitz A (2003) Non-diadromous recruitment in coastal populations of common bully (*Gobiomorphus cotidianus*). New Zealand Journal of Marine and Freshwater Research 37: 301–313.

Closs GP, Hicks AS and Jellyman PG (2013) Life histories of closely related amphidromous and non-migratory fish species: a trade-off between egg size and fecundity. Freshwater Biology 58: 1162–1177.

Cook BD, Bernays S, Pringle CM and Hughes JM (2009) Marine dispersal determines the genetic population structure of migratory stream fauna of Puerto Rico: evidence for island-scale population recovery processes. Journal of the North American Benthological Society 28: 709–718.

Cook BD, Pringle CM and Hughes JM (2010) Immigration history of amphidromous species on a Greater Antillean island. Journal of Biogeography 37: 270–277.

Cook RP (2004) Macrofauna of Laufuti Stream, Tau, American Samoa, and the role of physiography in its zonation. Pacific Science 58: 7–21.

Crandall ED, Taffel JR and Barber PH (2010) High gene flow due to pelagic larval dispersal among South Pacific archipelagos in two amphidromous gastropods (Neritomorpha: Neritidae). Heredity 104: 563–572.

Crook DA, MacDonald JI and Raadik TA (2008) Evidence of diadromous movements in a coastal population of southern smelts (Retropinninae: *Retropinna*) from Victoria, Australia. Marine & Freshwater Research 59: 638–646.

David B, Chadderton L, Closs G, Barry B and Markwitz A (2004) Evidence of flexible recruitment strategies in coastal populations of giant kokopu (*Galaxias argenteus*). Department of Conservation Science Internal Series 160. Department of Conservation, Wellington.

David BO and Hamer MP (2012) Remediation of a perched stream culvert with ropes improves fish passage. Marine & Freshwater Research 63: 440–449.

David BO, Closs GP, Crow SK and Hansen EA (2007) Is diel activity determined by social rank in a drift-feeding stream fish dominance hierarchy? Animal Behaviour 74: 259–263.

David BO, Hamer MP and Collier KJ (2009) Mussel spat ropes provide passage for banded kokopu (*Galaxias fasciatus*) in laboratory trials. New Zealand Journal of Marine and Freshwater Research 43: 883–888.

Debrot AO (2003a) A review of the freshwater fishes of Curacao, with comments on those of Aruba and Bonaire. Caribbean Journal of Science 39: 100–108.

Debrot AO (2003b) The freshwater shrimps of Curacao, West Indies (Decapoda, Caridea) Crustaceana. 76: 65–76.

Ebner BC and Thuesen P (2010) Discovery of stream-cling-goby assemblages (*Stiphodon* species) in the Australian Wet Tropics. Australian Journal of Zoology 58: 331–340.

Ebner BC, Thuesen PA, Larson HK and Keith P (2011) A review of distribution, field observations and precautionary conservation requirements for sicydiine gobies in Australia. Cybium 35: 397–414.

Elgar MA (1990) Evolutionary compromise between a few large and many small eggs: comparative evidence in teleost fish. Oikos 59: 283–287.

Ellien C, Valade P, Bosmans J, Taillebois L, Teichert N and Keith P (2011) Influence of salinity on larval development of *Sicyopterus lagocephalus* (Pallas, 1170) (Gobioidei). Cybium 35: 381–390.

Elsdon TS and Gillanders BM (2005) Alternative life-history patterns of estuarine fish: barium in otoliths elucidates freshwater residency. Canadian Journal of Fisheries and Aquatic Sciences 62: 1143–1152.

Elsdon TS, Wells BK, Campana SE, Gillanders BM, Jones CM, Limburg KE, Secor DH, Thorrold SR and Walther BD (2008) Otolith chemistry to describe movements and life-history parameters of fishes:

hypotheses, assumptions, limitations and inferences. Oceanography and Marine biology: An Annual Review 46: 297–330.

Feutry P, Valade P, Ovenden JR, Lopez PJ and Keith P (2012a) Pelagic larval duration of two diadromous species of Kuhliidae (Teleostei: Percoidei) from Indo-Pacific insular systems. Marine & Freshwater Research 63: 397–402.

Feutry P, Tabouret H, Maeda K, Pecheyran C and Keith P (2012b) Diadromous life cycle and behavioural plasticity in freshwater and estuarine Kuhliidae species (Teleostei) revealed by otolith microchemistry. Aquatic Biology 15: 195–204.

Fitzsimons JM, Parham JE and Nishimoto RT (2002) Similarities in behavioural ecology among amphidromous and catadromous fishes on the oceanic islands of Hawai'i and Guam. Environmental Biology of Fishes 65: 123–129.

Fitzsimons JM, McRae MG and Nishimoto RT (2007) Behavioral ecology of indigenous stream fishes in Hawai'i. Bishop Museum Bulletin in Cultural and Environmental Studies 3: 11–21.

Flevet E, Doledec S and Lim P (2001) Distribution of migratory fishes and shrimps along multivariate gradients in tropical island streams. Journal of Fish Biology 59: 390–402.

Gorbach KR, Benbow ME, McIntosh MD and Burky AJ (2012) Dispersal and upstream migration of an amphidromous neritid snail: implications for restoring migratory pathways in tropical streams. Freshwater Biology 57: 1643–1657.

Goto A (1981) Life history and distribution of a river sculpin *Cottus hangiogensis*. Bulletin of the Faculty of Fisheries Hokkaido University 32: 10–21.

Goto A (1986) Movement and population size of the river sculpin *Cottus hangiongensis* in the Daitobetsu River of southern Hokkaido. Ichthyological Research 32: 421–430.

Goto A (1990) Alternative life-history styles of Japanese freshwater sculpins revisited. Environmental Biology of Fishes 28: 101–112.

Goto A and Arai T (2003) Migratory histories of three types of *Cottus pollux* (small-egg, middle-egg, and large-egg types) as revealed by otolith microchemistry. Ichthyological Research 50: 67–72.

Goto A and Arai T (2006) Diverse migratory histories of Japanese *Trachidermus* and *Cottus* species (Cottidae) as inferred from otolith microchemistry. Journal of Fish biology 68: 1731–1741.

Goto A, Yokoyama R and Yamada M (2002) A fluvial population of *Cottus pollux* (middle-egg type) from the Honmyo River, Kyushu Island, Japan. Ichthyological Research 49: 318–323.

Ha PY and Kinzie RA (1996) Reproductive biology of *Awaous guamensis*, an amphidromous Hawaiian goby. Environmental Biology of Fishes 45: 383–396.

Hale R, Downes BJ and Swearer SE (2008) Habitat selection as a source of inter-specific differences in recruitment of two diadromous fish species. Freshwater Biology 53: 2145–2157.

Hale R, Swearer SE and Downes BJ (2009a) Is settlement at small spatial scales by diadromous fishes from the Family Galaxiidae passive or active in a small coastal river? Marine & Freshwater Research 60: 971–975.

Hale R, Swearer SE and Downes BJ (2009b) Separating natural responses from experimental artefacts: habitat selection by a diadromous fish species using odours from conspecifics and natural stream water. Oecologia 159: 679–687.

Hansen EA and Closs GP (2005) Diel activity and home range size in relation to food supply in a drift-feeding stream fish. Behavioral Ecology 16: 640–648.

Hickey MA and Closs GP (2006) Evaluating the potential of night spotlighting as a method for assessing species composition and brown trout abundance: a comparison with electrofishing in small streams. Journal of Fish Biology 69: 1513–1523.

Hickford MJH and Schiel DR (2011) Population sinks resulting from degraded habitats of an obligate life-history pathway. Oecologia 166: 131–140.

Hicks AS (2012) Facultative amphidromy in galaxiids and bullies: the science, ecology and management implications. Ph.D. thesis, University of Otago, Dunedin.

Hicks AS, Barbee NC, Swearer SE and Downes BJ (2010) Estuarine geomorphology and low salinity requirement for fertilisation influence spawning site location in the diadromous fish, *Galaxias maculatus*. Marine & Freshwater Research 61: 1252–1258.

Hicks BJ, West DW, Barry BJ, Markwitz A, Baker CF and Mitchell CP (2005) Chronosequences of strontium in the otoliths of two New Zealand migratory freshwater fish, inanga (*Galaxias maculatus*) and koaro (*G. brevipinnis*). International Journal of PIXE 15: 95–101.

Hoareau TB, Lecomte-Finiger R, Grondin H, Conand C and Berrebi P (2007) Oceanic larval life of La Réunion 'bichiques', amphidromous gobiid post-larvae. Marine Ecology Progress Series 333: 303–308.

Hoese DF and Allen GR (2011) A review of the amphidromous species of the *Glossogobius celebius* complex, with the description of three new species. Cybium 35: 269–284.

Humphries P (1989) Variation in the life history of diadromous and landlocked populations of the spotted galaxias, *Galaxias truttaceus* Valenciennes in Tasmania. Australian Journal of Marine and Freshwater Research 40: 501–518.

Iguchi K (2007) Limitation of early seaward migration success in amphidromous fishes. Bishop Museum Bulletin in Cultural and Environmental Studies 3: 75–85.

Iguchi K and Mizuno N (1990) Diel changes of larval drift among amphidromous gobies in Japan, especially *Rhinogobius brunneus*. Journal of Fish Biology 37: 255–264.

Iguchi K and Mizuno N (1999) Early starvation limits survival in amphidromous fishes. Journal of Fish Biology 54: 705–712.

Iguchi K and Takeshima H (2011) Effect of saline water on early success of amphidromous fish. Ichthyological Research 58: 33–37.

Iguchi K, Ito F, Yamaguchi M and Matsubara N (1998) Spawning downstream migration of ayu in the Chikuma River. Bulletin of the National Research Institute of Fisheries Science 11: 75–84.

Iguchi K, Iwata Y, Nishida M and Otake T (2005) Skip of the routine habitat in an amphidromous migration of ayu. Ichthyological Research 52: 98–100.

Iida M, Watanabe S and Tsukamoto K (2009) Life history characteristics of a Sicydiinae goby in Japan, compared with its relatives and other amphidromous fishes. American Fisheries Society Symposium 69: 355–373.

Iida M, Watanabe S, Yamada Y, Lord C, Keith P and Tsukamoto K (2010) Survival and behavioural characteristics of amphidromous goby larvae of *Sicyopterus japonicus* (Tanaka, 1909) during their downstream migration. Journal of Experimental Marine Biology and Ecology 383: 17–22.

Ikeda M, Nunokawa M and Taniguchi N (2003) Lack of mitochondrial gene flow between populations of the endangered amphidromous fish *Plecoglossus altivelis ryukyuensis* inhabiting Amami-oshima Island. Fisheries Science 69: 1162–1168.

Ishihara T and Tachihara K (2008) Reproduction and early development of a freshwater pipefish *Microphis leiapis* in Okinawa-jima Island, Japan. Ichthyological Research 55: 349–355.

Ito S and Yanagisawa Y (2003) Mate choice and mating pattern in a stream goby of the genus *Rhinogobius*. Environmental Biology of Fishes 66: 67–73.

Ito S, Kanebayashi M, Sato A, Iguchi K, Yanagisawa Y and Oomori K (2010) Changes in male physiological condition during brooding activities in a natural population of a stream goby, *Rhinogobius* sp. Environmental Biology of Fishes 87: 135–140.

Jellyman PG and McIntosh AR (2010) Recruitment variation in a stream galaxiid fish: multiple influences on fry dynamics in a heterogeneous environment. Freshwater Biology 55: 1930–1944.

Joy MK and Death RG (2001) Control of freshwater fish and crayfish community structure in Taranaki, New Zealand: dams, diadromy or habitat structure? Freshwater Biology 46: 417–429.

Kano Y (2009) Hitchhiking behaviour in the obligatory upstream migration of amphidromous snails. Biology Letters 5: 465–468.

Katoh M and Nishida M (1994) Biochemical and egg size evolution of freshwater fishes in the *Rhinogobius brunneus* complex (Pisces, Gobiidae) in Okinawa, Japan. Biological Journal of the Linnean Society 51: 325–335.

Kawakami T and Tachihara K (2011) Dispersal of land-locked larval Ryukyu-ayu, *Plecoglossus altivelis ryukyuensis*, in the Fukuji Reservoir, Okinawa Island. Cybium 35: 337–343.

Keith P (2003) Biology and ecology of amphidromous Gobiidae of the Indo-Pacific and the Caribbean regions. Journal of Fish Biology 63: 831–847.

Keith P and Marquet G (2007) *Stiphodon rubromaculatus*, a new species of freshwater goby from Futuna island (Teleostei: Gobioidei: Sicydiinae). Cybium 31: 45–49.

Keith P, Watson RE and Marquet G (2002) *Stenogobius (Insularigobius) yateiensis*, a new species of freshwater goby from New Caledonia (Teleostei: Gobioidei). Bulletin Français de Pêche et de Pisciculture 364: 187–196.

Keith P, Marquet G and Watson RE (2004a) *Schismatogobius vanuatuensis*, a new species of freshwater goby from Vanuatu, South Pacific. Cybium 28: 237–241.

Keith P, Watson RE and Marquet G (2004b) *Sicyopteris aiensis*, a new species of freshwater goby (Gobioidei) from Vanuatu, South Pacific. Cybium 28: 111–118.

Keith P, Hoareau T and Bosc P (2005a) A new species of freshwater goby (Pisces: Teleostei: Gobioidei) from Mayotte island (Comoros) and comments about the genus *Cotylopus* endemic to the Indian Ocean. Journal of Natural History 39: 1395–1405.

Keith P, Galewski T, Cattaneo-Berrebi G, Hoareau T and Berrebi P (2005b) Ubiquity of *Sicyopterus lagocephalus* (Teleostei: Gobioidei) and phylogeography of the genus *Sicyopterus* in the Indo-Pacific area inferred from mitochondrial cytochrome b gene. Molecular Phylogenetics and Evolution 37: 721–732.

Keith P, Marquet G and Watson RE (2007a) *Stiphodon kalfatak*, a new species of freshwater goby from Vanuatu (Teleostei: Gobioidei: Sicydiinae). Cybium 31: 33–37.

Keith P, Marquet G and Watson RE (2007b) *Akihito futuna*, a new species of freshwater goby from the South Pacific (Teleostei: Gobioidei: Sicydiinae). Cybium 31: 471–476.

Keith P, Hoareau TB, Lord C, Ah-Yane O, Gimonneau G, Robinet T and Valade P (2008) Characterisation of post-larval to juvenile stages, metamorphosis and recruitment of an amphidromous goby, *Sicyopterus lagocephalus* (Pallas) (Teleostei: Gobiidae: Sicydiinae). Marine & Freshwater Research 59: 876–889.

Keith P, Marquet G and Pouilly M (2009) *Stiphodon mele* n. sp., a new species of freshwater goby from Vanuatu and New Caledonia (Teleostei: Gobiidae: Sicydiinae), and comments about amphidromy and regional dispersal. Zoosystema 31: 471–483.

Keith P, Lord C and Taillebois L (2010) *Sicyopus* (*Smilosicyopus*) *pentecost*, a new species of freshwater goby from Vanuatu and New Caledonia (Gobioidei: Sicydiinae). Cybium 34: 303–310.

Keith P, Marquet G and Taillebois L (2011a) Discovery of the freshwater genus *Sicyopus* (Teleostei: Gobioidei: Sicydiinae) in Madagascar, with the description of a new species and comments on regional dispersal. Journal of Natural History 45: 2725–2746.

Keith P, Lord C, Lorion J, Watanabe S, Tsukamoto K, Couloux A and Dettal A (2011b) Phylogeny and biogeography of Sicydiinae (Teleostei: Gobiidae) inferred from mitochondrial and nuclear genes. Marine Biology 158: 311–326.

Kelly DW, Thomas H, Thieltges DW, Poulin R and Tompkins DM (2010) Trematode infection causes malformations and population effects in a declining New Zealand fish. Journal of Animal Ecology 79: 445–452.

Kinzie RA (1988) Habitat utilization by Hawaiian stream fishes with reference to community structure in oceanic island streams. Environmental Biology of Fishes 22: 179–192.

Kondo M, Maeda K, Hirashima K and Tachihara K (2013) Comparative larval development of three amphidromous *Rhinogobius* species making reference to their habitat preferences and migration biology. Marine & Freshwater Research 64: 249–266.

Leathwick JR, Elith J, Chadderton WL, Rowe D and Hastie T (2008) Dispersal, disturbance and the contrasting biogeographies of New Zealand's diadromous and non-diadromous fish species. Journal of Biogeography 35: 1481–1497.

Loiselle PV (2005) Fishes of the fresh waters of Nosy Be, Madagascar, with notes on their distribution and natural history. Ichthyoloigical Exploration of Freshwaters 16: 29–46.

Lord C, Brun C, Hautecoeur M and Keith P (2010) Insights on endemism: comparison of the duration of the marine larval phase estimated by otolith microstructural analysis of three amphidromous *Sicyopterus* species (Gobioidei: Sicydiinae) from Vanuatu and New Caledonia. Ecology of Freshwater Fish 19: 26–38.

Lord C, Lorion J, Dettai A, Watanabe S, Tsukamoto K, Cruaud C and Keith P (2012) From endemism to widespread distribution: phylogeography of three amphidromous *Sicyopterus* species (Teleostei: Gobioidei: Sicydiinae). Marine Ecology Progress Series 455: 269–285.

Luton CD, Brasher AMD, Durkin DC and Little P (2005) Larval drift of amphidromous shrimp and gobies on the island of Oahu, Hawai'i. Micronesica 38: 1–16.

Lyons J (2005) Distribution of *Sicydium* Valenciennes 1837 (Pisces: Gobiidae) in Mexico and Central America. Hidrobiológica 15: 239–243.

Lytle DA (2001) Disturbance regimes and life-history evolution. American Naturalist 157: 525–536.

MacKenzie RA (2008) Impacts of riparian forest removal on Palauan streams. Biotropica 40: 666–675.

Maeda K and Tachibara K (2010) Diel and seasonal occurrence patterns of drifting fish larvae in the Teima Stream, Okinawa Island. Pacific Science 64: 161–176.

Maeda K, Mukai T and Tachihara K (2011) A new species of amphidromous goby, *Stiphodon alcedo*, from the Ryukyu Archipelago (Gobiidae: Sicydiinae). Cybium 35: 285–298.

March JG, Benstead JP, Pringle CM and Scatena FN (1998) Migratory drift of larval freshwater shrimps in two tropical streams, Puerto Rico. Freshwater Biology 40: 261–273.

Maruyama A, Rusuwa B and Yuma M (2003) Interpopulation egg-size variation of a landlocked *Rhinogobius* goby related to the risk of larval starvation. Environmental Biology of Fishes 67: 223–230.

McDowall RM (1970) The galaxiid fishes of New Zealand. Bulletin of the Museum of Comparative Zoology, Harvard University 139: 341–432.

McDowall RM (1988) Diadromy in Fishes: Migration Between Freshwater and Marine Environments. Croom Helm, London.

McDowall RM (1990) New Zealand Freshwater Fishes. Heinemann Reed, Auckland.

McDowall RM (1992) Diadromy: origins and definitions of terminology. Copeia 1992: 248–251.

McDowall RM (1997) The evolution of diadromy in fishes (revisited) and its place in phylogenetic analysis. Reviews in Fish Biology and Fisheries 7: 443–462.

McDowall RM (1998) Fighting the flow: downstream-upstream linkages in the ecology of diadromous fish faunas in West Coast New Zealand rivers. Freshwater Biology 40: 111–122.

McDowall RM (2000) The Reed Field Guide to New Zealand Freshwater Fishes. Reed Books, Auckland.

McDowall RM (2004) Ancestry and amphidromy in island freshwater fish faunas. Fish and Fisheries 5: 75–85.

McDowall RM (2007) On amphidromy, a distinct form of diadromy in aquatic organisms. Fish and Fisheries 8: 1–13.

McDowall RM (2009) Early hatch: a strategy for safe downstream larval transport in amphidromous gobies. Reviews in Fish Biology and Fisheries 19: 1–8.

McDowall RM (2010a) Why be amphidromous: expatrial dispersal and the place of source and sink population dynamics? Reviews in Fish Biology and Fisheries 20: 87–100.

McDowall RM (2010b) New Zealand Freshwater Fishes. Springer, Dordrecht.

McDowall RM and Allibone RM (1994) Possible Competitive-Exclusion of Common River Galaxias (*Galaxias vulgaris*) by Koaro (*G. brevipinnis*) following impoundment of the Waipori River, Otago, New Zealand. Journal of the Royal Society of New Zealand 24: 161–168.

McDowall RM and Charteris SC (2006) The possible adaptive advantages of terrestrial egg deposition in some fluvial diadromous fishes (Teleostei: Galaxiidae). Fish and Fisheries 7: 153–164.

McDowall RM, Robertson DA and Saito R (1975) Occurrence of galaxiid larvae and juveniles in the sea. New Zealand Journal of Marine and Freshwater Research 9: 1–9.

McDowall RM, Mitchell CP and Brothers EB (1994) Age at migration from the sea of juvenile Galaxias in New Zealand (Pisces: Galaxiidae). Bulletin of Marine Sciences 54: 385–402.

McEwan AJ and Joy MK (2009) Differences in the distributions of freshwater fishes and decapod crustaceans in urban and forested streams in Auckland, New Zealand. New Zealand Journal of Marine and Freshwater Research 43: 1115–1120.

Michel C, Hicks BJ, Stölting KN, Clarke AC, Stevens MI, Tana R, Meyer A and van den Heuvel M (2008) Distinct migratory and non-migratory ecotypes of an endemic New Zealand eleotrid (*Gobiomorphus cotidianus*)—implications for incipient speciation in island freshwater fish species. BMC Evolutionary Biology 8: 49.

Miles NG, West RJ and Norman MD (2009) Does otolith chemistry indicate diadromous lifecycles for five Australian riverine fishes? Marine & Freshwater Research 60: 904–911.

Moriyama A, Yanagaisawa Y and Mizuno N (1998) Starvation of drifting goby larvae due to retention of free embryos in upstream reaches. Environmental Biology of Fishes 52: 321–329.

Murphy CA and Cowan JH (2007) Production, marine larval retention or dispersal, and recruitment of amphidromous Hawaiian gobioids: issues and implications. Bishop Museum Bulletin in Cultural and Environmental Studies 3: 63–74.

Myers G (1949) Usage of anadromous, catadromous and allied terms for migratory fishes. Copeia 1949: 89–97.

Nip THM (2010) First records of several sicydiine gobies (Gobiidae: Sicydiinae) from mainland China. Journal of Threatened Taxa 2: 1237–1244.

Nordlie FG (2012) Life-history characteristics of eleotrid fishes of the western hemisphere, and perils of life in a vanishing environment. Reviews in Fish Biology and Fisheries 22: 189–224.

O'Connor WG and Koehn JD (1998) Spawning of the broad-finned Galaxias, *Galaxias brevipinnis* Gunther (Pisces: Galaxiidae) in coastal streams of southeastern Australia. Ecology of Freshwater Fish 7: 95–100.

Ohara K, Hotta M, Takahashi D, Asahida T, Ida H and Umino T (2009) Use of microsatellite DNA and otolith Sr:Ca ratios to infer genetic relationships and migration history of four morphotypes of *Rhinogobius* sp. OR. Ichthyological Research 56: 373–379.

Page TJ, Baker AM, Cook BD and Hughes JM (2005) Historical transoceanic dispersal of a freshwater shrimp: the colonization of the South Pacific by the genus *Paratya* (Atyidae). Journal of Biogeography 32: 581–593.

Patten BG (1971) Spawning and fecundity of seven species of northwest American *Cottus*. American Midland Naturalist 85: 493–506.

Poff NL and Ward JV (1989) Implications of Streamflow Variability and Predictability for Lotic Community Structure: A Regional Analysis of Streamflow Patterns. Canadian Journal of Fisheries and Aquatic Sciences 46: 1805–1818.

Poulin R, Closs GP, Lill AWT, Hicks AS, Herrmann KK and Kelly DW (2012) Migration as an escape from parasitism in New Zealand galaxiid fishes. Oecologia 169: 955–963.

Ramirez A, Engman A, Rosas KG, Perez-Reyes O and Martino-Cardona DM (2012) Urban impacts on tropical island streams: some key aspects influencing ecosystem response. Urban Ecosystems 15: 315–325.

Riede K (2004) Global register of migratory species—from global to regional scales. Final report of the R&D Projekt 808 05-0811. Federal Agency for Nature Conservation, Bonn.

Rolls RJ (2011) The role of life-history and location of barriers to migration in the spatial distribution and conservation of fish assemblages in a coastal river system. Biological Conservation 144: 339–349.

Rowe DK, Saxton BA and Stancliff AG (1992) Species composition of whitebait (Galaxiidae) fisheries in 12 Bay of Plenty rivers, New Zealand: evidence for river mouth selection by juvenile *Galaxias brevipinnis* (Günther). New Zealand Journal of Marine and Freshwater Research 26: 219–228.

Schmidt DJ, Crook DA, O'Connor JP and Hughes JM (2010) Genetic analysis of threatened Australian grayling *Prototroctes maraena* suggests recruitment to coastal rivers from an unstructured marine larval source population. Journal of Fish Biology 78: 98–111.

Scrimgeour GJ and Eldon GA (1989) Aspects of the reproductive biology of torrentfish, *Cheimarrichthys fosteri*, in 2 braided rivers of Canterbury, New Zealand. New Zealand Journal of Marine and Freshwater Research 23: 19–25.

Shafer D (2000) Evaluation of periodic and aperiodic otolith structure and somatic-otolith scaling for use in retrospective life history analysis of a tropical marine goby, *Bathygobius coalitus*. Marine Ecology Progress Series 199: 217–229.

Smith CC and Fretwell SD (1974) The optimal balance between size and number of offspring. American Naturalist 108: 499–506.

Sorensen PW and Hobson KA (2005) Stable isotope analysis of amphidromous Hawaiian gobies suggests their larvae spend a substantial period of time in freshwater river plumes. Environmental Biology of Fishes 74: 31–42.

Suzuki T, Chen I and Senou H (2011) A new species of *Rhinogobius* Gill, 1859 (Teleostei: Gobiidae) from the Bonin Islands, Japan. Journal of Marine Science and Technology—Taiwan 19: 693–701.

Taillebois L, Maeda K, Vigne S and Keith P (2012) Pelagic larval duration of three amphidromous Sicydiinae gobies (Teleostei: Gobioidei) including widespread and endemic species. Ecology of Freshwater Fish 21: 552–559.

Takahashi D and Kohda M (2004) Courtship in fast water currents by a male stream goby (*Rhinogobius brunneus*) communicates the parental quality honestly. Behavioral Ecology and Sociobiology 55: 431–438.

Tamada K (2009) Variations in clutch and egg sizes in the amphidromous goby *Rhinogobius* sp. CB along a river course and within a spawning season. Ichthyological Research 56: 69–75.

Tamada K (2011) River bed features affect the riverine distribution of two amphidromous *Rhinogobius* species. Ecology of Freshwater Fish 20: 33–41.

Tamada K and Iwata K (2005) Intra-specific variation of egg size, clutch size and larval survival related to material size in amphidromous *Rhinogobius* goby. Environmental Biology of Fishes 73: 379–389.

Teichert N, Richarson M, Valade P and Gaudin P (2012) Reproduction and marine life history of an endemic gobiid fish of Reunion Island. Aquatic Biology 15: 225–236.

Thuesen PA, Ebner BC, Larson H, Keith P, Silcock RM, Prince J and Russell DJ (2011) Amphidromy links a newly documented fish community of continental Australian streams, to oceanic islands of the west Pacific. PLOS ONE 6: e26685.

Tsukagoshi H, Yokoyama R and Goto A (2011) Mitochondrial DNA analysis reveals a unique population structure of the amphidromous sculpin *Cottus pollux* middle-egg type (Teleostei: Cottidae). Molecular Phylogenetics and Evolution 60: 265–270.

Tsunagawa T and Arai T (2008) Flexible migration of Japanese freshwater gobies *Rhinogobius* spp. as revealed by otolith Sr:Ca ratios. Journal of Fish Biology 73: 2421–2433.

Tsunagawa T and Arai T (2011) Migratory history of the freshwater goby *Rhinogobius* sp., CB in Japan. Ecology of Freshwater Fish 20: 3–41.

Valade P, Lord C, Grondin H, Bosc P, Taillebois L, Iida M, Tsukamoto K and Keith P (2009) Early life history and description of larval stages of an amphidromous goby, *Sicyopterus lagocephalus* (Gobioidei: Sicydiinae). Cybium 33: 309–319.

Watson RE, Keith P and Marquet G (2001) *Sicyopus (Smilosicyopus) chloe*, a new species of freshwater goby of New Caledonia (Teleostei: Gobioidei: Sicydiinae). Cybium 25: 41–52.

Watson RE, Keith P and Marquet G (2002) *Lentipes kaaea*, a new species of freshwater goby from New Caledonia (Teleostei: Gobioidei: Sicydiinae). Bulletin Français de Pêche et de Pisciculture 364: 173–185.

Watson RE, Keith P and Marquet G (2005) *Stiphodon sapphirinus*, a new species of freshwater goby of New Caledonia (Teleostei: Gobioidei: Sicydiinae). Cybium 29: 339–345.

Watson RE, Keith P and Marquet G (2007) *Akihito vanuatu*, a new genus and new species of freshwater goby from the South Pacific (Teleostei: Gobioidei: Sicydiinae). Cybium 31: 341–349.

Walter RP, Hogan JD, Blum MJ, Gagne RB, Hain EF, Gilliam JF and McIntyre PB (2012) Climate change and conservation of endemic amphidromous fishes in Hawaiian streams. Endangered Species Research 16: 261–272.

Walther BD and Limburg KE (2012) The use of otolith chemistry to characterize diadromous migrations. Journal of Fish Biology 81: 796–825.

Waters JM and Wallis GP (2001) Cladogenesis and loss of the marine life-history phase in freshwater galaxiid fishes (Osmeriformes: Galaxiidae). Evolution 55: 587–597.

Waters JM, Dijkstra LH and Wallis GP (2000) Biogeography of a southern hemisphere freshwater fish: how important is marine dispersal? Molecular Ecology 9: 1815–1821.

Wylie MJ (2011) Reproductive Biology of the Giant Kokopu Galaxias argenteus. M.Sc. Thesis University of Otago, Dunedin.

Yamasaki N, Kondo M, Maeda K and Tachihara K (2011) Reproductive biology of three amphidromous gobies, *Sicyopterus japonicus*, *Awaous melanocephalus* and *Stenogobius* sp., on Okinawa Island. Cybium 35: 345–359.

Yokoyama R and Goto A (2005) Evolutionary history of freshwater sculpins, genus *Cottus* (Teleostei; Cottidae) and related taxa, as inferred from mitochondrial DNA phylogeny. Molecular Phylogenetics and Evolution 36: 654–668.

CHAPTER 8

Life Histories of
Oceanodromous Fishes

Ana Couto,[1,2] *Miguel Baptista,*[1,2] *Miguel Furtado,*[1,2] *Lara L. Sousa*[1,3,4] *and Nuno Queiroz*[1,*]

Overview on oceanodromous fish migration

In the ocean, population requirements often imply the use of different habitats, either due to variable environmental conditions (e.g., temperature, dissolved oxygen), or a change in the necessities of the population itself (e.g., foraging habitat vs. spawning habitat) (Binder et al. 2011). In such cases, individuals benefit from moving to an alternative habitat and as a result, many fish species develop a life history that involves migration (e.g., Block et al. 2001; Sims et al. 2003; Walli et al. 2009; Chiang et al. 2014). The term migration, although frequently used in biology, is difficult to define satisfactorily; this is because in different fields, or more specifically, when studying different species, it can have different meanings (Baker 1978; Aidley 1981; Smith 1985). Therefore, some definitions fail to include movements that would otherwise be classified as migrations in some specific cases (Smith 1985). For example, long-distance, directed movements such as wide-scale foraging trips are often not considered migratory movements. However, as long-term movements can have considerable biological importance for oceanodromous species (and the whole sequence is usually necessary for successful completion of the life cycle), such cases should be classified as migrations (Smith 1985). Using a broader definition, Baker

[1] CIBIO/InBIO - Universidade do Porto, Centro de Investigação em Biodiversidade e Recursos Genéticos, Campus Agrário de Vairão, Rua Padre Armando Quintas, 4485-668 Vairão, Portugal.
[2] MARE – Marine and Environmental Sciences Centre, Laboratório Marítimo da Guia, Faculdade de Ciências da Universidade de Lisboa, Av. Nossa Senhora do Cabo, 939, 2750-374 Cascais, Portugal.
[3] Marine Biological Association of the United Kingdom, The Laboratory, Citadel Hill, Plymouth PL1 2PB, UK.
[4] Ocean and Earth Science, National Oceanography Centre Southampton, University of Southampton, Waterfront Campus, European Way, Southampton SO14 3ZH, UK.
* Corresponding author: nuno.queiroz@cibio.up.pt

(1978) described migration as "the act of moving from one spatial unit to other", with this definition being later applied to oceanic fishes (Aidley 1981). Similarly, Jones (1984) when referring to open ocean migrators, applied the word 'migration' to the recurrent movement pattern of fish moving to and from a given location with the seasons. For tuna species, a group of well-studied oceanodromous fish, a more specific definition was proposed by Nakamura (1969) and later used by Humston et al. (2000), where (1) movements within a habitat induced by alterations in local abiotic or biotic conditions and/or (2) directed movements of fish between habitats due to changes in biological requirements, were regarded as migrations. Nonetheless, a common feature in all definitions is that migration involves the movement of individuals and populations between well-defined areas or habitats, usually on a cyclical or seasonal timescale (Metcalfe et al. 2008).

This chapter focuses on the migratory processes and patterns of oceanodromous fish, using as examples studies where movement data was obtained using archival and Pop-off Satellite Archival Tags (PSATs) (see Chapter 13). For the purpose of this chapter, migration will be classified as long-term horizontal movements between areas, mainly driven by alterations in species requirements, such as reproduction, feeding events or wide-scale foraging opportunities. In the first part of the chapter, we emphasize the importance of tracking, characterizing the migratory movement patterns of several teleost, such as tunas and swordfish, and elasmobranchs species. Subsequently, migration triggers and cues, such as environmental changes or reductions in feeding success, will be described; later followed by the definition of the main drivers behind oceanodromous fish migrations, focusing on feeding vs. reproduction dichotomy and wider scale foraging opportunities, while providing valuable examples. At the end of the chapter, navigational cues that fish may use in the open ocean will also be explored and the conservation status of the most emblematic oceanodromous migratory species will also be briefly examined.

Migratory species and tracking

Approximately 2.5% of all fish species display a migratory behaviour (Binder et al. 2011). In 1982, the United Nations Convention on the Law of the Sea (UNCLOS), created a new category of species named 'highly migratory species', acknowledging that some species move considerable distances in the open ocean and often between zones with different jurisdictions (Burke 1984). Due to the vast expanse of the oceanic environment and the difficulty of following animals, documenting the movements of marine migratory species has been challenging (Chapman et al. 2015). For example, conventional tagging (mark-recapture) of fish provides only two locations, one where the fish was released and other where the fish was caught (Priede and French 1991). As detailed in Chapter 13, unprecedented advances in archival and satellite tagging have been achieved during the last decade (Schaefer and Fuller 2016). Numerous technological innovations such as increased memory, sensor performance and improved component reliability were implemented and have allowed long-term tracking of movements and migrations of a variety of fish species. Hence, it has been possible to characterize the migratory movements of several teleost and elasmobranchs species. For example, pop-off archival tags have been successfully used to track the movements of large pelagic fishes, including tunas (e.g., Block et al. 1998; Wilson et al. 2005; Block 2011), billfishes (e.g., Takahashi et al. 2003; Kraus et al. 2011; Chiang et al. 2014) and sharks (e.g., Skomal et al. 2004; Bonfil et al. 2005; Sims et al. 2005; Weng et al. 2007; Bonfil et al. 2010; Hueter et al. 2013; Chapman et al. 2015). So far, 17 shark species, representing seven families from four orders have been satellite tagged (with migrations observed in 40% of the studied species), although 40% of the studies focused on as few as three different shark species; white shark *Carcharodon carcharias* (Linnaeus, 1758) (20%) (e.g., Bonfil et al. 2005; Bruce et al. 2006; Weng et al. 2007; Domeier and Nasby-Lucas 2008; Jorgensen et al. 2009), basking shark *Cetorhinus maximus* (Gunnerus, 1765) (16%) (e.g., Sims et al. 2003; Skomal et al. 2004; Sims et al. 2005; Skomal et al. 2009) and whale shark *Rhincodon typus* (Smith, 1828) (14%) (e.g., Wilson et al. 2006; Hammerschlag et al. 2011). Similarly, a recent review study (Braun et al. 2015) revealed that migrations were observed for swordfish *Xiphias gladius* (Linnaeus, 1758), blue marlin *Makaira nigricans* (Lacepède, 1802), and black marlin *Istiompax indica* (Cuvier, 1832), which accounted for 40% of the analyzed species. Finally, with the exception of bigeye tuna *Thunnus obesus* (Lowe, 1839), migrations were described for the Atlantic tuna *Thunnus thynnus* (Linnaeus, 1758), Pacific bluefin tuna *Thunnus orientalis* (Temminck and Schlegel, 1844), albacore tuna *Thunnus alalunga* (Bonnaterre, 1788) and yellowfin tuna *Thunnus albacares* (Bonnaterre, 1788) (De Metrio et al. 2002, Itoh et al. 2003, Block et al. 2005, Kitagawa et al. 2007, Schaefer et al. 2007, Teo et al. 2007, Childers et al. 2011). See Fig. 8.1 for spatial distribution of migratory archival- and satellite-based studies.

Migration triggers

The timing of key life history traits, such as migratory reproduction and/or feeding, can have significant fitness consequences (Baker 1978; McNamara et al. 2011). Migration timing determines the degree of spatio-temporal overlap with important resources or environmental conditions and ecological interactions that are vital to survival and/ or reproduction (Scheuerell et al. 2009), particularly in environments with strong seasonal variations. Many migratory species are known to use endogenous (e.g., hormone concentrations and body size) often coupled with environmental cues (e.g., temperature and food supply) to initiate migration. For example, migrations in sharks and tunas are frequently directly related to seasonal changes in water temperature (Kimura and Sugimoto 1997; Wilson et al. 2001; Wilson et al. 2006). Whale sharks are known to aggregate in the coastal waters off Ningaloo Reef, migrate to the North-eastern Indian Ocean (Wilson et al. 2006; pop-off archival tags) and return to the same location in subsequent years. This seasonal aggregation of whale sharks off Ningaloo Reef appears to be associated with a seasonal southerly movement of warm water masses down the coast of western Australia. Hence, the seasonal alteration in the current pattern may act as temporal signal triggering the southerly movements of whale sharks back to Ningaloo Reef. Also, the extension of warm waters down

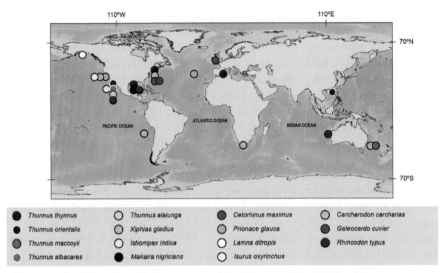

Figure 8.1. Worldwide spatial distribution of the tagging studies described in this chapter.

the coast of western Australia allows whale sharks to exploit abundant food sources in an expanded geographic range (Wilson et al. 2001). The relation between whale shark migration and shifts in the currents was also investigated in Maldives, where the seasonal movement of this species appeared related with monsoons (Anderson and Ahmed 1993; Kumari and Raman 2010). From May to November, the current flows from west to east, leading to the development of local upwelling and triggering a plankton bloom on the 'downstream' side of the Maldives. Whale sharks were most commonly found on the eastern side of the Maldives during this period, moving to the opposite side after the monsoon period.

Alterations in the water temperature can also indirectly influence the migration timing of apex predators by affecting prey distribution and movements. For example, changes in the Kuroshio (warm) and Oyashio (cold) currents likely influence the movements of the Japanese saury *Cololabis saira*, a major prey species of albacore tuna, and as a consequence triggered a migratory dispersion of the tuna population to different areas in order to satisfy food requirements (Kimura and Sugimoto 1997). As a result, an increase in the frequency of the trans-Pacific (west-east) migratory movements was also observed (Kimura and Sugimoto 1997). The migration of Pacific bluefin tuna from western to eastern Pacific was also linked with changes in prey abundance (Itoh et al. 2003; Kitagawa et al. 2009), although it might be also related with reproduction (see below reproduction vs. feeding dichotomy). In years when sardines, one of the known prey of Pacific bluefin tuna, are abundant off Japan, a higher proportion of this species stay in the western Pacific compared with years when sardines are scarce (Polovina 1996). The migration of this species along the coast of western North America, as demonstrated by satellite tags (Domeier et al. 2005; Kitagawa et al. 2007; Boustany et al. 2010), is also likely related with the seasonal shifts in prey availability induced by favourable environmental conditions (upwelling/downwelling events). The increase of water temperature in central

California (USA) during fall, accompanied by a northward movement or recruitment of prey, leads to a northerly movement of the Pacific bluefin tuna along the coast. Strong downwelling events in December cause a decrease in the productivity of the northern region, and consequent reduction of the concentration of preys in the zone, resulting in the return of tunas to the waters off southern California and further south (Kitagawa et al. 2007) (Fig. 8.2).

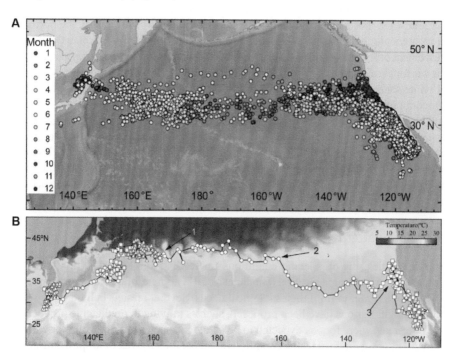

Figure 8.2. Trans-oceanic migratory paths of Pacific bluefin tuna *Thunnus orientalis* in the North Pacific Ocean [A—Boustany et al. (2010); B—Kitagawa et al. (2009)]. *Used with permission.*

Changes in the prey distribution and food availability due to shifts in the water temperature were also linked to the movements of swordfish, which migrated northwards to Peru, a region with year-round upwelling, and returned south to Chile during the onset of summer upwelling (Abascal et al. 2010). The influence of upwelling on the migration of marine animals was also observed for non-fish species. For example the migration of humpback whales to northerly areas coincides with coastal upwelling at target locations, resulting in cool surface waters (Rasmussen et al. 2007). The same pattern was observed for sooty shearwaters, where the arrival to wintering grounds in the Northern Hemisphere occurs when oceanic productivity is higher than in the South Pacific, which is due to coastal upwelling events (Shaffer et al. 2006). Although the motivation for white shark migrations is unknown, sharks migrations off California may be related with foraging success (Weng et al. 2007; Domeier and Nasby-Lucas 2008). Weng et al. (2007) proposed that the timing of departure of white sharks tagged with PSATs in the Farallones islands, off California, to offshore oceanic waters is connected with the decline of pinnipeds abundance, resulting in a decrease of hunting success and therefore causing the sharks to leave the area shortly after (Weng et al. 2007). In Australian waters, the time white sharks spent in a determined area also seems related with food resources availability, but when these disperse, sharks undertake rapid and directed movement away from such areas (Bruce et al. 2006). White sharks remained in an area off eastern Victoria (South-east Australia) where snappers, a known prey of white sharks in southern Australia, seasonally aggregate to spawn and left that area in autumn when the snappers were dispersing (Malcolm et al. 2001). Also, the northern movement along the eastern Australian coast during autumn may be related with the movement of several schooling species and these schools movements provide cues for the seasonal migration of white sharks (Bruce et al. 2006). Motivation for departure of white sharks from Guadalupe island (off the north-western Mexican coast) is also concomitant with the decline of prey, as yellowfin tuna (a possible prey item; Domeier and Nasby-Lucas 2007) and fur seals, which usually coincides with a decrease in the water temperature (Domeier and Nasby-Lucas 2008). Hence, individual white sharks possibly choose a departure time that is closely related with their specific hunting success and physiological condition (Domeier and Nasby-Lucas 2008). Migration as a response to declines in prey abundance has been observed in other marine species, like leatherback turtles (Sherrill-Mix et al. 2008) and humpback whales (Best 1995).

Migratory patterns

Marine species rely on specific conditions for feeding and reproduction that often cannot be met in a single location (Binder et al. 2011). Therefore, animals are required to migrate in order to fulfil their biological needs. Several reasons behind large-scale horizontal migrations have been proposed, such as moving to predictable productive oceanic areas where food is expected to be abundant, such as frontal zones or eddies, to overwintering grounds where conditions are ideal for survival, or to spawning areas where mating occurs or conditions for spawning are favourable (Nøttestad et al. 1999).

Reproduction vs. feeding dichotomy

Reproduction is one of the main drivers of fish migrations. Roshier and Reid (2003) refer to migration as the periodic, seasonally-driven movement to and from regular breeding and non-breeding grounds (generally foraging grounds), implying a strong philopatry [defined as "going to a place formerly occupied instead of equally probable places" (Gerking 1959; Grubbs et al. 2007)] for breeding locations. Therefore, long-distance migrations typically connect foraging and reproductive grounds (Dingle 2014). In some cases only mature adults perform migrations, while immature individuals display small-scale movements (e.g., Schaefer et al. 2007).

Satellite tracking of Atlantic bluefin tuna revealed rapid and ocean-wide migrations, ranging from cool sub-polar foraging grounds in the Atlantic Ocean to breeding grounds in warm subtropical waters. The western stock migrates to the Gulf of Mexico from April to June, where warm waters are favourable for the development of the eggs and larval stages (Block et al. 1998; Lutcavage et al. 1999; Block et al.

2005; Wilson et al. 2005; Teo et al. 2007; Block 2011). Similarly, mature individuals from the eastern stock move to productive, warm areas in the Mediterranean Sea from June to August, which are also feeding grounds for both pre- and post-spawning fish (De Metrio et al. 2002; Block et al. 2005; De Metrio et al. 2005; Rooker et al. 2007; Teo et al. 2007; Hilborn et al. 2010; Aranda et al. 2013; Cermeño et al. 2015) (Fig. 8.3). Pacific bluefin tuna also exhibit trans-Pacific migratory behaviour, likely driven by reproduction purposes. Juveniles migrate from the spawning grounds in the western Pacific Ocean to the eastern Pacific, and they only return back to the western Pacific as adults (Itoh et al. 2003). This trans-Pacific migration was originally observed using mark-recapture tags during the 1960s (Orange and Fink 1963; Clemens and Flittner 1969) and was confirmed by satellite tracking (Itoh et al. 2003; Kitagawa et al. 2009). Results from archival tagging of Pacific bluefin tunas off the coast of California (USA) and Baja California (Mexico) suggested that a higher number of mature individuals return to the spawning grounds in the west (Boustany et al. 2010). Southern bluefin tuna *Thunnus maccoyii* (Castelnau, 1872) display similar migration patterns to those of other bluefin species. Adults tagged in the western Tasman Sea remained in this foraging area from June to December, leaving as early as September and as late as December (Patterson et al. 2008) (Fig. 8.4). One individual travelled into the Indian Ocean spawning grounds south of Indonesia, corroborating previous assumptions that, during austral winter, adults would forage in the temperate waters of the Southern Hemispheres, migrating to the spawning grounds located in the North-western Indian Ocean from spring to autumn (Caton 1994). Similarly, reproduction-driven, cyclical migrations of mature individuals were observed for populations of yellowfin tuna (Schaefer et al. 2007), swordfish (Reeb et al. 2000, Takahashi et al. 2003, Neilson et al. 2009, Abascal et al. 2010) and black marlin *Istiompax indica* (Cuvier, 1832) (Chiang et al. 2014), with individuals moving from food-rich foraging grounds to warmer, southern spawning waters (Fig. 8.5).

Linking migratory movements with reproduction in elasmobranchs has been historically difficult, although a few studies have tried to associate wide-scale movement to parturition (Weng et al. 2008; Skomal et al. 2009; Hueter et al. 2013; Vandeperre et al. 2014). For example, juvenile females blue sharks *Prionace glauca*

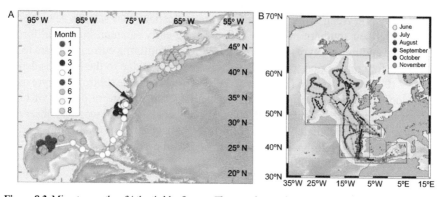

Figure 8.3. Migratory paths of Atlantic bluefin tuna *Thunnus thynnus* in northwest [A—Block et al. (2005)] and northeast [B—Aranda et al. (2013)] Atlantic Ocean. *Used with permission.*

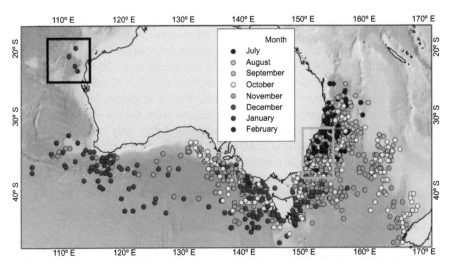

Figure 8.4. Geolocations of southern bluefin tuna *Thunnus maccoyii* in Australian waters (Patterson et al. 2008). The blue square represents the tagging locations and the black square the spawning grounds. *Used with permission.*

(Linnaeus, 1758) tagged in the Azorean Archipelago (North Atlantic, Portugal), undertook seasonal large scale latitudinal migrations, moving to northern latitudes during summer, while males remained in the area until January before moving west, to offshore areas in the northwestern Atlantic, probably to mate during spring (Vandeperre et al. 2014). By the end of spring, blue sharks returned to the Azores and then moved southwards during late summer, and thus performing a clear cyclic pattern of migration (Vandeperre et al. 2014). Furthermore, the world's two largest fish, whale and basking sharks, leave summer feeding grounds in the North Atlantic to southern tropical locations across the equator, likely to provide stable conditions for gestation and parturition, as well as suitable nursery habitat for new-borns (Skomal et al. 2009; Hueter et al. 2013) (Fig. 8.6). Lastly, female salmon sharks *Lamna ditropis* (Hubbs and Follet 1947), tagged in the Prince William Sound (Alaska, USA), migrated between subarctic foraging grounds (around 60° N) and subtropical reproductive areas (around 30° N) (Weng et al. 2008). There are indications that salmon sharks give birth during their southern migration in late spring-early summer, in the California current and subtropical gyre (Weng et al. 2008). The California current appears to be not only a foraging region, but also a parturition ground, due to the long residency times (~ 107 days) observed (Weng et al. 2008). In contrast, the subtropical gyre may be predominantly a parturition ground, as sharks spent short-term periods of time (~ 77 days) there, and migrated north sooner (Weng et al. 2008). It was also observed that some sharks immediately return north after arriving at the subtropical gyre area, which is consistent with having given birth and with no other functions to fulfil in the region (Weng et al. 2008).

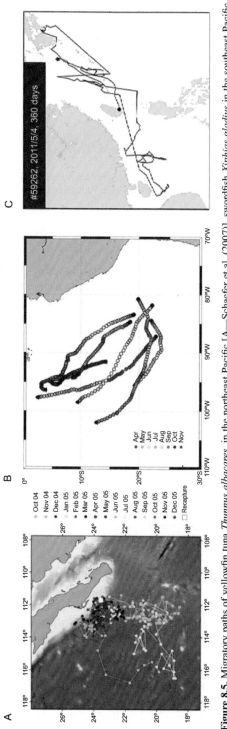

Figure 8.5. Migratory paths of yellowfin tuna *Thunnus albacares*, in the northeast Pacific [A—Schaefer et al. (2007)], swordfish *Xiphias gladius* in the southeast Pacific [B—Abascal et al. (2010)] and black marlin *Istiompax indica* in the southwest Pacific [C—Chiang et al. (2014)]. *Used with permission.*

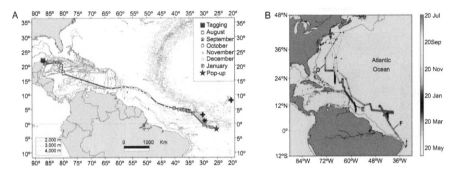

Figure 8.6. Trans-equatorial migration of whale sharks *Rhincodon typus* [A—Hueter et al. (2013)] and basking sharks *Cetorhinus maximus* [B—Skomal et al. (2009)] along the western Atlantic Ocean. *Used with permission.*

Wide-scale foraging opportunities

As mentioned, large-scale horizontal migrations can occur in response to wide-scale predictable productive oceanic locations (Nøttestad et al. 1999), even when prey are abundant in coastal areas (Jorgensen et al. 2009). Such large-scale movements were observed by several high trophic level pelagic species, such as tunas (e.g., Polovina et al. 2001; Block et al. 2011), billfish (e.g., Brill and Lutcavage 2001) and sharks (e.g., Sims et al. 2003; Domeier and Nasby-Lucas 2008; Gore et al. 2008; Nasby-Lucas et al. 2009; Block et al. 2011). For example, one of the probable reasons for trans-Pacific migrations performed by a albacore tuna (Childers et al. 2011) is to explore the various frontal zones and eddies that dominate the Kuroshio and Oyashio region off Japan (Okazaki et al. 2002; Shaffer et al. 2006). Strikingly, this area is the main spawning ground of pelagic fishes, such as anchovy and sardine, with a spawning peak in May (Oozeki et al. 2007), and coinciding with the arrival of albacore tuna. The migration of bluefin tuna from the northwest Atlantic to the northeast Atlantic has also been linked to wide-scale foraging success, with tuna movements associated with a productive upwelling off the Iberian Peninsula (western Europe) that attracts sardine and mackerel species to the area (Brill and Lutcavage 2001; Walli et al. 2009), from late May or early June to late September early October (Haynes et al. 1993) (Fig. 8.7).

The search for oceanic foraging areas is one of the hypothesis put forward to explain why white sharks perform extensive migrations. Several studies involving PSAT tagging of white sharks in North-east Pacific indicated that many individuals aggregate near coastal areas during autumn and winter (Weng et al. 2007; Domeier and Nasby-Lucas 2008; Jorgensen et al. 2009), before migrating to the same shared offshore foraging area, a region centred halfway between Baja California and the Hawaiian islands (Domeier and Nasby-Lucas 2008; Nasby-Lucas et al. 2009) (Fig. 8.8). This area is an oxygen minimum frontal zone, where the potential prey species aggregate and are targeted by white sharks (Domeier and Nasby-Lucas 2008; Jorgensen et al. 2009; Nasby-Lucas et al. 2009). Moreover, PSAT tagged white sharks in the Chatham Islands (New Zealand) made rapid and direct movements towards subtropical and tropical locations, being consistent with described behaviour ranging

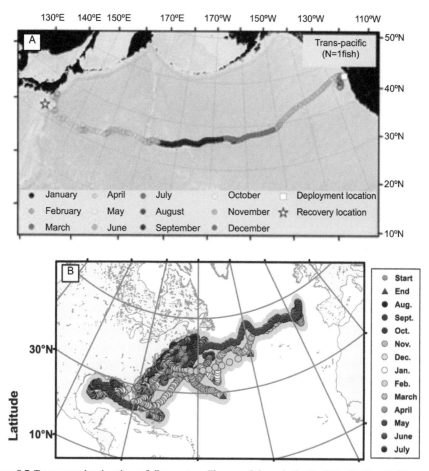

Figure 8.7. Trans-oceanic migrations of albacore tuna *Thunnus alalunga* in the North Pacific [A—Childers et al. (2011)] and Atlantic bluefin tuna *Thunnus thynnus* in the North Atlantic [B—Galuardi et al. 2010]. *Used with permission.*

from site fidelity to feeding related migrations (Bruce et al. 2006; Jorgensen et al. 2009; Bonfil et al. 2010). The migrations of two planktivorous sharks were also associated with wide-scale foraging opportunities, with individuals leaving coastal productive frontal regions (Sims et al. 2003; Skomal et al. 2004; Sims et al. 2005; Wilson et al. 2006) to offshore areas in order to explore different food-rich habitats (Eckert and Stewart 2001; Wilson et al. 2006; Gore et al. 2008). In a study that plainly illustrates this behaviour, tiger sharks *Galeocerdo cuvier* (Péron and Lesueur, 1822) tagged with satellite-linked high resolution tags in the northwest Atlantic, displayed an extensive space-use throughout the region, with evident long-distance north-south migrations. During winter, tiger sharks were mostly associated with coral reef-bound islands in the Bahamas, Turks and Caicos Islands, and Anguilla/Saint Martin in the Caribbean Sea, while during summer the majority of tiger sharks rapidly moved into temperate, oceanic, foraging grounds (Lea et al. 2015) (Fig. 8.9). The northern range expansion of tiger sharks during summer coincides with the northward extension of

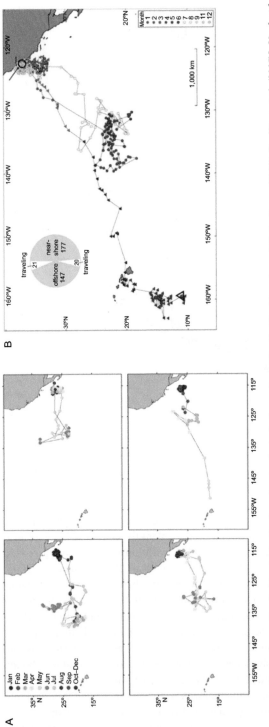

Figure 8.8. Migratory paths of white sharks *Carcharodon carcharias* in the northwest Pacific [A—Domeier and Nasby-Lucas (2008); B—Weng et al. (2007)]. *Used with permission.*

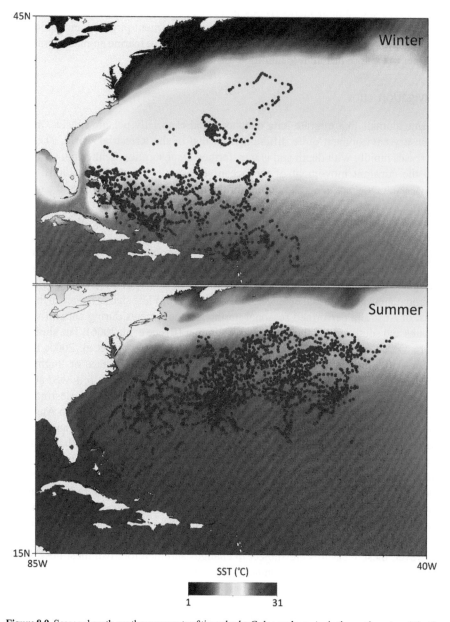

Figure 8.9. Seasonal north-south movements of tiger sharks *Galeocerdo cuvier* in the northwestern Atlantic Ocean in relation to sea surface temperature (SST) during winter and summer (Lea et al. 2015). *Used with permission.*

the Gulf Stream (Lea et al. 2015). Similar north-south migrations were observed for a range of species such as swordfish (Takahashi et al. 2003; Abascal et al. 2010), blue and yellowfin tunas (Block et al. 2011), black marlin (Chiang et al. 2014), sunfish *Mola mola* (Linnaeus, 1758) (Sims et al. 2009), white shark (Bruce et al. 2006), mako sharks, blue and salmon sharks (Block et al. 2011). Therefore, the seasonal increase of sea water temperature in northern areas and consequent physical and biological processes (e.g., increase in the productivity) seems to drive marine animals to migrate, in order to explore different locations.

Navigation cues

Marine animals that migrate long distances, and especially pelagic species, inhabit a sensory environment quite different from that of the terrestrial world. The light decreases rapidly with depth and is transformed by absorption, scattering, refraction and the constant movement of the ocean surface (Lohmann et al. 2008). Also, visual landmarks are generally absent for the majority of time. However, the marine environment provides animals with other cues that do not exist on land. Therefore animals navigating through the ocean also have access to a suite of navigational cues (Klimley et al. 2002; Lohmann et al. 2008). It seems unlikely that the oriented movement could occur without some form of environment reference (Klimley et al. 2002). There are four particular types of cues that may provide the required navigational information for ocean migrants, namely geomagnetic, chemical, hydrodynamic and celestial cues (Klimley et al. 2002). White sharks tagged in the east coast of Australia performed seasonal northward movements during autumn-winter months and southward in the spring-early summer (Bruce et al. 2006). The similarity in tracks recorded in the latter study and other studies (Bonfil et al. 2005; Bonfil et al. 2010) suggests that white sharks may follow common routes or 'highways' when moving between areas, indicating that individuals probably followed similar cues, or may have shared a common ability to navigate between destinations (Bruce et al. 2006). The mechanisms used by white sharks to navigate to Australia and back during a transoceanic migration (Bonfil et al. 2005) are also unknown; aside from a few shallow seamounts, there are no other topographic features that could be used for along-route orientation (Bonfil et al. 2005). However, by swimming near the surface during the oceanic travel, it was possible for the white shark to use visual cues such as celestial bodies as an important navigational mechanism (Bonfil et al. 2005). The use of solar cues for orientation during fish migrations in the open ocean has been previously described (Binder et al. 2011). Some species may use information obtained from changes in the sun's angle in the horizontal plane and/or vertical plane. However, these measures are subject to changes and therefore fish must possess an internal biological clock and calendar to compensate. Also, the sun is often obscured and, therefore, it is not always the best orienting cue. Nonetheless, it has been suggested that fish may continue to orient using polarized light as a directional cue (Binder et al. 2011). Klimley et al. (2002) also stated that swimming closer to the surface would facilitate the use of the Earth's main dipole field as a reference. There is a growing body of support for geomagnetic orientation when fishes undergo long-distance oceanic migrations (Gould 1998; Binder et al. 2011). When water moves across the Earth's magnetic field, it induces a weak electric current that may be detected by fishes (Binder et al. 2011). Studies have subsequently shown that, for example, yellowfin and Atlantic bluefin tuna precise navigation is mainly due to the ability to sense the various components of the Earth's magnetic field (Walker 2006). Bruce et al. (2006) observed that the migration periods of white sharks were characterized by frequent ascents and descents between the surface and depth, and linked the oscillatory diving behaviour with the ability of sharks to detect and follow cues in the environment, either olfactory, geomagnetic, or orientation relative to the sun. Similarly, Klimley et al. (2002) also suggested that oscillatory dives may be used to explore the water column to gain directional information during migrations, by detecting the specific source of a water mass with a unique chemical composition and odour. For example, Atlantic bluefin tuna are known to use olfactory cues to locate their respective spawning grounds, with fish diving deeper, probably in an attempt to locate the denser water masses typical of the Gulf of Mexico or the Mediterranean Sea (Teo et al. 2007). It has also been proposed that juvenile albacore use frontal areas to navigate in the North Pacific, by using the chlorophyll *a* transition zone as migration route and as a foraging habitat (Polovina et al. 2001). Furthermore, Cotton et al. (2005) suggested that basking sharks rely on thermal discontinuities to navigate during their migrations. This 'foraging or migration corridors' theory is supported by the correlation between movement patterns and sea surface temperature and corroborated by behavioural studies (Sims et al. 2003). However, it is not clear whether individuals use the fronts as a cue to move toward favourable feeding locations or as a 'simple' pathway, which are two different mechanisms that can have different implications for the distribution of fish in relation to a given front (Schick et al. 2004). Evidence of the use of oceanic fronts in movements was also observed for billfishes, with studies suggesting that that horizontal movements of striped marlin *Kajikia audax* (Philippi, 1887) are set predominately by currents, with tagged fish making continuous small-scale random movements in relation to mesoscale eddies (Brill et al. 1993; Brill and Lutcavage 2001).

Conservation status of oceanodromous species

The fisheries of highly migratory fish species have great economic importance in all oceans and semi-enclosed seas, except in polar regions. In 2006, around 30% of highly migratory tuna and tuna-like species, and more than 50% of highly migratory oceanic sharks were considered either overexploited or depleted (Maguire 2006). Tuna and tuna-like species are some of the most economically important species due to the intensive international trade (Maguire 2006), accounting for 4,8 million tonnes in 2004, which corresponded to approximately US$7,670 million (2004 values). The Atlantic bluefin tuna is currently considered as 'endangered' under the A2 criterion of the Red List of the International Union for Conservation of Nature (IUCN), mainly due to a 51% global decline in the Spawning Stock Biomass (SSB) over the past four decades (Collette et al. 2011c). However, recently, the SSB showed signs of increasing (Fromentin et al. 2014). The southern bluefin tuna, which has been intensively fished since the early 1950s, is considered to be 'critically endangered' by the IUCN Red List, with an estimated 85% SSB reduction between 1973 and 2009, but with no signs of recovery (Collette et al. 2011d). The Pacific bluefin tuna is listed as 'vulnerable' in the IUCN Red List (Collette et al. 2014), and recent stock assessment studies, based on several biological reference points, indicate that overfishing is occurring (Collette et al. 2014; ISC 2014). The albacore and yellowfin tuna are both considered 'near threatened' (Collette et al. 2011a,b) and are important species (albacore: ~2,500 USD/tonne in 2008; yellowfin tuna: ~1,900 USD/tonne in 2014) for the international trade (Collette et al. 2011a,b). The stocks of yellowfin tuna are, or close to being, fully exploited in all oceans (Collette et al. 2011b); the albacore stocks are fully exploited in the South Atlantic and Pacific and overexploited in the North Atlantic (Collette et al. 2011a). The swordfish is also an important commercial species (around 12,125 USD/tonne in 2012), caught mainly by longlining (e.g., Buencuerpo et al. 1998; Marín et al. 1998; Megalofonou 2005; Abascal et al. 2010). Studies have indicated that the spawning biomass was above the required levels to maintain maximum sustainable yield in the Pacific (Hinton and Maunder 2011) and Atlantic oceans (Neilson et al. 2013), and the species is currently considered of 'least concern' by the IUCN Red List. Yet, there are some indications, such as low average sizes/high catch rates, that have raised concerns about the sustainability of swordfish fisheries, particularly in the Southeast Pacific (Abascal et al. 2010). The worldwide stocks of tuna, as well as other tune-like species (including billfishes), are managed by various international commissions operating in each ocean, such as the Inter American Tropical Tuna Commission (IATTC; established in 1949), the International Commission for the Conservation of Atlantic Tunas (ICCAT; established in 1969), the Commission for Southern Bluefin Tuna (CCSBT; established in 1993) and the International Scientific Committee (ISC; established in 1995) for tuna and tuna-like Species in the North Pacific Ocean, among others (Majkowski 2007). All major tuna fisheries target tuna populations that migrate across multiple Economic Exclusive Zones (EEZ) and into the high seas, making them susceptible to fishing pressure from several nations, including those that are not parties in cooperative fishing agreements. In 1982, UNCLOS marked the beginning of international level fisheries regulation, followed by the 1995 Fish Stocks Agreement (FSA) (Dulvy et al. 2008; Techera and Klein 2011), which further elaborated the rights and obligations agreed in UNCLOS regarding fishing of highly migratory species on the high seas. The FSA attempted to improve the institutional basis for management of straddling and highly migratory fish stocks, by articulating principles for the conservation and management of such stocks, and by establishing rules for the creation and operation of regional fishery management organizations. This development spurred efforts to create new regional organizations to govern tuna harvests in the western and central Pacific and Indian oceans. It also strengthened the legal foundation for pre-existing international tuna management organizations, notably the IATTC and ICCAT, which previously only focused on the status of stocks using traditional fishery management tools (e.g., quotas and size limits for targeted species) (Fonteneau 2001). The newly emerging consensus is that management of pelagic stocks should be more cautious, in order to ensure, not only the long term health of tuna stocks, but also of other components of the pelagic ecosystems, thus highlighting the importance of an ecosystem-based fisheries management (Fonteneau 2001).

The total reported catches of shark species and families considered highly migratory was close to 100,000 tonnes in 2004 (Dulvy et al. 2008). However, it is likely that the total catch and mortality of sharks (for example as bycatch) is much higher than the reported values (Dulvy et al. 2008). Unfortunately, the state of many shark populations is currently unknown, or poorly known (Dulvy et al. 2008). Whale sharks, basking sharks, white sharks and tiger sharks are all considered 'vulnerable' by the IUCN Red List (Norman 2005; Fergusson 2009; Simpfendorfer 2009; Fowler 2005b). The basking shark has been exploited commercially for centuries in several parts of the world mainly for its liver oil, which was used as lighting fuel for lamps and presently as a source of chemical compounds. Basking shark fisheries have been declining in several areas but it is difficult to separate natural fluctuations from fishing effects (Maguire 2006). A reduction in the catches of basking shark was observed between 1960 and 1980 and 2004, from 8,000 tonnes to 239 tonnes, indicating that this species is probably globally overexploited (Maguire 2006). Populations of whale shark appear to have been depleted by harpoon fisheries in Southeast Asia and, perhaps, by incidental capture in other fisheries (Norman 2005). The vulnerability of whale sharks to commercial fishing derives mainly from their high value in the international fin trade, highly migratory nature and natural low abundance (Norman 2005). White sharks have no targeted fisheries, and thus the majority of worldwide captures are incidental through commercial fisheries operating surface longlines, setlines, gillnets, trawls, fish-traps and other gear (Fergusson 2009). Strikingly, pelagic sharks remain a low priority for fisheries management, and despite increasing concern about their conservation status, oceanic sharks are at risk of depletion (Dulvy et al. 2008). The International Plan of Action for the Conservation and Management of Sharks (IPOA-Sharks), which was adopted in 1999 by the FAO Committee on Fisheries, is a complement to the UNCLOS and to the FSA. In short, the IPOA-Sharks asked fishing nations and regional fishery management organizations to develop national action plans aiming the conservation and management of sharks, covering both target and bycatch species caught in their waters or elsewhere by nationals (Dulvy et al. 2008; Techera and Klein 2011). However, as it is a voluntary international instrument, and not a treaty or law, thus it acts as a framework for regulatory action (Dulvy et al. 2008; Techera and Klein 2011). Other conservation agreements may offer an alternative for pelagic shark conservation, including regional treaties such as the Barcelona and Bern conventions and international treaties such as the Convention on International Trade in Endangered Species (CITES) and the Convention on Migratory Species (CMS). Both white and basking shark are included in the Barcelona convention on annex II 'list of endangered or threatened species' and in the appendix II 'strictly protected fauna species' of the Bern convention. These two species, along with whale sharks, are also listed on CITES appendix II (whale and basking sharks in 2002, white shark in 2004) which intends to limit the international trade to achieve sustainable fishing levels (Fowler 2000; Bonfil et al. 2005; Fowler 2005a; Domeier and Nasby-Lucas 2007; Dulvy et al. 2008). Whale, white and basking sharks are completely protected under the CMS appendices (whale shark on appendix II in 1999; white shark on appendices I and II in 2002; basking shark on appendices I and II in 2005; Fowler 2000; Fowler 2005a; Dulvy et al. 2008), with the establishment of zero quotas for basking and white shark in the EU, and for white shark in New Zealand (Dulvy et al. 2008). Blue shark catches are regulated under annex III of the Bern and Barcelona conventions, which allow a certain level of exploitation depending on the population levels or require exploitation licences (Fowler 2005a; Dulvy et al. 2008). The need for a stronger protection of migratory sharks and the development of a global conservation agreement to mitigate bycatch has been identified in 2005 by the CMS, which concluded that 35 species of migratory elasmobranchs, including those species mentioned here, could benefit from the protection of CMS proposals (Dulvy et al. 2008).

A new paradigm shift has arisen worldwide (Hooker et al. 1999) regarding the protection of migratory ocean species, which consists in the creation of open ocean Marine Protected Areas (MPAs) that would include known seasonal breeding or feeding locations (as an example, MPAs for sharks were proposed by Kinney and Simpfendorfer 2009; Pendoley et al. 2014). However, species are also at risk during the migration between protected habitats (e.g., Shillinger et al. 2008; Womble and Gende 2013). Failure to successfully protect migratory corridors generally arises as the result of a combination of issues. For instance, the consistency of migratory routes may vary between individuals and populations, hindering the delineation of essential protection zones (Pendoley et al. 2014). With the exceptions detailed in this chapter, long-term movement information remains sparse for many large oceanic species, making it difficult to assess the potential efficacy of oceanic MPAs and the implementation of protection in the migration corridors for such highly mobile species (Lea et al. 2015). With the development of electronic tag data it is now known that several tuna, billfish and shark species have greater spatial distributions than previously assumed (e.g., Bonfil et al. 2005; Gore et al. 2008; Abascal et al. 2010; Block 2011; Childers et al. 2011). Incorporating spatial information into fisheries management and species protection policies will improve the effectiveness of management measures (Abascal et al. 2010; Costa et al. 2012). As an example, the information gathered with satellite tracking could provide the basis for large management schemes, using international policy vehicles. Furthermore, if conservation corridors were to be created, key ecological foraging hotspots and migratory paths could be sustainably managed (Block et al. 2011). Advances in electronic tracking technologies, together with new analytical techniques, have enhanced our understanding of oceanodromous fish migrations, by allowing scientists to follow their movements over great distances, during longer periods of time and, importantly, with greater detail (e.g., Teo et al. 2007; Weng et al. 2007; Skomal et al. 2009; Block et al. 2011; Childers et al. 2011). In the future, data from electronic tagging and bio-logging studies is expected to have an even stronger role in conservation management (Bograd et al. 2010), since the increased understanding of species migration is a crucial part of management and conservation plans (Costa et al. 2012; Lea et al. 2015).

Acknowledgements

We are grateful to Pedro Morais and Françoise Daverat for their helpful comments. Funding was provided by Fundação para a Ciência e a Tecnologia (FCT) through a Investigator Fellowship to N.Q. (ref.: IF/01611/2013), and Ph.D. scholarships to L.L.S. (ref.: SFRH/BD/68717/2010) and M.B. (ref.: SFRH/BD/88175/2012). A.C. was funded by national funds from FCT/Ministério da Educação e Ciência (MEC) through PROMAR ref.: 31-03-05-FEP-37.

References

Abascal FJ, Mejuto J, Quintans M and Ramos-Cartelle A (2010) Horizontal and vertical movements of swordfish in the southeast Pacific. ICES Journal of Marine Science 67: 466–474.

Aidley DJ (1981) Animal Migration. Cambridge University Press, Cambridge.

Anderson R and Ahmed H (1993) The shark fisheries of the Maldives. FAO, Rome, and Ministry of Fisheries, Male, Maldives.

Aranda G, Abascal FJ, Varela JL and Medina A (2013) Spawning behaviour and post-spawning migration patterns of atlantic bluefin tuna (*Thunnus thynnus*) ascertained from satellite archival tags. PLOS ONE 8: e76445.

Baker R (1978) Evolutionary Ecology of Animal Migration. Holmes & Meier Publishers, New York.

Best PB (1995) A suspended migration of humpback whales *Megaptera novaeangliae* on the west coast of South Africa. Marine Ecology Progress Series 118: 1–12.

Binder TR, Cooke SJ and Hinch SG (2011) The biology of fish migration. *In*: Farrell AP (ed.). Encyclopedia of Fish Physiology: From Genome to Environment. Academic Press, San Diego 3: 1921–1927.

Block B (2011) Tracking oceanic fish. *In*: Farrell AP (ed.). Encyclopedia of Fish Physiology: From Genome to Environment. Academic Press, San Diego. 3: 1928–1936.

Block BA, Dewar H, Farwell C and Prince ED (1998) A new satellite technology for tracking the movements of Atlantic bluefin tuna. Proceedings of the National Academy of Sciences USA 95: 9384–9389.

Block BA, Dewar H, Blackwell SB, Williams TD, Prince ED, Farwell CJ, Boustany AM, Teo SLH, Seitz A, Walli A and Fudge D (2001) Migratory movements, depth preferences, and thermal biology of the Atlantic bluefin tuna. Science 293: 1310–1314.

Block BA, Teo SL, Walli A, Boustany A, Stokesbury MJ, Farwell CJ, Weng KC, Dewar H and Williams TD (2005) Electronic tagging and population structure of Atlantic bluefin tuna. Nature 434: 1121–1127.

Block BA, Jonsen I, Jorgensen S, Winship A, Shaffer SA, Bograd S, Hazen E, Foley D, Breed G and Harrison A-L (2011) Tracking apex marine predator movements in a dynamic ocean. Nature 475: 86–90.

Bograd SJ, Block BA, Costa DP and Godley BJ (2010) Biologging technologies: new tools for conservation. Endangered Species Research 10: 1–7.

Bonfil R, Meÿer M, Scholl MC, Johnson R, O'Brien S, Oosthuizen H, Swanson S, Kotze D and Paterson M (2005) Transoceanic migration, spatial dynamics, and population linkages of white sharks. Science 310: 100–103.

Bonfil R, Francis M, Duffy C, Manning M and O'Brien S (2010) Large-scale tropical movements and diving behavior of white sharks *Carcharodon carcharias* tagged off New Zealand. Aquatic Biology 8: 115–123.

Boustany AM, Matteson R, Castleton M, Farwell C and Block BA (2010) Movements of Pacific bluefin tuna (*Thunnus orientalis*) in the eastern North Pacific revealed with archival tags. Progress in Oceanography 86: 94–104.

Braun CD, Kaplan MB, Horodysky AZ and Llopiz JK (2015) Satellite telemetry reveals physical processes driving billfish behavior. Animal Biotelemetry 3: 2.

Brill RW and Lutcavage ME (2001) Understanding environmental influences on movements and depth distributions of tunas and billfishes can significantly improve population assessments. Proceedings of the American Fisheries Society Symposium 25: 179–198.

Brill RW, Holts D, Chang R, Sullivan S, Dewar H and Carey F (1993) Vertical and horizontal movements of striped marlin (*Tetrapturus audax*) near the Hawaiian Islands, determined by ultrasonic telemetry, with simultaneous measurement of oceanic currents. Marine Biology 117: 567–574.

Bruce B, Stevens J and Malcolm H (2006) Movements and swimming behaviour of white sharks (*Carcharodon carcharias*) in Australian waters. Marine Biology 150: 161–172.

Buencuerpo V, Rios S and Morón J (1998) Pelagic sharks associated with the swordfish, *Xiphias gladius*, fishery in the eastern North Atlantic Ocean and the Strait of Gibraltar. Fishery Bulletin 96: 667–685.

Burke WT (1984) Highly migratory species in the new Law of the Sea. Ocean Development and International Law 14: 273–314.

Caton A (1994) Review of aspects of southern bluefin tuna biology, population, and fisheries. FAO Fisheries Technical Paper (FAO).

Cermeño P, Quílez-Badia G, Ospina-Alvarez A, Sainz-Trápaga S, Boustany AM, Seitz AC, Tudela S and Block BA (2015) Electronic tagging of Atlantic bluefin tuna (*Thunnus thynnus*, L.) reveals habitat use and behaviors in the Mediterranean Sea. PLOS ONE 10: e0116638.

Chapman DD, Feldheim KA, Papastamatiou YP and Hueter RE (2015) There and back again: a review of residency and return migrations in sharks, with implications for population structure and management. Annual Review of Marine Science 7: 547–570.

Chiang W-C, Musyl MK, Sun C-L, DiNardo G, Hung H-M, Lin H-C, Chen S-C, Yeh S-Z, Chen W-Y and Kuo C-L (2014) Seasonal movements and diving behaviour of black marlin (*Istiompax indica*) in the northwestern Pacific Ocean. Fisheries Research 166: 92–102.

Childers J, Snyder S and Kohin S (2011) Migration and behavior of juvenile North Pacific albacore (*Thunnus alalunga*). Fisheries Oceanography 20: 157–173.

Clemens A and Flittner GA (1969) Bluefin tuna migrate across the Pacific Ocean. California Fish and Game 55: 132–135.

Collette B, Acero A, Amorim AF, Boustany A, Canales Ramirez C, Cardenas G, Carpenter KE, Chang S-K, de Oliveira Leite N, Jr., Di Natale A, Die D, Fox W, Fredou FL, Graves J, Guzman-Mora A, Viera Hazin FH, Hinton M, Juan Jorda M, Minte Vera C, Miyabe N, Montano Cruz R, Masuti E, Nelson R, Oxenford H, Restrepo V, Salas E, Schaefer K, Schratwieser J, Serra R, Sun C, Teixeira Lessa RP, Pires Ferreira Travassos PE, Uozumi Y and Yanez E (2011a) *Thunnus alalunga*. The IUCN Red List of Threatened Species. Assessed 18 March 2015.

Collette B, Acero A, Amorim AF, Boustany A, Canales Ramirez C, Cardenas G, Carpenter KE, Chang S-K, de Oliveira Leite N, Jr., Di Natale A, Die D, Fox W, Fredou FL, Graves J, Guzman-Mora A, Viera Hazin FH, Hinton M, Juan Jorda M, Minte Vera C, Miyabe N, Montano Cruz R, Masuti E, Nelson R, Oxenford H, Restrepo V, Salas E, Schaefer K, Schratwieser J, Serra R, Sun C, Teixeira Lessa RP, Pires Ferreira Travassos PE, Uozumi Y and Yanez E (2011b) *Thunnus albacares*. The IUCN Red List of Threatened Species. Assessed 15 March 2015.

Collette B, Amorim AF, Boustany A, Carpenter KE, de Oliveira Leite N, Jr., Di Natale A, Die D, Fox W, Fredou FL, Graves J, Viera Hazin FH, Hinton M, Juan Jorda M, Kada O, Minte Vera C, Miyabe N, Nelson R, Oxenford H, Pollard D, Restrepo V, Schratwieser J, Teixeira Lessa RP, Pires Ferreira Travassos PE and Uozumi Y (2011c) *Thunnus thynnus*. The IUCN Red List of Threatened Species. Assessed 15 March 2015.

Collette B, Chang S-K, Di Natale A, Fox W, Juan Jorda M, Miyabe N, Nelson R, Uozumi Y and Wang S (2011d) *Thunnus maccoyii*. The IUCN Red List of Threatened Species. Assessed 15 March 2015.

Collette B, Fox W, Juan Jorda M, Nelson R, Pollard D, Suzuki N and Teo S (2014) *Thunnus orientalis*. The IUCN Red List of Threatened Species. Assessed 15 March 2015.

Costa DP, Breed GA and Robinson PW (2012) New insights into pelagic migrations: implications for ecology and conservation. Annual Review of Marine Science 43: 73–96.

Cotton PA, Sims DW, Fanshawe S and Chadwick M (2005) The effects of climate variability on zooplankton and basking shark (*Cetorhinus maximus*) relative abundance off southwest Britain. Fisheries Oceanography 14: 151–155.

De Metrio G, Arnold G, Block B, De la Serna J, Deflorio M, Cataldo M, Yannopoulos C, Megalofonou P, Beemer S and Farwell C (2002) Behaviour of post-spawning Atlantic bluefin tuna tagged with pop-up satellite tags in the Mediterranean and eastern Atlantic. ICCAT Collective Volume of Scientific Papers 54: 415–424.

De Metrio G, Arnold G, De la Serna J, Block B, Megalofonou P, Lutcavage M, Oray I and Deflorio M (2005) Movements of bluefin tuna (*Thunnus thynnus* L.) tagged in the Mediterranean Sea with pop-up satellite tags. ICCAT Collective Volume of Scientific Papers 58: 1337–1340.

Dingle H (2014) Migration: The Biology of Life on the Move. Oxford University Press, Oxford.

Dodson and Neill WH. Mechanisms of Migration in Fishes. Springer, New York.

Domeier ML and Nasby-Lucas N (2007) Annual re-sightings of photographically identified white sharks (*Carcharodon carcharias*) at an eastern Pacific aggregation site (Guadalupe Island, Mexico). Marine Biology 150: 977–984.

Domeier ML and Nasby-Lucas N (2008) Migration patterns of white sharks *Carcharodon carcharias* tagged at Guadalupe Island, Mexico, and identification of an eastern Pacific shared offshore foraging area. Marine Ecology Progress Series 370: 221–237.

Domeier ML, Kiefer D, Nasby-Lucas N, Wagschal A and O'Brien F (2005) Tracking Pacific bluefin tuna (*Thunnus thynnus orientalis*) in the northeastern Pacific with an automated algorithm that estimates latitude by matching sea-surface-temperature data from satellites with temperature data from tags on fish. Fishery Bulletin 103: 292–306.

Dulvy NK, Baum JK, Clarke S, Compagno LJ, Cortes E, Domingo A, Fordham S, Fowler S, Francis MP and Gibson C (2008) You can swim but you can't hide: the global status and conservation of oceanic pelagic sharks and rays. Aquatic Conservation 18: 459–482.

Eckert SA and Stewart BS (2001) Telemetry and satellite tracking of whale sharks, *Rhincodon typus*, in the Sea of Cortez, Mexico, and the north Pacific Ocean. Environmental Biology of Fishes 60: 299–308.

Fergusson I, Compagno LJV and Marks M (2009) *Carcharodon carcharias*. The IUCN Red List of Threatened Species. Assessed 15 March 2015.

Fonteneau A (2001) Potential use of marine protected areas applied to tuna fisheries and offshore pelagic ecosystems. pp. 55–66. *In*: Thiel H and Koslow A (eds.). Managing Risks to Biodiversity and the Environment on the High Seas, including Tools such as Marine Protected Areas - Scientific Requirements and Legal Aspects. Proceedings of the Expert Workshop held at the International Academy for Nature Conservation Isle of Vilm, Germany, 27 February–4 March 2001.

Fowler SL (2000) Whale shark *Rhincodon Typus*: policy and research scoping study: June–September 2000. Nature Conservation Bureau.

Fowler SL (2005a) Sharks, rays and chimaeras: the status of the Chondrichthyan fishes: status survey, Vol. 63. IUCN.

Fowler SL (2005b) *Cetorhinus maximus*. The IUCN Red List of Threatened Species. Assessed 15 March 2015.

Fromentin J-M, Bonhommeau S, Arrizabalaga H and Kell LT (2014) The spectre of uncertainty in management of exploited fish stocks: The illustrative case of Atlantic bluefin tuna. Marine Policy 47: 8–14.

Gerking SD (1959) The restricted movement of fish populations. Biological Reviews 34: 221–242.

Gore MA, Rowat D, Hall J, Gell FR and Ormond RF (2008) Transatlantic migration and deep mid-ocean diving by basking shark. Biology Letters 4: 395–398.

Gould JL (1998) Sensory bases of navigation. Current Biology 8: R731–R738.

Grubbs RD, Musick JA, Conrath CL and Romine JG (2007) Long-term movements, migration, and temporal delineation of a summer nursery for juvenile sandbar sharks in the Chesapeake Bay region. American Fisheries Society Symposium 50: 87.

Hammerschlag N, Gallagher A and Lazarre D (2011) A review of shark satellite tagging studies. Journal of Experimental Marine Biology and Ecology 398: 1–8.

Haynes R, Barton ED and Pilling I (1993) Development, persistence, and variability of upwelling filaments off the Atlantic coast of the Iberian Peninsula. Journal of Geophysical Research 98: 22681–22692.

Hilborn R, Galuardi B, Royer F, Golet W, Logan J, Neilson J and Lutcavage M (2010) Complex migration routes of Atlantic bluefin tuna (*Thunnus thynnus*) question current population structure paradigm. Canadian Journal of Fisheries and Aquatic Sciences 67: 966–976.

Hinton MG and Maunder MN (2011) Status of swordfish in the Eastern Pacific Ocean in 2010 and outlook for the future. Inter-American Tropical Tuna Commission, Stock Assessment Report 12: 133–177.

Hooker SK, Whitehead H and Gowans S (1999) Marine protected area design and the spatial and temporal distribution of cetaceans in a submarine canyon. Conservation Biology 13: 592–602.

Hueter RE, Tyminski JP and de la Parra R (2013) Horizontal movements, migration patterns, and population structure of whale sharks in the Gulf of Mexico and northwestern Caribbean Sea. PLOS ONE 8: e71883.

Humston R, Ault JS, Lutcavage M and Olson DB (2000) Schooling and migration of large pelagic fishes relative to environmental cues. Fisheries Oceanography 9: 136–146.

ISC (2014) Report of the Pacific bluefin tuna working group workshop. International scientific committee for tuna and tuna-like species in the North Pacific Ocean.

Itoh T, Tsuji S and Nitta A (2003) Migration patterns of young Pacific bluefin tuna (*Thunnus orientalis*) determined with archival tags. Fishery Bulletin 101: 514–534.

Jones FH (1984) A view from the ocean. *In*: McCleave JD (ed.). Mechanisms of Migration in Fishes 1: 26. Springer, New York.

Jorgensen SJ, Reeb CA, Chapple TK, Anderson S, Perle C, Van Sommeran SR, Fritz-Cope C, Brown AC, Klimley AP and Block BA (2009) Philopatry and migration of Pacific white sharks. Proceedings of the Royal Society B: Biological Sciences: 10.1098/rspb.2009.1155.

Kimura S and Sugimoto MN (1997) Migration of albacore, *Thunnus alalunga*, in the North Pacific Ocean in relation to large oceanic phenomena. Fisheries Oceanography 6: 51–57.

Kinney MJ and Simpfendorfer CA (2009) Reassessing the value of nursery areas to shark conservation and management. Conservation Letters 2: 53–60.

Kitagawa T, Boustany AM, Farwell CJ, Williams TD, Castleton MR and Block BA (2007) Horizontal and vertical movements of juvenile bluefin tuna (*Thunnus orientalis*) in relation to seasons and oceanographic conditions in the eastern Pacific Ocean. Fisheries Oceanography 16: 409–421.

Kitagawa T, Kimura S, Nakata H, Yamada H, Nitta A, Sasai Y and Sasaki H (2009) Immature Pacific bluefin tuna, *Thunnus orientalis*, utilizes cold waters in the Subarctic Frontal Zone for trans-Pacific migration. Environmental Biology of Fishes 84: 193–196.

Klimley AP, Beavers SC, Curtis TH and Jorgensen SJ (2002) Movements and swimming behavior of three species of sharks in La Jolla Canyon, California. Environmental Biology of Fishes 63: 117–135.

Kraus RT, Wells RD and Rooker JR (2011) Horizontal movements of Atlantic blue marlin (*Makaira nigricans*) in the Gulf of Mexico. Marine Biology 158: 699–713.

Kumari B and Raman M (2010) Whale shark habitat assessments in the northeastern Arabian Sea using satellite remote sensing. International Journal of Remote Sensing 31: 379–389.

Lea JSE, Wetherbee BM, Queiroz N, Burnie N, Aming C, Sousa LL, Mucientes GR, Humphries NE, Harvey GM, Sims DW and Shivji MS (2015) Repeated long-distance migratory targeting of highly contrasting ecosystems and philopatry by a large marine predator. Scientific Reports 5: 11202.

Lohmann KJ, Lohmann CM and Endres CS (2008) The sensory ecology of ocean navigation. Journal of Experimental Biology 211: 1719–1728.

Lutcavage ME, Brill RW, Skomal GB, Chase BC and Howey PW (1999) Results of pop-up satellite tagging of spawning size class fish in the Gulf of Maine: do North Atlantic bluefin tuna spawn in the mid-Atlantic? Canadian Journal of Fisheries and Aquatic Sciences 56: 173–177.

Maguire J-J (2006) The state of world highly migratory, straddling and other high seas fishery resources and associated species. FAO.

Majkowski J (2007) Global fishery resources of tuna and tuna-like species. FAO.

Marín YH, Brum F, Barea LC and Chocca JF (1998) Incidental catch associated with swordfish longline fisheries in the south-west Atlantic Ocean. Marine & Freshwater Research 49: 633–639.

McNamara JM, Barta Z, Klaassen M and Bauer S (2011) Cues and the optimal timing of activities under environmental changes. Ecology Letters 14: 1183–1190.

Megalofonou P (2005) Incidental catch and estimated discards of pelagic sharks from the swordfish and tuna fisheries in the Mediterranean Sea. Fishery Bulletin 103: 620–634.

Metcalfe JD, Righton D, Eastwood P and Hunter E (2008) Migration and habitat choice in marine fishes. pp. 187–223. *In*: Magnhagen C, Braithwaite VA, Forsgren E and Kapoor BG (eds.). Fish Behaviour. Science Publishers, New Hampshire.

Nakamura H (1969) Tuna: Distribution and Migration. Fishing News Ltd., London. 76p.

Nasby-Lucas N, Dewar H, Lam CH, Goldman KJ and Domeier ML (2009) White shark offshore habitat: a behavioral and environmental characterization of the eastern Pacific shared offshore foraging area. PLOS ONE 4: e8163.

Neilson JD, Smith S, Royer F, Paul SD, Porter JM and Lutcavage M (2009) Investigations of horizontal movements of Atlantic swordfish using pop-up satellite archival tags. Tagging and tracking of marine animals with electronic devices. Springer, New York.

Neilson J, Arocha F, Cass-Calay S, Mejuto J, Ortiz M, Scott G, Smith C, Travassos P, Tserpes G and Andrushchenko I (2013) The recovery of Atlantic swordfish: the comparative roles of the regional fisheries management organization and species biology. Reviews in Fisheries Science 21: 59–97.

Norman B (2005) *Rhincodon typus*. The IUCN Red List of Threatened Species. Assessed 15 March 2015.

Nøttestad L, Giske J, Holst JC and Huse G (1999) A length-based hypothesis for feeding migrations in pelagic fish. Canadian Journal of Fisheries and Aquatic Sciences 56: 26–34.

Okazaki Y, Nakata H and Kimura S (2002) Effects of frontal eddies on the distribution and food availability of anchovy larvae in the Kuroshio Extension. Marine & Freshwater Research 53: 403–410.

Oozeki Y, Takasuka A, Kubota H and Barange M (2007) Characterizing Spawning Habitats of Japanese Sardine, *Sardinops Melanostictus*, Japanese Anchovy, *Engraulis Japonicus*, and Pacific Round Herring, *Etrumeus Teres*, in the Northwestern Pacific. California Cooperative Oceanic Fisheries Investigations Report 48: 191.

Orange CJ and Fink BD (1963) Migration of a tagged bluefin tuna across the Pacific Ocean. California Fish and Game 49: 307–308.

Patterson TA, Evans K, Carter TI and Gunn JS (2008) Movement and behaviour of large southern bluefin tuna (*Thunnus maccoyii*) in the Australian region determined using pop-up satellite archival tags. Fisheries Oceanography 17: 352–367.

Pendoley KL, Schofield G, Whittock PA, Ierodiaconou D and Hays GC (2014) Protected species use of a coastal marine migratory corridor connecting marine protected areas. Marine Biology 161: 1455–1466.

Polovina JJ (1996) Decadal variation in the trans-Pacific migration of northern bluefin tuna (*Thunnus thynnus*) coherent with climate-induced change in prey abundance. Fisheries Oceanography 5: 114–119.

Polovina JJ, Howell E, Kobayashi DR and Seki MP (2001) The transition zone chlorophyll front, a dynamic global feature defining migration and forage habitat for marine resources. Progress in Oceanography 49: 469–483.

Priede I and French J (1991) Tracking of marine animals by satellite. International Journal of Remote Sensing 12: 667–680.

Rasmussen K, Palacios DM, Calambokidis J, Saborío MT, Dalla Rosa L, Secchi ER, Steiger GH, Allen JM and Stone GS (2007) Southern Hemisphere humpback whales wintering off Central America: insights from water temperature into the longest mammalian migration. Biology Letters 3: 302–305.

Reeb C, Arcangeli L and Block B (2000) Structure and migration corridors in Pacific populations of the swordfish *Xiphius gladius*, as inferred through analyses of mitochondrial DNA. Marine Biology 136: 1123–1131.

Rooker JR, Alvarado Bremer JR, Block BA, Dewar H, De Metrio G, Corriero A, Kraus RT, Prince ED, Rodríguez-Marín E and Secor DH (2007) Life history and stock structure of Atlantic bluefin tuna (*Thunnus thynnus*). Reviews in Fisheries Science 15: 265–310.

Roshier D and Reid J (2003) On animal distributions in dynamic landscapes. Ecography 26: 539–544.

Schaefer KM and Fuller DW (2016) Methodologies for investigating oceanodromous fish movements: archival and pop-up satellite archival tags. pp. 14–19. *In*: Morais P and Daverat F (eds.). An Introduction to Fish Migration. CRC Press, Boca Raton, FL, USA.

Schaefer KM, Fuller DW and Block BA (2007) Movements, behavior, and habitat utilization of yellowfin tuna (*Thunnus albacares*) in the northeastern Pacific Ocean, ascertained through archival tag data. Marine Biology 152: 503–525.

Scheuerell MD, Zabel RW and Sandford BP (2009) Relating juvenile migration timing and survival to adulthood in two species of threatened Pacific salmon (*Oncorhynchus* spp.). Journal of Applied Ecology 46: 983–990.

Schick R, Goldstein J and Lutcavage M (2004) Bluefin tuna (*Thunnus thynnus*) distribution in relation to sea surface temperature fronts in the Gulf of Maine (1994–96). Fisheries Oceanography 13: 225–238.

Shaffer SA, Tremblay Y, Weimerskirch H, Scott D, Thompson DR, Sagar PM, Moller H, Taylor GA, Foley DG and Block BA (2006) Migratory shearwaters integrate oceanic resources across the Pacific Ocean in an endless summer. Proceedings of the National Academy of Sciences of the United States of America 103: 12799–12802.

Sherrill-Mix SA, James MC and Myers RA (2008) Migration cues and timing in leatherback sea turtles. Behavioral Ecology 19: 231–236.

Shillinger GL, Palacios DM, Bailey H, Bograd SJ, Swithenbank AM, Gaspar P, Wallace BP, Spotila JR, Paladino FV and Piedra R (2008) Persistent leatherback turtle migrations present opportunities for conservation. PLOS Biology 6: e171.

Simpfendorfer C (2009) *Galeocerdo cuvier*. The IUCN Red List of Threatened Species. Assessed 15 March 2015.

Sims DW, Southall EJ, Richardson AJ, Reid PC and Metcalfe JD (2003) Seasonal movements and behaviour of basking sharks from archival tagging: no evidence of winter hibernation. Marine Ecology Progress Series 248: 187–196.

Sims DW, Southall EJ, Tarling GA and Metcalfe JD (2005) Habitat-specific normal and reverse diel vertical migration in the plankton-feeding basking shark. Journal of Animal Ecology 74: 755–761.

Sims DW, Queiroz N, Doyle TK, Houghton JD and Hays GC (2009) Satellite tracking of the World's largest bony fish, the ocean sunfish (*Mola mola* L.) in the North East Atlantic. Journal of Experimental Marine Biology and Ecology 370: 127–133.

Skomal GB, Wood G and Caloyianis N (2004) Archival tagging of a basking shark, *Cetorhinus maximus*, in the western North Atlantic. Journal of the Marine Biological Association of the United Kingdom 84: 795–799.

Skomal GB, Zeeman SI, Chisholm JH, Summers EL, Walsh HJ, McMahon KW and Thorrold SR (2009) Transequatorial migrations by basking sharks in the western Atlantic Ocean. Current Biology 19: 1019–1022.

Smith RJF (1985) The Control of Fish Migration. Zoophysiology Monograph Series, Springer, New York.

Takahashi M, Okamura H, Yokawa K and Okazaki M (2003) Swimming behaviour and migration of a swordfish recorded by an archival tag. Marine & Freshwater Research 54: 527–534.

Techera EJ and Klein N (2011) Fragmented governance: reconciling legal strategies for shark conservation and management. Marine Policy 35: 73–78.

Teo SL, Boustany A, Dewar H, Stokesbury MJ, Weng KC, Beemer S, Seitz AC, Farwell CJ, Prince ED and Block BA (2007) Annual migrations, diving behavior, and thermal biology of Atlantic bluefin tuna, *Thunnus thynnus*, on their Gulf of Mexico breeding grounds. Marine Biology 151: 1–18.

Vandeperre F, Aires-da-Silva A, Fontes J, Santos M, Santos RS and Afonso P (2014) Movements of Blue Sharks (*Prionace glauca*) across their life history. PLOS ONE 9: e103538.

Walker MM (2006) Magnetoreception. *In*: Zielinski B and Hara T (eds.). Sensory Systems Neuroscience: Fish Physiology. Academic Press, San Diego 25: 335–374.

Walli A, Teo SLH, Boustany A, Farwell CJ, Williams T, Dewar H, Prince E and Block BA (2009) Seasonal movements, aggregations and diving behavior of Atlantic bluefin tuna (*Thunnus thynnus*) revealed with archival tags. PLOS ONE 4: e6151.

Weng KC, Boustany AM, Pyle P, Anderson SD, Brown A and Block BA (2007) Migration and habitat of white sharks (*Carcharodon carcharias*) in the eastern Pacific Ocean. Marine Biology 152: 877–894.

Weng KC, Foley DG, Ganong JE, Perle C, Shillinger GL and Block BA (2008) Migration of an upper trophic level predator, the salmon shark *Lamna ditropis*, between distant ecoregions. Marine Ecology Progress Series 372: 253–264.

Wilson S, Lutcavage M, Brill R, Genovese M, Cooper A and Everly A (2005) Movements of bluefin tuna (*Thunnus thynnus*) in the northwestern Atlantic Ocean recorded by pop-up satellite archival tags. Marine Biology 146: 409–423.

Wilson S, Polovina J, Stewart B and Meekan M (2006) Movements of whale sharks (*Rhincodon typus*) tagged at Ningaloo Reef, Western Australia. Marine Biology 148: 1157–1166.

Wilson SG, Taylor JG and Pearce AF (2001) The seasonal aggregation of whale sharks at Ningaloo Reef, Western Australia: currents, migrations and the El Nino/Southern Oscillation. Environmental Biology of Fishes 61: 1–11.

Womble JN and Gende SM (2013) Post-breeding season migrations of a top predator, the harbor seal (*Phoca vitulina richardii*), from a marine protected area in Alaska. PLOS ONE 8: e55386.

PART 3

Technical Apparatus, Analytical Techniques and Data Analyses Used in Fish Migration Research

CHAPTER 9

Microchemical and Schlerochronological Analyses Used to Infer Fish Migration

*Françoise Daverat** and *Jean Martin*

Otoliths as tracers of fish migration

The use of calcified structures of fish (fin rays, scales, otoliths) to retrieve movements and migration is based on two major hypotheses. The first hypothesis is that calcified structures grow continuously throughout the life of the fish and hence record the entire life chronology. The second hypothesis is that the chemical composition of the calcified structure reflects the chemical composition of the ambient water. In this chapter, the case study of otoliths will be treated as priority, and some aspects on the use of fin rays, fin spines, vertebrae and scales will be discussed in a later chapter.

Otoliths are part of teleost fish skeleton, as bones of the inner ear. Otoliths are calcium carbonate concretions, usually crystallizing in the form aragonite, but also as vaterite or calcite, and are embedded in a structural protein matrix made of otolin and rich in aspartate and glutamate residues. Otoliths allow fish to have a perception of position in the water column (pressure, gravity), angular motion (rotation, translation) and hearing. There are three pairs of otoliths (sagittae, lapilli, asterisci), which are formed during embryonic development, and grow continuously throughout the fish's life. In this regard, otoliths meet the first hypothesis. In the same way as trees, otoliths record the growth rhythms of fish, with rings of contrasting opacity depending on growth velocity, either at the daily scale or at the seasonal scale. Hence, the different layers of otolith material record the entire life of the fish.

The second interest of otoliths is their chemical tracer property: otolith composition partially reflects ambient water composition (Kalish 1989; Campana 1999; Daverat et al. 2005). As mentioned, otoliths are formed by a calcium carbonate matrix, and since some alkaline elements (e.g., strontium, barium, magnesium) have a similar

IRSTEA-Institut National de Recherché en Sciences et Technologies pour l'Environnement et l'Agriculture.
50 av. de Verdun, Cestas Cedex 33612, France.
* Corresponding author: francoise.daverat@irstea.fr

ionic radius to calcium, these elements might replace calcium in the formation of the otolith matrix. These elements, and others, can be used as proxies to trace fish migration, or fish movements, due to the heterogeneous composition of the different water masses between which a fish has moved. In consequence, the ability to detect migrations using otolith structure and composition is based on a third hypothesis: the heterogeneous composition of the different water masses between which a fish has moved. The validation of this third hypothesis should be a pre-requisite of each study, but is a difficult issue. Predictable heterogeneity of water masses guides the choice towards relevant microchemical tracers. The choice of relevant chemical tracers, revealing the spatial heterogeneities of fish habitats, is central to infer fish migration.

One of the most obvious water mass heterogeneity features is salinity. Chemical differences between seawater and freshwater provides a basis for the investigation of fish movements through salinity gradients. As a consequence, otolith microchemistry is well suited to track diadromous fish migration (Walther and Limburg 2012). Seawater has a rather stable composition in Strontium (Sr), Barium (Ba) and $^{87}Sr{:}^{86}Sr$ isotopes ratio. Unlike seawater, freshwater has a large diversity of composition, mostly due to the geological ground where the water is flowing. Depending on bedrock geology, the Sr:Ca ratio in freshwater can be either lower or higher (Brown and Severin 2009) than in marine water. However, in a large number of cases, Sr concentration in freshwater is much lower than in seawater (only 12% of the analyzed streams and rivers in the U.S. had higher Sr concentration than adjacent coastal areas (Kraus and Secor 2004)), while Ba concentration is much higher in freshwater than in seawater (Gaillardet et al. 2003). The Ba:Ca ratio in otoliths have also been proven to be a useful proxy of habitat salinity (Elsdon et al. 2008). Water chemical composition was found to be the primary factor influencing Sr:Ca and Ba:Ca ratios in most fish species studied (Webb et al. 2012).

The Sr isotope ratio is very stable in seawater, so if the ratio in freshwater is very different from seawater, a gradient in Sr isotope values are expected. Sr and Ba concentrations have been used extensively to trace movements though gradients, especially for diadromous species (Elsdon et al. 2008). Measurement of Sr:Ca ratio is of particular interest for tracing migratory histories between marine and freshwater environments, and even within freshwaters but not within marine environments (Brown and Severin 2009).

Geology and hydrology provide clues on heterogeneity of water composition between different water bodies. In river basins where tributaries drain different bedrocks (limestone, metamorphic, basaltic, granitic), the water composition between tributaries varies markedly (Palmer and Edmond 1992). For example, in streams and rivers with low Ca concentration, an element to calcium ratio analysis can bias the interpretation of fish life histories (Elfman et al. 1999). In freshwater, water Sr isotopes ratio may discriminate different rivers (Kennedy et al. 2002; Martin et al. 2013), and are used to trace water origin. The other advantage of Sr isotopes is that there is no fractionation between water and the otolith, since Sr isotopes ratio are the same in the water and in the otolith (Kennedy et al. 2000). Sr and Ba concentrations may also differ between tributaries draining different bedrocks (Bowen 1956).

The origin of the water flowing into a river, either superficial or groundwater, will influence its chemical composition, since upwelled groundwater has a specific composition (Brown and Severin 2009). For example, the little Colorado River was found to have a significantly different $\delta^{13}C$ compared to the main Colorado River (Hayden et al. 2012). Other elements, as Li, Cd, may also differ between rivers (De Pontual et al. 2000; Tomas et al. 2005). In a similar manner, coastal areas may also have diverse water chemical composition, either due to river plumes (Dorval et al. 2007; Elsdon et al. 2008), upwelling (Woodson et al. 2013) or groundwater discharge (Encarnação et al. 2013). Other examples have described the influence of a hurricane which marked a specific signature in the otoliths of a marine fish (Elfman et al. 1999). The composition of water chemical tracers varies continuously (Sturrock et al. 2012), so it is also important to investigate the time-scale variation of those chemical tracers used to infer migrations and movements, since variation patterns might change weekly/monthly/seasonally. Therefore the heterogeneity of water masses have to be constant over time to successfully retrieve movements between water masses with high confidence, or the chemical variability of water masses must be investigated simultaneously to the time covered by the life of the fishes that are being studied. Complementarily, a geochemical atlas based on basins' geological features can provide basis for expected water composition heterogeneity (Hegg et al. 2013).

Stable isotopes such as $\delta^{13}C$, $\delta^{15}N$ and $\delta^{18}O$ measured in calcified structures are very useful to track movements. Spatial heterogeneity of $\delta^{13}C$, $\delta^{15}N$ and $\delta^{18}O$ across oceanic water masses and across freshwater gradients, or isoscapes are available worldwide (McMahon et al. 2013), and provide a basis for the interpretation of movements. $\delta^{18}O$ modern values can be predicted by salinity and temperature values (Epstein and Mayeda 1953) so that $\delta^{18}O$ measured in fish calcified structures was used as a temperature proxy (Thorrold et al. 1997). Just as in soft tissues, $\delta^{13}C$ and $\delta^{15}N$ measures in calcified structures are affected by the diet of the fish and reflect partly the habitat used by the fish with the same limitations of physiological fractionation. Stable isotopes are particularly useful in oceanic water masses where elemental ratios do not offer spatial heterogeneity. However, the physiology of the fish was found to bias the signals given either by stable isotopes and by elemental ratios, as found for $\delta^{18}O$ for plaice (Darnaude et al. 2014).

Hence, the relation between water composition and otolith composition can be biased by the physiological regulation of elements during their incorporation into the aragonite matrix (Sturrock et al. 2014). The incorporation of chemical tracers into the otolith is driven by complex physiology processes that are not yet fully understood, similar to the otoliths' biomineralization (Fablet et al. 2011). A simple illustration of this complexity is the species specific relation between ambient water composition and the fish otolith composition, which can also be specific to a group of related species (Chang and Geffen 2012) or even stock specific (Barnes and Gillanders 2013). The difference between Allis shad *Alosa alosa,* and European eel *Anguilla anguilla* otolith composition measured during the same years, within the same saline gradient of the Gironde Estuary (SW France), illustrates this species specific difference (Daverat and Tomas 2006; Lochet et al. 2008). In this estuary, the Sr:Ca ratio in the otoliths of Allis shad ranged from 1×10^{-2} to 3×10^{-2}, while the Sr:Ca ratio in the otoliths of the European eel ranged from 1×10^{-2} to 10×10^{-2}.

Validation of the ability of the otolith composition to reflect ambient water composition is required when addressing migration of a species using otolith microchemistry, even for broadly used tracers such as Sr and Ba. Numerous validation experiments consisted in rearing different fish species in controlled elemental water composition to decipher the physiological effects from the environmental effects (Kawakami et al. 1998; Kraus and Secor 2004; Daverat et al. 2005; Collingsworth et al. 2010; Yokouchi et al. 2011; Reis-Santos et al. 2013). Several exogenous and endogenous factors can affect the incorporation of chemical tracers into fish otoliths. Temperature affects otolith's biomineralization and fish metabolism (Thresher 1999). Starvation, or more generally a poor body condition, affects Sr:Ca ratio along otoliths, as revealed by a caging experiment where growth rate was found to marginally affect Sr:Ca ratio (Daverat et al. 2012a). The genetical background of a stock might also affect the elemental incorporation into the otoliths, significantly for Sr and Ba but not for Magnesium (Mg), as observed for the mulloway *Argyrosomus japonicus* (Barnes and Gillanders 2013).

Ontogeny can also bias the environmental signature of a species otolith composition. Only a few experimental studies revealed strong ontogenetic effects (Sturrock et al. 2014). For example, metamorphosis and development physiology occurring at early life stages can significantly affect otolith composition. For example, metamorphosis might cause a shift in the metabolic regulation of free Sr^{2+} (Kalish 1989), registered as a drastic decrease of Sr:Ca ratio in the otoliths of some species that undergo metamorphosis during early ontogeny, as the European eel (Arai et al. 1997), Japanese eel *Anguilla japonica* (Otake et al. 1994), whitespotted conger *Conger myriaster* (Arai et al. 2002), daggertooth pike conger *Muraenesox cinereus* (Ling et al. 2005), reticulated moray *Gymnothorax reticularis* (Ling et al. 2005) and Dover sole *Solea solea* (De Pontual et al. 2003). Thus, the Sr:Ca ratio decrease in the otolith could never be attributed to the onset of a migration. This ontogenetic effect was also observed in otoliths of Atlantic salmon *Salmo salar* juveniles (Martin et al. 2013). Although a significant relationship between water and otolith chemistry was found, otoliths did not register the seasonal variations of water Sr:Ca and Ba:Ca ratios (Martin et al. 2013). The otolith profiles exhibited a similar Ba:Ca ratio peak pattern, following the yolk sac absorption mark, which could not be explained as a result of Ba:Ca ratio changes in the ambient water, but rather reflected an ontogenetic signal (Martin et al. 2013). High elemental concentrations found at the core of the otoliths suggested an internal control of elemental incorporation during the early stages of juvenile ontogeny (Ruttenberg et al. 2005; Warner et al. 2005). Strong physiological control of metal incorporation was found for other elements, such as Zinc (Zn), Manganese (Mn) and Copper (Cu); yet with confusing results, partly explained by the presence of Cu and Zn metal-binding proteins in otoliths (Miller et al. 2006). The metabolism of early stages of fish is likely to induce stage specific signal in otolith composition as the otolith core microchemistry revealed strong metabolic effects on otolith composition (Ruttenberg et al. 2005; Warner et al. 2005).

Several studies have shown variable concentrations of Mn and Zn in fish otoliths throughout life history, with species differences (Campbell et al. 2002; Ranaldi and Gagnon 2008; Friedrich and Halden 2010). Fluctuations were assigned to differences in metabolic rates and to diet uptake. Zn:Ca ratio in otoliths of several species exhibited seasonal oscillatory distribution (Friedrich and Halden 2010).

Contrasting results were also found for Mn, with elevated Mn:Ca ratio found in the core regions of otoliths of a wide range of species (Brophy et al. 2004; Barbee and Swearer 2007). Outside core regions, Mn:Ca ratio exhibited highly variable patterns, with habitat independent variations (Milton et al. 2008), or variations coupled with the hypoxic conditions of ambient water (Limburg et al. 2011). Additional knowledge on what mainly controls the incorporation of transition elements (e.g., Sr, Ba, Zn, Mn) and other elements present in polluted areas (e.g., Cadmium (Cd), lead (Pb)) are mandatory to establish sound ecological relationships with chemical signals recorded in fish otoliths. Diet was not found, so far, to significantly affect the otolith elemental composition compared to water (Walther and Thorrold 2006). This could be explained by the fact that food composition is most often at equilibrium with the ambient water composition. Other experimental studies looking at the origin of otolith Sr isotope composition origin also showed that food contributed up to 70% to otolith Sr, when marine food was provided to salmonids reared in freshwater (Kennedy et al. 2002; Martin et al. 2013). Diet was found to affect $\delta^{13}C$ and $\delta^{15}N$ in calcified structures as well as in soft tissues (see Chapter 10, Hoffman 2016). Measuring $\delta^{13}C$ and $\delta^{15}N$ in essential amino acids instead of all the amino acids overcomes this bias (McMahon et al. 2011). Essential amino acids are directly taken from the food and not synthesized by the fish, reflecting without bias the habitat isotopic signature (McMahon et al. 2011).

The relation between otolith composition and water composition is not straightforward, as fish metabolism imposes a delay in the response of habitat change, depending on the tracer chosen. For example, two studies mentioned a delay of two weeks before the otolith composition fully reflects ambient water composition in Ba and Sr (Miller 2011; Yokouchi et al. 2011). Nevertheless both studies showed that the habitat transition could be detected within a few days. This delay is shorter as far as Sr or C isotopes are concerned (McMahon et al. 2011; Hayden et al. 2012; Martin et al. 2013).

Analytical techniques and equipment used to infer fish migrations and movements

In the present chapter, the necessary steps to achieve a study using calcified structures chemistry to infer migrations are presented. The first step is the choice of relevant tracer of migration across the entire range of habitats used by the fish. This first step requires a spatial information on the heterogeneity of habitats and therefore on the subsequent heterogeneity of tracer. The second step is the knowledge of possible bias of relationship between fish calcified structure composition and ambient water composition. The third step is the strategy of preparation, analysis and data treatment to retrieve the final migration information.

Isoscapes of seawater $\delta^{13}C$, $\delta^{18}O$ and organic $\delta^{15}N$ provide a great basis for the geochemical discrimination of oceanic habitats. A large review addressed the isoscapes of the North Atlantic ocean, showing how large-scale geochemical processes will drive stable isotopes heterogeneity across water masses of different temperature and salinity (McMahon et al. 2013).

Fine-scale geographic discrimination of rivers could be achieved by the combined use of elemental and Sr isotope ratios. A geological map provides an idea of expected heterogeneity of water composition, because the rivers draining different bedrock will have different water composition (Bataille et al. 2014). Several characteristics of Sr isotopes make them ideal spatial markers for characterizing fish movements or natal sources in freshwater environments. First, the ^{87}Sr:^{86}Sr ratio in rivers arises from bedrock geology (Kennedy et al. 2000). Second, Sr isotopes are not trophic fractionated and the ^{87}Sr:^{86}Sr ratio in otoliths closely matched that of stream water (Kennedy et al. 2000). The ^{87}Sr:^{86}Sr ratio in river water is largely controlled by the age of rocks, which influence the amount of ^{87}Sr produced by the radiogenic decay of ^{87}Rb, and flow variations that modulate water mixtures from tributaries with differing ratios (Walther and Limburg 2012). As a result, considerable variations of ^{87}Sr:^{86}Sr ratio can be found among and within tributaries based on the rock type and age (Beard and Johnson 2000). It is useful to collect all knowledge sources that can provide a clue on the heterogeneity of chemical composition of water. By placing spatial variation in Sr isotope ratios in the context of bedrock geology, it could be predicted in advance (e.g., from geological maps) whether sites are likely to show sufficient geochemical variation for the technique to be useful.

Hydrology is also an important feature to take into account. The mixing degree of marine water and freshwater along an estuary determines the gradient of water composition, which is determined by the degree of marine water incursion and, to a lesser extent, to the distance of tidal penetration. Therefore, estuaries are highly dynamic mixing zones that are influenced by the biogeochemical processes occurring along the salinity gradient (Church 1986; Zwolsman and van Eck 1999). The load of dissolved elements changes along the salinity gradient through adsorption and desorption processes (Zwolsman and van Eck 1999; Dorval et al. 2005).

So, the validation of the relationship between water composition and otolith composition is essential to avoid bias during data analyses. Therefore, the collection of water in different locations, and the subsequent analysis of chemical composition, is essential to provide a 'chemical map' of the region of interest which greatly diminishes any possible biases (Dorval et al. 2007). However, there are two drawbacks, (1) water samples are by essence instantaneous and hence do not integrate chemical composition at the same time resolution as fish otoliths do, and (2) there are differences between water composition and otolith composition solely explained by the fish physiology which might be species-specific, as explained before.

To date, studies examining links between environmental variables and otolith chemistry have been largely restricted to laboratory experiments (Secor et al. 1995; Bath et al. 2000; Elsdon and Gillanders 2003). Although controlled experiments have provided notable information on how the physical and chemical environment can influence otolith chemistry, their direct application to wild fishes is more limited (Elsdon and Gillanders 2005). In such studies, the conditions (diet, oxygen, light, temperature and ambient elemental concentrations) experienced by laboratory fish deviate from natural conditions. Field caging studies have provided a method of validating the chemical signals produced by a specific habitat, but the exact relationships between otolith chemistry and ambient water differ among species and elements (Kraus and Secor 2004; Forrester 2005; Fodrie and Herzka 2008; Mohan et al. 2012).

Fish have to be kept and in most cases fed in cages in all the habitats studied for a long period of time, because the accretion of otoliths is slow (Kline 1990), making caging a technically and financially very demanding approach. Besides, most wild fish species are not adapted to live in cages. The stress induced by caging introduces a bias in otolith's chemical composition, which prevents the correct interpretation of otolith composition in terms of habitats occupied (Daverat et al. 2012a).

Alternatively, the otolith composition of different fish from distinct water locations can be mimicked by rearing fish in experimental setups with water collected in different locations and renewed as often as necessary. This approach has been used in a large number of studies to obtain the range of tracer concentration expected in otoliths (Elsdon and Gillanders 2003; Kraus and Secor 2003; Daverat et al. 2005). Another interesting protocol consists in transferring the same batches of fish from one water source to another, to account for inter-individual variability in the elemental incorporation into otoliths (Elsdon and Gillanders 2003; Kraus and Secor 2003; Daverat et al. 2005; Elsdon and Gillanders 2005; Yokouchi et al. 2011).

The composition of otoliths from resident fish of the same species will provide a baseline of expected otolith chemical composition of a defined habitat. However, in most cases, resident fish are not available and there is no insurance that a fish has remained permanently at the site of collection.

When there is no literature support on the incorporation of elements and isotopes for a given species, there is a need to undertake controlled laboratory experiments to validate the exact relationships between otolith chemistry and ambient water (Bath et al. 2000; Milton and Chenery 2001; Martin and Thorrold 2005). The minimum requirement is an experimental set up where fish of the age and life stage of interest are reared under controlled water composition.

Collection and preparation of samples

First, it is imperative to check if the accretion of otolith material is continuous for the species of interest. Second it is of major importance to know how to interpret the structure of the otolith in terms of life history traits. Are there metamorphosis marks? Are there any life stage transition marks? What is the rhythm of ring deposition (daily rings, annual rings)? What is the average rate of otolith accretion?

Otolith tags can differ with life history stages (larval, juvenile, sub-adult and adult otolith growth) and metamorphosis (Elsdon et al. 2008). If profiles of chemical tags bridge life-history stages, then otolith composition differences may represent ontogenetic effects and not changes in chemical environment. This implies a perfect knowledge and control of the otolith structure in order to sample and analyze the part, or the different parts, of the otolith corresponding to the time periods of interest.

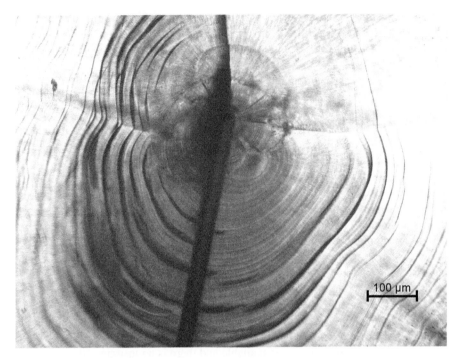

Figure 9.1. Picture of the core of a flounder otolith, showing a laser ablation trajectory starting from the core, going through the metamorphosis area (accessory primordia) down to juvenile phase area (Daverat et al. 2012b).

Preparation of otoliths or scales for microchemistry

General methods of calcified structures preparation for microchemistry are using the same protocol as for age estimation, except that there is no contamination with exogenous trace elements. For the extraction of otoliths or collection of scales, or fin ray, metal forceps or metal devices have to be avoided and plastic, glass or ceramic forceps should be used. If possible, sample preparations have to be undertaken in a clean room under a laminar flow hood or cabinet. If possible, calcified structures have to be analyzed shortly after extraction and after fish sample collection. Some authors recommend storing fish samples in alcohol rather than freezing for further otolith microchemistry analyses (Milton and Chenery 1998).

To avoid contamination with a media (e.g., alcohol, body fluids), we recommend extracting the otoliths shortly after fish collection. The steps of preparation can be undertaken in a clean room to avoid contamination. All instruments used to prepare the calcified structures can be cleaned with diluted ultrapure nitric acid and rinsed with ultrapure water (miliQ water). The plastic vials used to store otoliths, or other calcified structures, also have to be cleaned with diluted ultrapure nitric acid and rinsed with miliQ water.

The chemical composition of the otolith, to provide accurate information about migration and/or habitat use, should be defined according to its location within the otolith or to a specific life stage. This is why, it is often required to section the otoliths to access to a plane including the core and the edge, which corresponds to the whole life chronology of the fish. Otolith section preparation requires embedding the calcified structure in resin. To avoid contamination, the composition of the resin must be trace elements free. Similarly, the steps of preparation consisting in sectioning and polishing the surface of the otoliths require the use of metal free devices (diamond, silicium or carbide saw, silicium paper, diamond suspension with ultrapure water as a lubricant).

Methods of analysis

The spatial resolution in otolith microchemistry is of importance because spatial resolution means temporal resolution. Depending on the research question, targeting one or more life stages, the spatial resolution of the analysis has to be adapted to the

Table 9.1. Most common methods used to analyze otolith chemistry tracers for fish migration.

Tracer	Analytical technique	Sample requirements	Advantages/disadvantages/ reference
$\delta^{13}C$, $\delta^{18}O$, $\delta^{15}N$	Micro-Milling + IRMS Spectro μ sampler	A minimum of 20 μg of aragonite required for best detection IRMS	Micro-milling spatial precision (up to 100 μm wide) can limit usage of micro-milling to large otoliths, destructive.
Stable isotopes, elements	NanoSIMS, SIMS	Spots of 10 μm	Non-destructive (Matta et al. 2013).
Compound specific Amino acid $\delta^{13}C$, $\delta^{15}N$	Micro-Milling + GC-IRMS	Minimum of 10 mg of otolith powder required per sample	Bulk analysis, loss of life stage information, only suitable for large juvenile otoliths (destructive).
all elements	SXFM	Surface analysis 10 μm x 20 μm (3s)	Provide maps of elements, long acquisition time (6–24 hours per sample), data on elements difficult to access by other methods, non-destructive (Limburg et al. 2007).
all elements	PIXE		Provide maps of elements, non-destructive (Elfman et al. 1999).
Sr:Ca ratio	WDS	Surface analysis, spots as low as a few μm	Acquisition time is 3 minutes per spot, limited to Sr and Na, non-destructive.
all elements	LA ICPMS	In the order of 20 μm width A few μm for high resolution	Destructive method (the powder of otolith ablated to acquire measure is lost after measure).
$\delta^{87}Sr$, $\delta^{34}S$	MC ICPMS	Depends on Sr and S concentration, up to 100 μm wide beam	Destructive method (the powder of otolith ablated to acquire measure is lost after measure).

Legend: IRMS- Isotope Ratio Mass Spectrometer; NanoSIMS- Nano Secondary Ion Mass Spectrometry; GC-IRMS- thermo trace Gas Chromatograph coupled with Isotope Ratio Mass Spectrometer; WDS- Wavelength Dispersive Spectrometry; SXFM- Scanning X-ray Fluorescence Microscopy; PIXE- Proton Induced X-ray Emission; LA ICPMS- Laser Ablation Inductively Coupled Mass Spectrometry; MC-ICPMS- Quadrupole MultiCollector Inductively Coupled Mass Spectrometry.

structure of the otolith, such as size of daily increments, size of annual increments or marks of specific life history stages.

The choice of the analysis method will depend on the trace element chosen, and on the spatial resolution required for the research issue.

Data acquired with any method requires a post treatment, and a fitting curve obtained with the regular analysis of standards. The final results are concentrations of elements located in the otolith, either along a transect (e.g., from nucleus to the edge) or in specific points of the otolith (e.g., nucleus, edge, specific ontogeny marks in the otolith). Then, image analysis is required to assign chemical data with the corresponding position along the otolith, which can be a specific life stage or to assign data according to fish age or growth period.

Investigating fish migration: data treatment

Raw microchemistry data is the chemical composition of one or more tracers, in one or more areas of interest in the otolith (or in other calcified structure). The treatment of microchemistry data consists in two assignment steps, one assigns the different life stages along each otolith, and the second assigns one habitat, or several habitats, to each life stage depending on otolith's chemical composition. The correct habitat assignment to a specific life stage is facilitated if otolith samples were prepared to insure that otolith structures (or other calcified structure) are visible and interpretable after chemical analysis. When a destructive analytical technique is used, like laser ablation, then pictures of samples taken before the chemical analysis can be useful to locate chemical measurements afterwards.

Models of habitat-stage assignment

The unsupervised assignment of the habitat where a fish lives, using the chemical composition of a calcified structure, may rely on different sources of information. Thus, a direct assignment can rely on water composition when there is a straightforward link between water and otolith composition (Kennedy et al. 2000). However, more complex assignment models can take into account any source of *a priori* knowledge to make the correspondence of a chemical composition with a habitat.

The environmental fluctuations of water chemical composition, such as seasonal variation of river discharge and tide fluctuations, can also be added to assignment models to increase the accuracy of habitat specific signature. Moreover, all *a priori* knowledge on fish physiology influence on chemical signature can be added into assignment models (Darnaude et al. 2014). Some authors include individual growth models into habitat assignment models to consider stage-specific growth influence on chemical composition (Hoover et al. 2012), like metamorphosis (Daverat et al. 2012a).

Fish ecology can also be used as an *a priori* knowledge, since the probability of a fish to use one habitat at a specific life stage, and the probability to move between habitats, can be constrained (e.g., a fish cannot move between freshwater and the sea without passing through brackish environments) and thus providing more reliable habitat assignment (Fablet et al. 2007; Daverat et al. 2012b).

There are several statistical methods used in habitat assignment, and the choice of the most appropriate depends on the amount of available data and on the ecological problem. Origin assignment can be achieved using a wide range of methods (Mercier et al. 2010), as Bayesian framework models and neural networks, or even more conventional statistical methods (e.g., Quadratic Discriminant Analysis, Linear Discriminant Analysis). The main drawback of using *a priori* assignment is that the funding assumption is that all putatively used habitats signatures are known, and as a consequence the sampling strategy allows sampling all the required habitat signatures. Quadratic Discriminant Analysis and Linear Discriminant Analysis are particularly dependant on this hypothesis, because the habitats are defined *a priori* and they do not allow the assignment to unknown sources (Swan et al. 2003; Brazner et al. 2004; Pruell et al. 2010). The best results, or increased confidence in model outputs, are achieved if the water chemical composition of all the putative origins/habitats of fish were sampled. Bayesian tools are increasingly being used in the analysis of samples of mixed origin, in part, because they enable different sources of data to be combined in one model (e.g., water composition, geologic heterogeneity, fish ecology), and also because it allow practitioners to account for and analyze several sources of uncertainty at one time (Pella and Masuda 2006; Pflugeisen and Calder 2013). In particular, Bayesian methods have been employed in an attempt to provide an answer to the problem of an unknown number of origins in a population (Neubauer et al. 2013).

The interpretation of otolith composition transects as life histories can be undertaken using zoning which is the segmentation of a data-series into zones on the basis of similarity or dissimilarity metrics (Hedger et al. 2008). Local zoning involves passing a window across the sequence of otolith composition to obtain information on the local pattern, allowing the identification of discontinuities, with these discontinuities being the breakpoints between zones. Global zoning divides the sequence into a series of zones, usually through a recursive procedure, which are internally homogeneous and different from adjacent zones. Although this method may be easy to apply and a good tool for a rapid observation of the data, it has several disadvantages. One problem is that the temporal aspect of the life history of the fish is not addressed with this tool. Another drawback is that there is no model of habitat assignment in this type of analysis, the different zones being interpreted directly for each tracer and for each fish based on expert advice. It is not well suited to analyze multi-tracer composition otolith sequences.

The full reconstruction of fish trajectories, from otolith composition sequences, is a little more complex. Using signal analysis techniques, such as more commonly used to reconstruct fish trajectories from data storage tags, it is possible to acquire fish life histories of habitat use (Woilliez et al. 2014). The first step to reconstruct trajectories is to construct a model of habitat assignment (Fablet et al. 2007). In Fablet et al. (2007), the authors have used the distribution of the tracers value over the whole data set (all the points of measures of all the individual fish otolith sequences), and knowledge about known habitat signatures and fish physiology bias. The second step is to convert each individual otolith composition sequence, as a function of otolith distance, into individual otolith composition sequence as a function of fish age, based on the interpretation of otolith structure along the otolith composition transect. Then, when these individual time series are obtained, a model of transition, from one habitat to another habitat, is fitted on the whole data and finally each individual fish trajectory are reconstructed (Fablet et al. 2007; Daverat et al. 2011; Daverat et al. 2012b). Another study based on the same approach have used Fourier analysis to reconstruct oceanodromous fish larvae trajectories (Hoover et al. 2012). The advantage of analyzing otolith composition and otolith age sequences to retrieve trajectories is that useful statistics, at the whole sample level, can be acquired, as the average age at habitat shift, the average habitat used at a specific life stage and the diversity of habitat use patterns.

Coupling otolith natal origin and genetic to infer population structure

The coupling of natal origin inferred from otolith composition and population genetic tools to identify migration can be useful, because the first tool provides an instantaneous view of the reproductive groups (the fish born at the same location), and the second tool provides population structure (population *sensu stricto*) (Bradbury et al. 2008; Perrier et al. 2011; Collins et al. 2013; Longmore et al. 2014; Martin et al. 2014).

Case studies

Potamodromous migrations—Humpback chum in the Colorado River Basin— (Limburg et al. 2013)

The habitat use and the migration patterns of humpback chum *Gila cypha* in the Colorado River Basin, between the main river and tributaries, were inferred using otolith microchemistry (Limburg et al. 2013). The determination of several tracers, usually used to track fish movements, was achieved after a careful study on local geology, water geochemistry and by quantifying the variation of these tracers in the study area. The literature review and water analysis confirmed that Sr and δ^{13}C were the most relevant tracers in this context. Otoliths (lapillus in the case of humpback chum) were prepared, embedded in epoxy resin and polished to expose the core and the edge on the same plane. Two different methods of analysis were performed, µPIXE analysis was used to measure Sr:Ca ratio along the otoliths, while SIMS was used to quantify δ^{13}C at the core and at the edge of the otolith. Fish migrations were inferred by examination of water composition at the different locations, by looking at ruptures in the transect, discrepancy or accordance between core and edge δ^{13}C together with growth history. This revealed that humpback chum migrate between the main river and the tributary to comply their life cycle.

Anadromous migrations—Natal origin and homing of the Atlantic salmon (Martin et al. 2013)

Otolith microchemistry was used to identify the natal origin of the Atlantic salmon *Salmo salar*, and investigate the relative contribution of stocked fish (obtained from artificial reproduction) to the returning adult-stock, and to investigate the level of homing in a vast area, 12 tributaries of the Adour Basin (Southwestern France) (Martin et al. 2013). A literature review of geochemistry data served as a basis to determine relevant tracers. The final decision on the tracer's choice was based on an extensive water sampling strategy, with a large spatial coverage (sampling of all tributaries), and a large temporal coverage (three years and two seasons in each year). Based on contrasting geology among freshwater habitats, that generate unique geochemical signatures, Sr:Ca, Ba:Ca and $^{87}Sr:^{86}Sr$ ratios were used successfully as natural tags for determining individuals' natal origins.

Fish sampling strategy consisted in an exhaustive sampling of wild juveniles in 12 rivers, throughout the entire native range in the Adour Basin, and sampling juveniles in two hatcheries where stocked fish originated from. Adult salmon were sampled during their reproduction migration. Sagittal otoliths were imbedded in epoxy resin and polished to expose the core. Otolith microchemistry analysis was performed in two steps. A fine transect, 30 µm wide, was used on a femtosecond laser ablation coupled to an ICPMS device to measure elemental ratios. In a second step, a transect of 100 µm wide was used to measure Sr isotopes using a femtosecond laser ablation coupled to a MC-ICP-MS. Mean otolith Sr:Ca, Ba:Ca and $^{87}Sr:^{86}Sr$ ratios were calculated in the region of the otolith accreted while in the natal tributary, but after yolk absorption (avoiding any maternally derived material accreted at the core), and prior to outmigration. Quadratic Discriminant Function Analysis (QDFA) was successful at classifying juveniles according to their natal rivers. Adults of unknown natal origin were assigned to their natal rivers using the juvenile fingerprints from the QDFA approach.

Statistical analysis showed that juvenile otolith signatures were matching water signatures. Although most of the adults showed a marked homing instinct, some wild fish strayed into non-natal spawning areas, but they were originated from the neighboring rivers. Few adults originated from unknown rivers. The hatchery-reared fish as adult spawners represented only 10% of the fish sampled during spawning migration.

Catadromous migrations—(Daverat et al. 2012b)

European flounder *Platichthys flesus* life history patterns were investigated at three basins along a latitudinal gradient (Minho- N-Portugal, Gironde- SW-France, Seine- N-France) (Daverat et al. 2012b). Sr:Ca and Ba:Ca otolith signatures and microstructure were used to retrospectively determine habitats occupied by flounders during their life, including during early larval ontogeny. Trajectories of habitat use were reconstructed based on the method of Fablet et al. (2007). Flounders exhibited high life history plasticity among and even within basins, noticed by the diversity of habitats used along larval ontogeny and throughout their lives, and by the age at which flounders migrated to freshwater. Egg signatures probably had a strong maternal influence, and our interpretation suggests that flounders spawned and/or hatched predominantly in brackish waters in Minho, while in Gironde and Seine, flounders spawned and/or hatched either in coastal, brackish or freshwater environments. The freshwater egg signature was most frequent in the Seine. These interpretations contradict the current general assumption that flounders spawn exclusively in coastal waters. During pre-metamorphosis and metamorphosis, flounders were predominantly in brackish waters in Minho, while in Gironde and Seine they were mainly in coastal and freshwater environments, respectively. The diversity of flounder life histories (LH) (i.e., sequence of habitat residence—freshwater, brackish and coastal), after metamorphosis was greater in Minho (LH = 18), than in Seine (LH = 15) or Gironde (LH = 13). The age at which flounders migrated to freshwater also varied between sites, at an earlier age in Minho and Gironde (< 0.5 years old) than in the Seine, where flounders migrating from the coast into freshwater reached maximum frequencies at the ages 1.3 years old. Thus, catadromy in European flounder may be facultative, and the reasons influencing flounders life history high plasticity deserves thorough research.

Amphidromous migrations—barramundi life histories (Walther et al. 2011)

Movements of migratory barramundi *Lates calcarifer* were quantified in two large unregulated rivers in northern Australia, using both elemental (Sr/Ba) and isotope ($^{87}Sr/^{86}Sr$) ratios in aragonitic otoliths (Walther et al. 2011). Chemical life history profiles indicated significant individual variation in habitat use, particularly among chemically distinct freshwater habitats within a catchment. A global zoning algorithm was used to quantify distinct changes in chemical signatures across profiles. This algorithm identified between two and six distinct chemical habitats in individual profiles, indicating variable movement among habitats. Profiles of $^{87}Sr/^{86}Sr$ ratios were notably distinct among individuals, with highly radiogenic values recorded in some otoliths. This variation suggested that fish made full use of habitats across the entire catchment basin.

Oceanodromous migrations—(Longmore et al. 2014)

Otolith trace element and stable isotope analyses were combined with microsatellite data to investigate population structure and connectivity in the migratory deep-sea black scabbardfish *Aphanopus carbo*, sampled along a latitudinal gradient spanning much of the known species range in the Northeast Atlantic (Longmore et al. 2014). In each sampled life stage, otolith trace element and oxygen isotope compositions are similar among fish from different capture locations, but otolith compositions vary greatly between life stages. Oxygen isotope compositions indicate ontogenetic migrations from relatively warm water conditions, during larval growth, to cooler waters with increasing age. Analysis of microsatellite DNA also suggests lack of genetic structure among the areas sampled. The multidisciplinary approach employed collectively suggests that *A. carbo* individuals undergo an ocean-scale ontogenetic migration, beginning with spawning in southern warm-water Macaronesian areas (potentially dominated by Madeira), followed by a large proportion of immature fish moving to and feeding on the continental slope in northern areas. The results lend the first conclusive evidence for defining the life-history circuit of this species, and the perception of its stock structure across the North Atlantic.

Scales, fin rays and other calcified structures used to infer fish migration

Fish scales, rays and spines have also been used successfully as alternatives to otoliths in inferring fish migrations and movements (Clarke et al. 2007; Smith 2010; Davies et al. 2011; Phelps et al. 2012; Woodcock and Walther 2014). The greatest advantage of using these three structures lies in the fact that they are non-lethal techniques (Wolff et al. 2013; Woodcock and Walther 2014), which is particularly important when studying endangered species. Scales of teleost fish are composed of two layers, the upper osseous layer and the lower fibrillary plate. The fibrillary plate is known to be composed of multiple layers of lamellae, each of which is filled with parallel collagen fibers, hydroxyapatite and an organic matrix (Onozato and Watabe 1979). Fin rays have the same composition as fish bones, with an hydroxyapatite matrix (Phelps et al. 2012). The hydroxyapatite of fish scales and fin rays, in the same manner as otolith aragonite, record ambient water composition to some extent, but they might be regenerated when the fish is in calcium deficit (Yasuaki et al. 1989; Witten and Villwock 1997) and, as a consequence, scales and fin rays may not record the entire life of a fish. There are other limitations, scales can be lost and then regenerated (Bereiter-Hahn and Zylberberg 1993) and, sometimes, scales are not formed very early in life (Onozato and Watabe 1979); therefore, scales might not record the entire life of the fish. As a consequence, appropriate sampling of scales on the fish body, can increase the chance to collect scales that were formed at the early stage of the fish. This area varies with the fish species. Growing evidences are showing that scales composition in some cases can reflect ambient water composition (Wells et al. 2000), as well as fin rays (Phelps et al. 2012). Indeed, some studies used scales to trace fish movement with success for species with distinct life histories, as the potadromous westslope cutthroat trout *Oncorhyncus clarkii lewisi* (Wells et al. 2003), the anadromous brown trout *Salmo trutta* (Ramsay et al. 2011), the amphidromous Atlantic tarpon (Woodcock and Walther 2014) and barramundi *Lates calcarifer* (Pender and Griffin 1996), and the oceanodromous albacore tuna *Thunnus alalunga* (Davies et al. 2011).

Vertebrae were also used to infer fish movements but with lower use than scales and fin rays (Radtke and Shepherd 1991; Smith and Whitledge 2010).

References

Arai T, Otake T and Tsukamoto K (1997) Drastic changes in otolith microstructure and microchemistry accompanying the onset of metamorphosis in the Japanese eel *Anguilla japonica*. Marine Ecology Progress Series 161: 17–22.

Arai T, Ikemoto T, Kunito T, Tanabe S and Miyazaki N (2002) Otolith microchemistry of the conger eel, *Conger myriaster*. Journal of the Marine Biological Association of the UK 82: 303–305.

Barbee NC and Swearer SE (2007) Characterizing natal source population signatures in the diadromous fish *Galaxias maculatus*, using embryonic otolith chemistry. Marine Ecology Progress Series 343: 273–282.

Barnes TC and Gillanders BM (2013) Combined effects of extrinsic and intrinsic factors on otolith chemistry: implications for environmental reconstructions. Canadian Journal of Fisheries and Aquatic Sciences 70: 1159–1166.

Bataille CP, Brennan SR, Hartmann J, Moosdorf N, Wooller M and Bowen GJ (2014) A geostatistical framework for predicting variations in strontium concentrations and isotope ratios in Alaskan rivers Chemical Geology 389: 1–15.

Bath GE, Thorrold SR, Jones CM, Campana SE, McLaren JW and Lam JWH (2000) Strontium and barium uptake in aragonitic otoliths of marine fish. Geochimica et Cosmochimica Acta 64: 1705–1714.

Bereiter-Hahn J and Zylberberg L (1993) Regeneration of teleost fish scale. Comparative Biochemistry and Physiology Part A: Physiology 105: 625–641.

Bowen H (1956) Strontium and barium in sea water and marine organisms. Journal of the Marine Biological Association of the United Kingdom 35: 451–460.

Bradbury I, Campana S and Bentzen P (2008) Estimating contemporary early life-history dispersal in an estuarine fish: integrating molecular and otolith elemental approaches. Molecular Ecology 17: 1438–1450.

Brazner JC, Campana SE and Tanner DK (2004) Habitat fingerprints for Lake Superior coastal wetlands derived from elemental analysis of yellow perch otoliths. Transactions of the American Fisheries Society 133: 692–704.

Brophy D, Jeffries TE and Danilowicz BS (2004) Elevated manganese concentrations at the cores of clupeid otoliths: possible environmental, physiological, or structural origins. Marine Biology 144: 779–786.

Brown RJ and Severin KP (2009) Otolith chemistry analyses indicate that water Sr:Ca is the primary factor influencing otolith Sr:Ca for freshwater and diadromous fish but not for marine fish. Canadian Journal of Fisheries and Aquatic Sciences 66: 1790–1808.

Campana SE (1999) Chemistry and composition of fish otoliths: pathways, mechanisms and applications. Marine Ecology Progress Series 188: 263–297.

Campbell JL, Babaluk JA, Cooper M, Grime GW, Halden NM, Nejedly Z, Rajta I and Reist JD (2002) Strontium distribution in young-of-the-year Dolly Varden otoliths: Potential for stock discrimination. Nuclear Instruments and Methods in Physics Research Section B: Beam Interactions with Materials and Atoms 189: 185–189.

Chang M-Y and Geffen AJ (2012) Taxonomic and geographic influences on fish otolith microchemistry. Fish and Fisheries 14: 458–492.

Church TM (1986) Biogeochemical factors influencing the residence time of microconstituents in a large tidal estuary, Delaware Bay. Marine Chemistry 18: 393–406.

Clarke AD, Telmer KH and Shrimpton JM (2007) Elemental analysis of otoliths, fin rays and scales: a comparison of bony structures to provide population and life-history information for the Arctic grayling (*Thymallus arcticus*). Ecology of Freshwater Fish 16: 354–361.

Collingsworth PD, Van Tassell JJ, Olesik JW and Marschall EA (2010) Effects of temperature and elemental concentration on the chemical composition of juvenile yellow perch (*Perca flavescens*) otoliths. Canadian Journal of Fisheries and Aquatic Sciences 67: 1187–1196.

Collins SM, Bickford N, McIntyre PB, Coulon A, Ulseth AJ, Taphorn DC and Flecker AS (2013) Population structure of a neotropical migratory fish: contrasting perspectives from genetics and otolith microchemistry. Transactions of the American Fisheries Society 142: 1192–1201.

Darnaude AM, Sturrock A, Trueman CN, Mouillot D, Campana SE and Hunter E (2014) Listening in on the past: what can otolith $\delta18O$ values really tell us about the environmental history of fishes? PLOS ONE 9: e108539.

Daverat F and Tomas J (2006) Tactics and demographic attributes of the European eel (*Anguilla anguilla*): the case study of the Gironde watershed (Southwest France). Marine Ecology Progress Series 307: 247–257.

Daverat F, Tomas J, Lahaye M, Palmer M and Elie P (2005) Tracking continental habitat shifts of eels using otolith Sr/Ca ratios: validation and application to the coastal, estuarine and riverine eels of the Gironde-Garonne-Dordogne watershed. Marine & Freshwater Research 56: 619–627.

Daverat F, Martin J, Fablet R and Pécheyran C (2011) Colonisation tactics of three temperate catadromous species, eel *Anguilla anguilla*, mullet *Liza ramada* and flounder *Platichthys flesus*, revealed by Bayesian multielemental otolith microchemistry approach. Ecology of Freshwater Fish 20: 42–51.

Daverat F, Lanceleur L, Pécheyran C, Eon M, Dublon J, Schäfer J, Baudrimont M and Renault S (2012a) Accumulation of Mn, Co, Zn, Rb, Cd, Sn, Ba, Sr, Pb in the otoliths and tissues of eel (*Anguilla anguilla*) following long-term exposure in an estuarine environment. Science of the Total Environment 437: 323–330.

Daverat F, Morais P, Dias E, Martin J, Babaluk J, Fablet R, Pécheyran C and Antunes C (2012b) Plasticity of European flounder life history patterns discloses alternatives to catadromy. Marine Ecology Progress Series 465: 269–282.

Davies C, Brophy D, Jeffries T and Gosling E (2011) Trace elements in the otoliths and dorsal spines of albacore tuna (*Thunnus alalunga*, Bonnaterre, 1788): An assessment of the effectiveness of cleaning procedures at removing postmortem contamination. Journal of Experimental Marine Biology and Ecology 396: 162–170.

De Pontual H, Lagardere F, Troadec H, Batel A, Desaunay Y and Koutsikopoulos C (2000) Otoliths imprinting of sole (*Solea solea*) from the Bay of Biscay: a tool to discriminate individuals from nursery origins? Oceanologica Acta 23: 497–513.

De Pontual H, Lagardere F, Amara R, Bohn M and Ogor A (2003) Influence of ontogenetic and environmental changes in the otolith microchemistry of sole (*Solea solea*). Journal of Sea Research 50: 199–211.

Dorval E, Jones CM, Hannigan R and van Montfrans J (2005) Can otolith chemistry be used for identifying essential seagrass habitats for juvenile spotted seatrout, *Cynoscion nebulosus*, in Chesapeake Bay? Marine & Freshwater Research 56: 645–653.

Dorval E, Jones CM, Hannigan R and Montfrans JV (2007) Relating otolith chemistry to surface water chemistry in a coastal plain estuary. Canadian Journal of Fisheries and Aquatic Sciences 64: 411–424.

Elfman M, Limburg K, Kristiansson P, Malmqvist K and Pallon J (1999) Application of micro-PIXE to fish life history analyses: trace element analysis of otoliths. Nuclear Instruments and Methods in Physics Research Section B: Beam Interactions with Materials and Atoms 150: 272–276.

Elsdon TS and Gillanders BM (2003) Relationship between water and otolith elemental concentrations in juvenile black bream *Acanthopagrus butcheri*. Marine Ecology Progress Series 260: 263–272.

Elsdon TS and Gillanders BM (2005a) Consistency of patterns between laboratory experiments and field collected fish in otolith chemistry: an example and applications for salinity reconstructions. Marine & Freshwater Research 56: 609–617.

Elsdon TS and Gillanders BM (2005b) Strontium incorporation into calcified structures: separating the effects of ambient water concentration and exposure time. Marine Ecology Progress Series 285: 233–243.

Elsdon TS, Wells BK, Campana S, Gillanders BM, Jones CM, Limburg KE, Secor DH, Thorrold SR and Walther B (2008) Otolith chemistry to describe movements and life history parameters of fishes: hypotheses, assumptions, limitations and inferences. Oceanography and Marine Biology: An Annual Review 46: 297–330.

Encarnação J, Leitão F, Range P, Piló D, Chícharo MA and Chícharo L (2013) The influence of submarine groundwater discharges on subtidal meiofauna assemblages in south Portugal (Algarve). Estuarine, Coastal and Shelf Science 130: 202–208.

Epstein S and Mayeda T (1953) Variation of O^{18} content of waters from natural sources. Geochimica et Cosmochimica Acta 4: 213–224.

Fablet R, Daverat F and Pontual HD (2007) Unsupervised Bayesian reconstruction of individual life histories from otolith signatures: case study of Sr:Ca transects of European eel (*Anguilla anguilla*) otoliths. Canadian Journal of Fisheries and Aquatic Sciences 64: 152–165.

Fablet R, Pecquerie L, De Pontual H, Høie H, Millner R, Mosegaard H and Kooijman SA (2011) Shedding light on fish otolith biomineralization using a bioenergetic approach. PLOS ONE 6: e27055.

Fodrie FJ and Herzka SZ (2008) Tracking juvenile fish movement and nursery contribution within arid coastal embayments via otolith microchemistry. Marine Ecology Progress Series 361: 253–265.

Forrester GE (2005) A field experiment testing for correspondence between trace elements in otoliths and the environment and for evidence of adaptation to prior habitats. Estuaries 28: 974–981.

Friedrich LA and Halden NM (2010) Determining exposure history of northern pike and walleye to tailings effluence using trace metal uptake in otoliths. Environmental Science & Technology 44: 1551–1558.

Gaillardet J, Viers J and Dupré B (2003) Trace elements in river waters. Treatise on Geochemistry 5: 225–272.

Hayden TA, Limburg KE and Pine WE (2013) Using otolith chemistry tags and growth patterns to distinguish movements and provenance of native fish in the Grand Canyon. River Research and Applications 29: 1318–1329.

Hedger RD, Atkinson PM, Thibault I and Dodson JJ (2008) A quantitative approach for classifying fish otolith strontium: Calcium sequences into environmental histories. Ecological Informatics 3: 207–217.

Hegg JC, Kennedy BP and Fremier AK (2013) Predicting strontium isotope variation and fish location with bedrock geology: Understanding the effects of geologic heterogeneity. Chemical Geology 360: 89–98.

Hoffman JC (2016) Tracing the origins, migrations, and other movements of fishes using stable isotopes. pp. 169–196. *In*: Morais P and Daverat F (eds.). An Introduction to Fish Migration. CRC Press, Boca Raton, FL.

Hoover RR, Cynthia JM and Grosch CE (2012) Estuarine ingress timing as revealed by spectral analysis of otolith life history scans. Canadian Journal of Fisheries and Aquatic Sciences 69: 1266–1277.

Kalish JM (1989) Otolith microchemistry: validation of the effects of physiology, age and environment on otolith composition. Journal of Experimental Marine Biology and Ecology 132: 151–178.

Kawakami Y, Mochioka N, Morishita K, Toh H and Nakazono A (1998) Determination of the freshwater mark in otoliths of Japanese eel elvers using microstructure and Sr/Ca ratios. Environmental Biology of Fish 53: 421–427.

Kennedy BP, Blum JD, Folt CL and Nislow KH (2000) Using natural strontium isotopic signatures as fish markers: Methodology and application. Canadian Journal of Fisheries and Aquatic Sciences 57: 2280–2292.

Kennedy BP, Klaue A, Blum JD, Folt CL and Nislow KH (2002) Reconstructing the lives of fish using Sr isotopes in otoliths. Canadian Journal of Fisheries and Aquatic Sciences 59: 925–929.

Kline LL (1990) Population dynamics of young-of-the-year striped bass, *Morone saxatilis*, populations, based on daily otolith increments. Doctoral dissertation. University of Virginia, College of William and Mary, Gloucester Point. 253p.

Kraus RT and Secor DH (2003) Response of otolith Sr:Ca to a manipulated environment in young American eels. American Fisheries Society Symposium 33: 79–85.

Kraus RT and Secor DH (2004) Incorporation of strontium into otoliths of an estuarine fish. Journal of Experimental Marine Biology and Ecology 302: 85–106.

Limburg KE, Huang R and Bilderback DH (2007) Fish otolith trace element maps: new approaches with synchrotron microbeam x-ray fluorescence. X-Ray Spectrometry 36: 336–342.

Limburg KE, Olson C, Walther Y, Dale D, Slomp CP and Høie H (2011) Tracking Baltic hypoxia and cod migration over millennia with natural tags. Proceedings of the National Academy of Sciences 108: E177–E182.

Limburg KE, Hayden TA, Pine WE III, Yard MD, Kozdon R and Valley JW (2013) Oftravertine and time: otolith chemistry and microstructure detect provenance and demography of endangered humpback chub in Grand Canyon, USA. PLOS ONE 8: e84235.

Ling Y, Iizuka Y and Tzeng W (2005) Decreased Sr/Ca ratios in the otoliths of two marine eels, Gymnothorax reticularis and Muraenesox cinereus, during metamorphosis. Marine Ecology Progress Series 304: 201–206.

Lochet A, Jatteau P, Tomás J and Rochard E (2008) Retrospective approach to investigating the early life history of a diadromous fish: Allis shad *Alosa alosa* (L.) in the Gironde-Garonne-Dordogne watershed. Journal of Fish Biology 72: 946–960.

Longmore C, Trueman CN, Neat F, Jorde PE, Knutsen H, Stefanni S, Catarino D, Milton JA and Mariani S (2014) Oceanic scale connectivity and life cycle reconstruction in a deep-sea fish. Canadian Journal of Fisheries and Aquatic Sciences 71: 1312–1323.

Martin GB and Thorrold SR (2005) Temperature and salinity effects on magnesium, manganese, and barium incorporation in otoliths of larval and early juvenile spot *Leiostomus xanthurus*. Marine Ecology Progress Series 293: 223–232.

Martin J, Bareille G, Berail S, Pécheyran C, Daverat F, Bru N, Tabouret H and Donard O (2013) Spatial and temporal variations in otolith chemistry and relationships with water chemistry: a useful tool to distinguish Atlantic salmon Salmo salar parr from different natal streams. Journal of Fish Biology 82: 1556–1581.

Martin J, Daverat F, Rougemont Q, Jatteau P, Drouineau H, Bareille G, Berail S, Pécheyran C, Launey S, Feunteun E, Clavé D, Roques S, Nachón DJ and Mota M (submitted) Dispersal capacities of Allis Shad (*Alosa alosa*) under global change: insights of innovative otolith microchemistry analysis. Canadian Journal of Fisheries and Aquatic Sciences.

Matta ME, Orland IJ, Ushikubo T, Helser TE, Black BA and Valley JW (2013) Otolith oxygen isotopes measured by high-precision secondary ion mass spectrometry reflect life history of a yellowfin sole (*Limanda aspera*). Rapid Communications in Mass Spectrometry 27: 691–699.

McMahon KW, Berumen ML, Mateo I, Elsdon TS and Thorrold SR (2011) Carbon isotopes in otolith amino acids identify residency of juvenile snapper (Family: Lutjanidae) in coastal nurseries. Coral Reefs 30: 1135–1145.

McMahon KW, Hamady LL and Thorrold SR (2013) A review of ecogeochemistry approaches to estimating movements of marine animals. Limnology and Oceanography 58: 697–714.

Mercier L, Darnaude AM, Bruguier O, Vasconcelos RP, Cabral HN, Costa MJ, Lara M, Jones DL and Mouillot D (2010) Selecting statistical models and variable combinations for optimal classification using otolith microchemistry. Ecological Applications 21: 1352–1364.

Miller JA (2011) Effects of water temperature and barium concentration on otolith composition along a salinity gradient: Implications for migratory reconstructions. Journal of Experimental Marine Biology and Ecology 405: 42–52.

Miller MB, Clough AM, Batson JN and Vachet RW (2006) Transition metal binding to cod otolith proteins. Journal of Experimental Marine Biology and Ecology 329: 135–143.

Milton D, Halliday I, Sellin M, Marsh R, Staunton-Smith J and Woodhead J (2008) The effect of habitat and environmental history on otolith chemistry of barramundi *Lates calcarifer* in estuarine populations of a regulated tropical river. Estuarine, Coastal and Shelf Science 78: 301–315.

Milton DA and Chenery SR (1998) The effect of otolith storage methods on the concentrations of elements detected by laser-ablation ICPMS. Journal of Fish Biology 53: 785–794.

Milton DA and Chenery SR (2001) Can otolith chemistry detect the population structure of the shad hilsa *Tenualosa ilisha*? Comparison with the results of genetic and morphological studies. Marine Ecology Progress Series 222: 239–251.

Mohan JA, Rulifson RA, Corbett DR and Halden NM (2012) Validation of oligohaline elemental otolith signatures of striped bass by use of *in situ* caging experiments and water chemistry. Marine and Coastal Fisheries 4: 57–70.

Neubauer P, Shima JS and Swearer SE (2013) Inferring dispersal and migrations from incomplete geochemical baselines: analysis of population structure using Bayesian infinite mixture models. Methods in Ecology and Evolution 4: 836–845.

Onozato H and Watabe N (1979) Studies on fish scale formation and resorption. Cell and Tissue Research 201: 409–422.

Palmer M and Edmond J (1992) Controls over the strontium isotope composition of river water. Geochimica et Cosmochimica Acta 56: 2099–2111.

Pella J and Masuda M (2006) The gibbs and split merge sampler for population mixture analysis from genetic data with incomplete baselines. Canadian Journal of Fisheries and Aquatic Sciences 63: 576–596.

Pender PJ and Griffin RK (1996) Habitat history of barramundi *Lates calcarifer* in a North Australian river system based on barium and strontium levels in scales. Transactions of the American Fisheries Society 125: 679–689.

Perrier C, Daverat F, Evanno G, Pécheyran C, Bagliniere J-L, Roussel J-M and Campana S (2011) Coupling genetic and otolith trace element analyses to identify river-born fish with hatchery pedigrees in stocked Atlantic salmon (*Salmo salar*) populations. Canadian Journal of Fisheries and Aquatic Sciences 68: 977–987.

Pflugeisen BM and Calder CA (2013) Bayesian hierarchical mixture models for otolith microchemistry analysis. Environmental and Ecological Statistics 20: 179–190.

Phelps QE, Whitledge GW, Tripp SJ, Smith KT, Garvey JE, Herzog DP, Ostendorf DE, Ridings JW, Crites JW and Hrabik RA (2012) Identifying river of origin for age-0 Scaphirhynchus sturgeons in the Missouri and Mississippi rivers using fin ray microchemistry. Canadian Journal of Fisheries and Aquatic Sciences 69: 930–941.

Pruell RJ, Taplin BK and Karr JD (2010) Stable carbon and oxygen isotope ratios of otoliths differentiate juvenile winter flounder (*Pseudopleuronectes americanus*) habitats. Marine & Freshwater Research 61: 34–41.

Radtke RL and Shepherd BS (1991) Current methodological refinements for the acquisition of life history information in fishes: paradigms from pan-oceanic billfishes. Comparative Biochemistry and Physiology Part A: Physiology 100: 323–333.

Ramsay AL, Milner NJ, Hughes RN and McCarthy ID (2011) Comparison of the performance of scale and otolith microchemistry as fisheries research tools in a small upland catchment. Canadian Journal of Fisheries and Aquatic Sciences 68: 823–833.

Ranaldi MM and Gagnon MM (2008) Zinc incorporation in the otoliths of juvenile pink snapper (*Pagrus auratus* Forster): The influence of dietary versus waterborne sources. Journal of Experimental Marine Biology and Ecology 360: 56–62.

Reis-Santos P, Tanner SE, Elsdon TS, Cabral HN and Gillanders BM (2013) Effects of temperature, salinity and water composition on otolith elemental incorporation of *Dicentrarchus labrax*. Journal of Experimental Marine Biology and Ecology 446: 245–252.

Ruttenberg BI, Hamilton SL, Hickford MJH, Paradis GL, Sheehy MS, Standish JD, Ben-Tzvi O and Warner RR (2005) Elevated levels of trace elements in cores of otoliths and their potential for use as natural tags. Marine Ecology Progress Series 297: 273–281.

Secor DH, Henderson-Arzapalo A and Piccoli PM (1995) Can otolith microchemistry chart patterns of migration and habitat utilization in anadromous fishes? Journal of Experimental Marine Biology and Ecology 192: 15–33.

Smith K and Whitledge G (2010) Fin ray chemistry as a potential natural tag for smallmouth bass in northern Illinois rivers. Journal of Freshwater Ecology 25: 627–635.

Smith KT (2010) Evaluation of fin ray and fin spine chemistry as indicators of environmental history for five fish species. M.Sc. thesis. Southern Illinois University Carbondale. 89p.

Sturrock AM, Trueman CN, Darnaude AM and Hunter E (2012) Can otolith elemental chemistry retrospectively track migrations in fully marine fishes? Journal of Fish Biology 81: 766–795.

Sturrock AM, Trueman CN, Milton JA, Waring CP, Cooper MJ and Hunter E (2014) Physiological influences can outweight environmental signals in otolith microchemistry research. Marine Ecology Progress Series 500: 245–264.

Swan SC, Gordon JD, Morales-Nin B, Shimmield T, Sawyer T and Geffen AJ (2003) Otolith microchemistry of *Nezumia aequalis* (Pisces: Macrouridae) from widely different habitats in the Atlantic and Mediterranean. Journal of the Marine Biological Association of the UK 83: 883–886.

Thorrold SR, Campana SE, Jones CM and Swart PK (1997) Factors determining $\delta^{13}C$ and $\delta^{18}O$ fractionation in aragonitic otoliths of marine fish. Geochimica et Cosmochimica Acta 61: 2909–2919.

Thresher RE (1999) Elemental composition of otoliths as a stock delineator in fishes. Fisheries Research 43: 165–204.

Tomas J, Augagneur S and Rochard E (2005) Discrimination of the natal origin of young-of-the-year Allis shad (*Alosa alosa*) in the Garonne-Dordogne basin (south-west France) using otolith chemistry. Ecology of Freshwater Fish 14: 185–190.

Walther BD and Thorrold SR (2006) Water, not food, contributes the majority of strontium and barium deposited in the otoliths of a marine fish. Marine Ecology Progress Series 311: 125–130.

Walther BD and Limburg KE (2012) The use of otolith chemistry to characterize diadromous migrations. Journal of Fish Biology 81: 796–825.

Walther BD, Dempster T, Letnic M and McCulloch MT (2011) Movements of diadromous fish in large unregulated tropical rivers inferred from geochemical tracers. PLOS ONE 6: e18351.

Warner RR, Swearer SE, Caselle JE, Sheehy M and Paradis G (2005) Natal trace-elemental signatures in the otoliths of an open-coast fish. Limnology and Oceanography 50: 1529–1542.

Webb SD, Woodcock SH and Gillanders BM (2012) Sources of otolith barium and strontium in estuarine fish and the influence of salinity and temperature. Marine Ecology Progress Series 453: 189–199.

Wells BK, Bath GE, Thorrold SR and Jones CM (2000) Incorporation of strontium, cadmium, and barium in juvenile spot (*Leiostomus xanthurus*) scales reflects water chemistry. Canadian Journal of Fisheries and Aquatic Sciences 57: 2122–2129.

Wells BK, Rieman BE, Clayton JL, Horan DL and Jones CM (2003) Relationships between water, otolith, and scale chemistries of westslope cutthroat trout from the Coeur d'Alene River, Idaho: The potential application of hard-part chemistry to describe movements in freshwater. Transactions of the American Fisheries Society 132: 409–424.

Witten P and Villwock W (1997) Growth requires bone resorption at particular skeletal elements in a teleost fish with acellular bone (*Oreochromis niloticus*, Teleostei: Cichlidae). Journal of Applied Ichthyology 13: 149–158.

Woilliez M, Fablet R, Ngo T-T, Lalire M, Lazure P and Garren F (2014) A HMM-based model to geolocate pelagic fish from high-resolution individual temperature and depth histories: European sea bass as a case study. Proceedings of BLS5 2014: 5th Bio-logging Scientific Symposium.

Wolff BA, Johnson BM and Landress CM (2013) Classification of hatchery and wild fish using natural geochemical signatures in otoliths, fin rays, and scales of an endangered catostomid. Canadian Journal of Fisheries and Aquatic Sciences 70: 1775–1784.

Woodcock SH and Walther BD (2014) Trace elements and stable isotopes in Atlantic tarpon scales reveal movements across estuarine gradients. Fisheries Research 153: 9–17.

Woodson LE, Wells BK, Grimes CB, Franks RP, Santora JA and Carr MH (2013) Water and otolith chemistry identify exposure of juvenile rockfish to upwelled waters in an open coastal system. Marine Ecology Progress Series 473: 261–273.

Yasuaki T, Tetsuya H and Juro Y (1989) Scale regeneration of tilapia (*Oreochromis niloticus*) under various ambient and dietary calcium concentrations. Comparative Biochemistry and Physiology Part A: Physiology 92: 605–608.

Yokouchi K, Fukuda N, Shirai K, Aoyama J, Daverat F and Tsukamoto K (2011) Time lag of the response on the otolith strontium/calcium ratios of the Japanese eel, *Anguilla japonica* to changes in strontium/calcium ratios of ambient water. Environmental Biology of Fish 92: 469–478.

Zwolsman JJ and van Eck G (1999) Geochemistry of major elements and trace metals in suspended matter of the Scheldt estuary, southwest Netherlands. Marine Chemistry 66: 91–111.

CHAPTER 10

Tracing the Origins, Migrations, and Other Movements of Fishes Using Stable Isotopes

Joel C. Hoffman

Introduction

Stable isotope analysis of fish tissue is now an established tool to identify the origin or to trace migration patterns of freshwater, estuarine, and marine fishes. Studies of fish origins, migrations, and other movements are undertaken to better understand the ecology and ecological connectivity of fishes, provide habitat use information for conservation, assess potential exposure to ecotoxicological threats, and support environmental assessments. In early studies of migration and movement, spatial patterns were inferred from marking and tagging of fish, or else by the use of intrinsic markers found in meristic characters, scales, or otoliths that were location- or stock-specific (Jones 1968). Such techniques are still used and effective; findings from these studies, however, are limited by the ability to recapture tagged fish or to those stocks for which intrinsic markers have been studied and determined reliable. Recent developments in electronic tagging technologies (e.g., archival tags, acoustic and radio telemetry tags; see Schaefer 2016, this book) have greatly expanded our ability to track individual fish over varied spatial domains. Nevertheless, these tagging methods are not feasible or economical for many fishes.

The stable isotope composition of fish tissue is an intrinsic marker. As such, it has some advantages compared to tagging and marking fish. Populations or stocks can be tracked because the marker is inherent to all members of the group, many individuals can be analyzed for relatively little cost (typically, US$10–$30 per sample),

US Environmental Protection Agency, Office of Research and Development, National Health and Environmental Effects Research Lab, Mid-Continent Ecology Division, 6201 Congdon Blvd., Duluth, Minnesota 55804 USA.
Email: hoffman.joel@epa.gov

and there are no concerns about obtaining acceptable recapture rates or tag-induced changes in behavior. Whereas radio, acoustic, or archival tags allow an individual fish to be followed through space and time, stable isotope studies provide comparatively little information about a specific individual's behavior. With stable isotope studies, variation among individuals, however, can provide insight into the range of behaviors exhibited within the population (Bearhop et al. 2004). Further, the approach is broadly applicable to various fish sizes, life stages, and environmental settings, and can be applied without euthanizing fish. Because a fish's stable isotope ratio is derived from the fish's diet, the marker also integrates movements with feeding ecology. Combining various tagging methods (e.g., radio telemetry, otolith microchemistry, tissue stable isotope analysis) is an increasingly popular approach that can offer advantages of both intrinsic and extrinsic markers.

The basis of the stable isotope studies discussed here is that the isotopic composition of muscle, liver, blood, or other soft tissues (i.e., not otoliths or spines) sampled is primarily derived from the fish's diet. That is, the isotopic 'signature' of a fish is acquired as the stable isotope composition of the tissues change in response to a change in the isotopic composition of the diet; thus, the signature is a time-integrated marker of the diet (Fry and Arnold 1982; Tieszen et al. 1983). The stable isotope ratios of carbon ($^{13}C{:}^{12}C$), nitrogen ($^{15}N{:}^{14}N$), and sulfur ($^{34}S{:}^{32}S$) are the most commonly used for studies of fish movements (when analyzing soft tissues; Hobson 1999). To trace the origin, migration pattern, or movement of a fish, a stable isotope-based study exploits naturally-occurring differences in one or more element's isotopic composition between local food webs, specific to the different habitats or ecosystems of interest (Hobson 1999). These isotopic differences arise where there are habitat- or ecosystem-specific differences in either the underlying biogeochemical processes or the food web inputs (Peterson and Fry 1987). Even where there is not a marked difference in the isotopic composition of the local food webs, it may still be possible to trace movements of fish. A fish may have a distinct isotopic composition between two habitats or systems if either the fish of interest or its prey has different trophic niches between the two locations. In either instance, the fish's movement between locations must be quicker than the rate of change in the isotopic composition of its tissues. Thus, the fish arrives at the new location with an isotopic composition indicative of its previous location. Once the fish is established in the new location, the fish's tissues will begin to acquire an isotopic composition specific to its new habitat (Fig. 10.1). The time required for the change to be detectable in a fish's tissue and for isotopic equilibrium to be reached depends on the isotope turnover rate for the tissue of interest.

The focus of this chapter is to introduce the methods and analytical techniques for studying movements of fish, inclusive of elasmobranchs, by stable isotope analysis of soft tissues such as muscle, liver, and blood. For a discussion of stable isotope analysis of otoliths and scales, see Daverat and Martin (2016, this book). In the past decade, stable isotope analysis has become increasingly accessible to fishery scientists and ecologists as the number of commercial laboratories performing this analytical service has increased, analytical throughput has improved, and analytical costs have declined. The topics presented in this chapter include basic terminology in stable isotope science, sample analysis, and preparation; stable isotope turnover

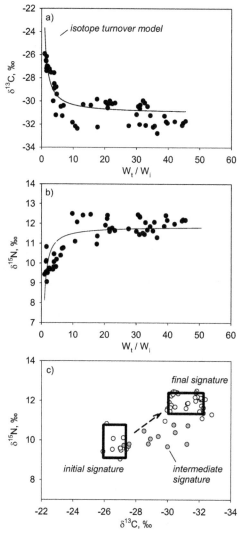

Figure 10.1. Somatic growth-dependent change in fish tissue $\delta^{13}C$ and $\delta^{15}N$ values (a, b) and habitat-specific $\delta^{13}C$ and $\delta^{15}N$ tissue signatures (c) of a fish establishing in a new location where the food web has a distinct stable isotope composition from the food web in the fish's location of origin (e.g., migration of a young-of-year fish from offshore to inshore habitat). The data set would be obtained by sampling the fish population repeatedly in time or space (or both). W_t is the weight at capture and W_i is the initial weight of fish migrating from their location of origin. The circles represent samples from individual fish. The solid line is the expected isotope turnover, assuming somatic growth exclusively contributes to isotope turnover (Eq. 2, $c = -1$; Fry and Arnold 1982). In panel c), the boxes define the 10th and 90th percentiles of fish with an isotopic signature derived from its initial location (defined as those fish with $W_t/W_i < 2$, open circles) and after settling into the new location (defined as those fish $W_t/W_i > 10$, open circles). The arrow indicates the direction of change and the fish with an intermediate value are indicated (gray circles). Note that there is a ± 2‰ variation among individuals approaching isotopic equilibrium, and thus, in this example, having a large difference between locations in both $\delta^{13}C$ values (maximum difference 6.9‰) and $\delta^{15}N$ values (maximum difference 3.4‰) is essential. Also note that the maximum difference illustrated in (a) and (b) is greater than the mean difference illustrated in (c).

and trophic fractionation; stable isotope distributions in ecosystems; study design; and case studies that demonstrate how stable isotope analysis can be used to study fish migration and movement.

Stable isotope terminology, analysis, and sample preparation

Stable isotopes are atoms of the same element (i.e., same number of protons) that have different masses because the number of neutrons varies. Stable isotopes, unlike radioisotopes, do not decay. The most abundant isotope of carbon (C) has six protons and six neutrons. It has an atomic mass of about 12, designated ^{12}C. The atomic mass is essentially the sum of the masses of the protons and neutrons (each of a weight equal to about one atomic mass unit) because electrons have a relatively insignificant mass. Carbon has a stable isotope with seven neutrons, designated ^{13}C. Ecosystem differences in the natural abundance of ^2H, ^{13}C, ^{15}N, and ^{34}S have proved useful for studies of fish origin, migration, and movement (based on analysis of soft tissues). The elements H, C, N, and S are all comprised of an assemblage of different isotopes. The natural abundance of stable isotopes varies by element. For example, ^{14}N comprises 99.63% of all nitrogen on the planet, whereas ^{15}N comprises only 0.37%. In contrast, ^{12}C comprises 98.9% of all carbon on the planet, whereas ^{13}C comprises 1.1%. The natural abundance in any given sample varies from the planetary mean, and systematic variation in the environment, reflected in the food web, is exploited to determine the origin of a given fish.

Ecological studies of fish movement generally report the isotopic composition of a sample in δ (delta) notation, which expresses the value as a ratio of heavy: light stable isotopes (e.g., ^2H:^1H, ^{13}C:^{12}C, ^{15}N:^{14}N, ^{34}S:^{32}S) in relation to an international standard in parts per thousand (‰, or "per mil"):

$$\delta X = (R_{sample}/R_{standard} - 1) \times 10^3, \hspace{3cm} \text{(Eq. 1)}$$

where X is the heavy stable isotope (e.g., ^{13}C), R is the atomic ratio of heavy:light stable isotopes (^{13}C:^{12}C), and the standard is internationally recognized (for C, it was established as Pee Dee Belemnite (PDB), a limestone formation in South Carolina, USA; see Sharp (2007) for detailed information on stable isotope standards and international references). For example, the ^{13}C:^{12}C (R) for the PDB standard is 0.0112372, or about 1.12 ^{13}C atoms for every 100 ^{12}C atoms. If a sample has 1.11 ^{13}C atoms for every 100 ^{12}C, then the $\delta^{13}C$ value would be –21.1‰. In ecological studies, the difference in $\delta^{13}C$ values between two locations is often 5‰ to 15‰, which is a small difference in the natural abundance of ^{13}C.

These small differences in natural abundance are measured using a high precision Isotope Ratio Mass Spectrometer (IRMS), which is designed to quantitatively measure the abundance of ions of different masses. Prior to quantitative measurement, the element of interest first must be converted into a form that can readily be ionized. For light elements, including H, C, N, and S, this is achieved by combusting the sample completely to convert it into a gas (e.g., for C, it would be converted to CO_2). Thus, the IRMS is generally coupled to an elemental analyzer, in which the sample is converted to a gas. The isotopic composition of the pure gas is measured using the IRMS (for more information on measurement, see Criss 1999; Sharp 2007).

For quality assurance, the two most important factors to consider when working with an IRMS laboratory is the calibration and standards used. Calibration should occur regularly and the laboratory should be able to demonstrate that their current measures match either international or laboratory standards (or both) to within ± 1‰. To demonstrate this, laboratories generally incorporate standards at the start or finish (or both) of each 'run' (i.e., a discrete set of samples analyzed continuously). The instrument is calibrated with samples of a standard mass; when samples are prepared, an effort should be made to conform all sample masses to the amount indicated by the laboratory or the results can be biased. When analyzing samples, the analytical facility should introduce replicate laboratory standards at regular intervals (e.g., every 10–12 samples in the run) for quality assurance and to determine if substantial 'drift' in the instrument occurred. Where substantial 'drift' does occur, quality controls can be used to apply a mathematical correction, though overall precision may be affected. For acceptable precision, replicate reference material (whether a laboratory or international standard) should have an error ≤ 0.2‰ and this should be maintained across runs.

When analyzing samples for hydrogen isotope abundance (δD), a standardized equilibration method should be used to correct for water vapor contamination (Wassenaar and Hobson 2003; Doucett et al. 2007; Chessen et al. 2009). Uncontrolled hydrogen exchange between ambient water vapor and organic hydrogen in samples presents an obstacle to reporting δD values that are comparable among sampling location and times, tissue types, and laboratories. In complex molecules, possibly 12–22% of the hydrogen can freely exchange with ambient water vapor (Wassenaar and Hobson 2000).

When analyzing fish tissue, the most commonly used tissue is white muscle from the dorsal region, which has low isotopic variability relative to other tissues (Pinnegar and Polunin 1999). Muscle samples can be acquired without euthanizing the fish using biopsy tools, such as a biopsy needle or dermal tissue punch, to obtain small tissue samples (e.g., Baker et al. 2004; Schielke and Post 2010). Fin clips can also be used and generally show strong agreement with white muscle tissue (Tronquart et al. 2012), probably because they have a similar isotope turnover time (Suzuki et al. 2005). Liver and blood are also analyzed (e.g., McIntyre and Flecker 2006; Buchheister and Latour 2011) because these tissues have faster isotope turnover rates than muscle and therefore could add information about the timing of movement. In birds, blood plasma has been shown to isotopically turn over much more rapidly than red blood cells, which have a similar isotope turnover rate to muscle (Hobson 1999). Mucus has similarly been shown to have a much faster $\delta^{13}C$ and $\delta^{15}N$ turnover rate than muscle in fish (Church et al. 2009). Investigators have also used the organic fraction in bone (i.e., collagen; Schoeninger and DeNiro 1984), and more recently in scales (Ramsay et al. 2012) and otoliths (McMahon et al. 2011).

When collecting samples, either tissue samples or the whole fish must be returned to the laboratory for processing. If samples must be stored, freezing at –20°C will not alter the stable isotope composition of tissues (Bosley and Wainright 1999; Sarakinos et al. 2002). To avoid decomposition, samples should be stored on ice in the field and frozen as soon as possible. In remote locations, however, some preservation will be required. Preservation of tissues in ethanol or formalin significantly alters the stable isotope composition of tissue, so investigators should proceed with these methods with caution (Sarakinos et al. 2002; Kelly et al. 2006; Arrington and Winemiller 2002). Preservation by salt alters the composition only slightly (< 1‰ $\delta^{13}C$ and $\delta^{15}N$), but the shift should be quantified and arithmetically corrected to remove sample bias (Arrington and Winemiller 2002; Xu et al. 2011).

To prepare the samples for analysis, tissue samples should be isolated (i.e., incidental material such as skin or bones removed), rinsed in deionized water, thoroughly dried (at least 48 hours at 45°C or freeze-drying), and homogenized by grinding to minimize any within-tissue variation. Tissue grinding tools include a spatula (the sample is crushed inside the original container), mortar and pestle, and ball mill. To avoid cross-contamination, clean and sterilize (i.e., wipe down thoroughly with alcohol-soaked wipe) all grinding utensils thoroughly between samples. For commonly analyzed stable isotopes, only a small amount of dried tissue is required, about 2 mg or less. The precise sample weight required for analysis varies with respect to material (e.g., fish tissue or plant matter) and tissue (due to varying H, C, N, and S percent composition), element, and laboratory, and so it is necessary to discuss preferred sample weights with the IRMS laboratory prior to sample preparation. Different sample weights are required, in part, because IRMS analysis quality depends on the mass of the element analyzed. The ground sample is then packed into a foil capsule (either tin [$\delta^{13}C$, $\delta^{15}N$, $\delta^{34}S$] or silver [δD]), following instructions provided by the IRMS laboratory. For sampling storage and handling, use clean storage containers such as combusted glass vials to prevent contamination by other organic materials.

During preparation, samples are sometimes subject to acidification or lipid extraction. Acidification removes inorganic carbonates, which would contaminate the tissue's carbon isotope composition. Generally, it will not be necessary to acidify the sample if muscle, fins, blood, or some other organic animal tissue is analyzed. If in question, a sub-sample of ground tissue may be treated directly with a small amount of 10% hydrochloric acid to determine whether inorganic carbonates are present (if the sample bubbles after acidification, there are carbonates present). The treatment is used sparingly because it can inadvertently alter the $\delta^{15}N$ value of the sample (Pinnegar and Polunin 1999). If samples are acidified, it is advisable to analyze a second, untreated sample to measure the $\delta^{15}N$ value (Bunn et al. 1995; Carabel 2006).

Lipid extraction removes variability in $\delta^{13}C$ values associated with varying lipid content in samples. Kinetic isotope effects during lipid synthesis result in the depletion of ^{13}C in lipids compared to carbohydrates and proteins (DeNiro and Epstein 1977). Thus, independent of a fish's diet or environment, tissues with a higher lipid content will have a lower $\delta^{13}C$ value compared to tissues with a lower lipid content, potentially confounding comparisons among or within tissues (McConnaughey and McRoy 1979). Correcting $\delta^{13}C$ values for lipid content is advisable when working with samples such as whole organisms, muscle samples, fin clips, or liver that are a variable mixture of lipids and other organic molecules and that have a lipid content greater than 5–10%. For these samples, the lipid-effect will be higher than 1‰ and thus could affect data interpretation (Post et al. 2007; for aquatic organism muscle tissue, this is equivalent to an atomic C:N > 4.0). To extract the lipid, a polar solvent, generally a mix of chloroform and methanol, is introduced to the bulk tissue sample (Schlechtriem et al. 2003). Unfortunately, lipid extraction can enrich the ^{15}N of the sample, as well (Murry et al. 2006). Investigators have used a variety of mathematical approaches to correct for the effect of lipids on untreated tissue $\delta^{13}C$ values using data from lipid-extracted muscle, liver, and whole organism samples, thereby avoiding the need to extract every sample and analyze both a treated and untreated sample (Logan et al. 2008). The two most common approaches are normalization based on lipid content (McConnaughey and McRoy 1979; Kiljunen et al. 2006; Mintenbeck et al. 2008) and arithmetic mass balance using tissue C:N as a proxy for lipid content (Alexander et al. 1996; Fry et al. 2003). In a comparative analysis of correction methods, Logan et al. (2008) found that the correction method was less important than the level of specificity of the data from which the correction was derived (taxa and tissue type). Hoffman and Sutton (2010) compared four common corrections developed using a wide variety of fish species and found all yield similar results, within ± 1‰.

With respect to ^2H, a number of studies have found that lipid content does affect δD values of both whole organisms and muscle tissue, and that lipid-extracted samples much more closely resemble the consumer's diet (Jardine et al. 2009). An arithmetic correction has not been determined, however.

Stable isotope turnover and trophic fractionation

The ability to study movements between habitats is dependent on the isotope turnover rate in the tissue being analyzed (Tieszen et al. 1983; Hobson and Clark 1992; Hobson et al. 2010). Isotope turnover is a function of somatic growth and metabolism. Somatic growth-based isotope turnover occurs as the fish adds new tissue, diluting the pool of tissue derived from the diet in its previous location. Metabolic isotope turnover occurs as tissue is broken down and new tissue is synthesized. Whether a fish has a fast or slow growth rate, somatic growth is the primary contributor to isotope turnover (Hesslein et al. 1993; Herzka and Holt 2000; Bosley et al. 2002; Sakano et al. 2005). Metabolic turnover can accelerate the isotope turnover rate beyond that exerted by growth alone (Vander Zanden et al. 1998; Herzka et al. 2001; Trueman et al. 2005). However, other physiological processes, such as routing of particular components of diet to particular tissues ('isotopic routing') and progressive enrichment in tissue ^{15}N during starvation (owing to the same physiological mechanisms that cause trophic enrichment), can influence isotope turnover and cause certain tissues to not isotopically resemble a consumer's diet (Gannes et al. 1997). The isotope turnover time is the time it takes a fish (or its tissues) to approach isotopic equilibrium with its food sources. That is, a consumer's isotopic composition is integrated over the isotope turnover time. When a migratory fish moves to a new location and begins feeding, its isotopic composition will take some time to reach equilibrium with the isotopic composition of its prey (the faster the isotope turnover rate, the sooner the fish arrives at its new isotopic equilibrium).

It is critical to find a life stage and tissue (or tissues) that will yield an isotope turnover rate of appropriate temporal scale to study the type of movement of interest. Because mass-specific growth and metabolism rates are size-dependent, the time it takes for a fish to approach isotopic equilibrium with its diet generally should increase with increasing size (Martínez del Rio et al. 2009). Fish larvae have been found to approach isotopic equilibrium in 10–20 days (Herzka 2005; Hoffman et al.

2011), juvenile fishes in two–four months (Suzuki et al. 2005; Logan et al. 2006; Buchheister and Latour 2010) and adults in six months to two years (Hesslein et al. 1993; Weidel et al. 2011). Local climate may also be relevant due to a temperature effect; indeed, McIntyre and Flecker (2006) found that fish from tropical freshwaters had faster isotope turnover rates than species of similar size from temperate freshwaters.

Different tissues have different isotope turnover rates. For example, liver tissue consistently has been shown to have a faster isotope turnover rate than muscle tissue in fish (Logan et al. 2006; Suzuki et al. 2005; Buchheister and Latour 2010; Weidel et al. 2011). This is useful for migration studies because the liver provides more recent information on diet than muscle tissue; the two tissues in tandem could provide greater temporal resolution than either tissue alone regarding the timing of movement (Phillips and Eldridge 2006).

Modeling isotope turnover can aid movement studies. Two different models commonly are used for aquatic organisms. Fry and Arnold (1982) described isotope turnover as an empirical dilution equation that expresses the stable isotope value as a weight-dependent mixture between its initial and final isotopic values, represented by a power function with a constant metabolic decay rate as the fish approaches an asymptotic final value:

$$\delta_t = \delta_f + (\delta_i - \delta_f) \times (W_i/W_t)^c \tag{Eq. 2}$$

where δ_t is the stable isotope ratio (e.g., $\delta^{13}C$) of the animal or tissue at time t; δ_f and δ_i are the final (equilibrium) and initial stable isotope ratios, respectively; W_i and W_t are the initial weight and weight at time t, respectively; and c is the metabolic coefficient. The model simplifies to dilution-only isotope turnover (i.e., somatic growth only) when $c = -1$ and can be used to date ecologically significant events associated with changes in isotopic ratios (Fry and Arnold 1982), such as age at settlement of fish larvae (Herzka et al. 2001).

Hesslein et al. (1993) used an exponential model with terms for a specific growth rate (k; where $k = \ln(W_t/W_i)/t$) and a metabolic turnover rate (m).

$$\delta_t = \delta_f + (\delta_i - \delta_f) \times e^{-(k + m)t} \tag{Eq. 3}$$

The advantages of this model are that (1) time, t, is an explicit term; (2) m can be estimated using a curve-fitting procedure if k is known; and (3) it can be used with slow-growing fishes for whom isotope turnover is slow because it is primarily the result of metabolic turnover (m, i.e., k is much less than m). If δ_f and δ_i are both known and represent the final and initial location of a fish, and if k and m are empirically measured or can be estimated, then the equation can be solved for the time it would theoretically require for the isotopic signature of the population in the first habitat (δ_i) to resemble that in the second (δ_f) upon moving into the second habitat.

Isotopic fractionation occurs when there is partial separation of the light isotopes from the heavy isotopes; the processes that cause fractionation have been characterized as either non equilibrium effects (e.g., diffusion, evaporation, other kinetic isotope effects) or equilibrium effects (Criss 1999). Trophic fractionation is the difference between the isotopic composition of a consumer (whole body) and its diet; it is a metabolic effect that incorporates the effects of several non equilibrium fractionations.

In contrast, trophic discrimination is the difference between the isotopic composition of a single tissue (e.g., muscle) and the diet (Martínez del Rio et al. 2009). Martínez del Rio et al. (2009) argue that the distinction arises because there are many processes, including fractionation, that influence the isotopic composition of a tissue, whereas it is the effect of all physiological processes that result in the difference between the whole organism and its diet. The stable isotope ratios used in migration studies are useful because the δD, $\delta^{13}C$, $\delta^{15}N$, and $\delta^{34}S$ values of consumers generally reflect their diet along with a predictable trophic fractionation. For example, the carbon stable isotope ratio of a consumer closely resembles its diet, within ± 1‰ (DeNiro and Epstein 1978; Fry and Sherr 1984). Based on reviews that incorporated estimates from many animals, consumers have an average trophic fractionation per trophic level of + 0.4‰ $\delta^{13}C$ and + 3.4‰ $\delta^{15}N$ and + 0.5‰ $\delta^{34}S$ (for whole organisms and muscle tissue) (Vander Zanden and Rasmussen 2001; Post 2002; McCutchan et al. 2003). The trophic fractionation of ^2H, particularly in soft tissues, has received less attention than that of the other stable isotopes. Estep and Dabrowski (1980) found little trophic discrimination of ^2H in laboratory-reared mouse tissue. Similarly, Solomon et al. (2009) found little trophic discrimination of ^2H in laboratory-reared fish muscle tissue, though Birchall et al. (2005) reported a relatively large trophic enrichment (ca. + 90‰) of bone collagen in fish. A number of factors that are not well quantified can contribute to δD values of tissues (Doucett et al. 2007), including environmental water (Solomon et al. 2009), and thus caution is recommended when interpreting δD values from tissues.

Although numerous reviews have independently derived similar estimates of trophic fractionation (Vander Zanden and Rasmussen 2001; Post 2002; McCutchan et al. 2003), trophic fractionation can vary with respect to diet quality, trophic position, taxon, environment (e.g., marine versus freshwater), and tissue type (Vander Zanden and Rasmussen 2001; McCutchan et al. 2003; Caut et al. 2009). This potential variation should be considered when analyzing data because small variations in the trophic fractionation estimate may alter the interpretation of diet sources contributing to fish production (Vander Zanden and Rasmussen 2001). Other biochemical processes such as isotopic routing can alter stable isotope composition independent of diet and merit some consideration prior to undertaking a study (Gannes et al. 1997; Martínez del Rio 2009). With respect to elasmobranchs, there have been concerns that tissue urea content may present a concern for $\delta^{15}N$ analysis; however, isotope turnover studies of muscle, liver, and blood $\delta^{15}N$ and $\delta^{13}C$ values have shown that isotope turnover rates and trophic discrimination estimates are consistent with values measured for teleost fishes (Logan and Lutcavage 2010).

Stable isotope distribution in ecosystems

Different elemental pools in different aquatic ecosystems vary in their isotopic composition owing to both fractionation and differences in biogeochemical cycling. It is these differences that are often exploited in studies of fish origin and migration. For example, the $\delta^{13}C$ value of total dissolved CO_2 (DIC) in the ocean is about 0‰ (Mook et al. 1974), due to the equilibrium fractionation of atmospheric CO_2

($\delta^{13}C = -7‰$) (Deines 1980) across the atmosphere-ocean boundary. The fractionation by C3 plants during C fixation is about $-21‰$; thus, marine phytoplankton generally have a $\delta^{13}C$ value of about $-21‰$. In contrast, the $\delta^{13}C$ value of DIC in freshwaters varies widely depending on the DIC source. In rivers, where important sources can include ^{13}C-depleted weathered carbonate or DIC respired from ^{13}C-depleted terrigenous organic matter (Mook and Tan 1991), the $\delta^{13}C$ values for DIC are generally lower than $-10‰$. Owing to the difference in DIC source ^{13}C abundance, freshwater phytoplankton in rivers often have a $\delta^{13}C$ value of $< -30‰$ (Peterson and Fry 1987). Fractionation by algae, however, is not constant and can vary with DIC concentration, nutrient availability, phytoplankton growth rate, and cell size (Goericke et al. 1994; Goericke and Fry 1994). Thus, it is advisable to carefully measure $\delta^{13}C$ baseline values in the locations of interest, as they can vary. It is also possible to create a stable isotope label by enriching the environment by isotope addition; these studies, however, generally have not been designed to address movement in fishes (but see Caudill 2003 for an application to the mayfly *Callibaetis ferrugineus hageni*).

The purpose here is to provide a review of those ecosystems, or habitats, between which sufficiently large isotopic differences exist, and thus allowing to trace fish movements or migrations between them. This is not an exhaustive review of the differences that could exist between ecosystems or habitats, nor is it meant to review the many sources of variability that might obscure or confound such differences. The interested reader is encouraged to document available habitat types, determine relevant spatial and temporal scales, consider possible geochemical processes, and evaluate the food web characteristics as they pertain to the research question at hand.

Marine versus freshwater. Gradients in ^{13}C, ^{15}N, and ^{34}S abundance have been useful for tracking migrations of marine fishes into freshwater (e.g., Hesslein et al. 1991; Limburg 1998; MacAvoy et al. 1998). Movement between fresh and marine waters can be tracked using the carbon stable isotope ratio because the respective food webs in these waters are fueled by vegetation that utilize carbon pools with different isotopic compositions (Peterson and Fry 1987). Marine phytoplankton have a $\delta^{13}C$ value of ca. $-24‰$ to $-19‰$ because the $\delta^{13}C$ of total dissolved inorganic CO_2 (DIC) in the ocean is about $0‰$ and the incorporation of carbon by C_3 plants proceeds with a fractionation of about $-21‰$. Riverine sources, such as matter derived from terrestrial vegetation (C_3 plants, ca. $-28‰$ $\delta^{13}C$), soils (ca. $-26‰$ $\delta^{13}C$), and phytoplankton ($< -30‰$ $\delta^{13}C$), generally are depleted in ^{13}C compared to the estuary and marine systems (Peterson and Fry 1987).

Freshwater organisms may be ^{15}N-depleted compared to marine organisms because terrestrial derived organic matter and freshwater algae often have a $\delta^{15}N$ value of $-2‰$ to $7‰$ (Peterson and Fry 1987; Cloern et al. 2002), which is generally less than estuarine and coastal marine phytoplankton ($7‰$ to $10‰$) (Wainright and Fry 1994; Ostrom et al. 1997; Deegan and Garritt 1997). As a spatial tracer, $\delta^{15}N$ values should be interpreted cautiously for numerous reasons. First, an ecosystem's baseline $\delta^{15}N$ value, often represented by the particulate organic matter (POM), can be altered indirectly and directly by biological and biogeochemical processes (Miyake and Wada 1971; Wada and Hattori 1978; Mariotti et al. 1981; Cifuentes et al. 1988; Altabet 1988). This variability in the baseline $\delta^{15}N$ value will propagate up the food chain to primary consumers and then fishes, potentially confounding spatial differences. Second, temporal changes in baseline $\delta^{15}N$ values can occur faster than a consumer's isotope turnover and result in a mis-match between a consumer's $\delta^{15}N$ value and its environment (O'Reilly et al. 2004). Third, the relatively large ^{15}N trophic fractionation (3‰ to 4‰ $\delta^{15}N$) means that while tissue $\delta^{15}N$ values are useful for determining the trophic level, small changes in the trophic level of a fish can substantially alter the tissue $\delta^{15}N$ value and confound its interpretation as a spatial tracer.

The sulfur stable isotope ratio, $\delta^{34}S$, can also identify movements between marine habitats and freshwater. Marine fishes generally are more enriched in ^{34}S than freshwater fishes because marine sulfate ($\delta^{34}S$ ca. 21‰) is generally ^{34}S-enriched compared to freshwater sulfate ($\delta^{34}S < 20$‰) (Peterson and Fry 1987). Fractionation of sulfate by plants is small (Trust and Fry 1992), as is trophic fractionation, thereby preserving this difference in the $\delta^{34}S$ value of consumers. Marine and polyhaline estuarine fishes have been measured with $\delta^{34}S$ values ranging from 10‰ to 22‰, whereas freshwater fishes have been measured with values generally < 5‰ (Fry 1988; Hesslein et al. 1993; MacAvoy et al. 2001; Hoffman et al. 2007). Indeed, sulfur isotopes are perhaps the most powerful light isotope tracer of movement between marine and freshwaters, because of the large difference in $\delta^{34}S$ values between these ecosystems.

Estuaries and coastal habitats. Within estuaries, marked changes in the abundance of stable isotopes, particularly ^{13}C and ^{34}S, will occur from the upper estuary to the ocean, reflecting the different geochemistry of source waters (freshwater versus marine) and the physical mixing dynamics (Fry 2002; Fry and Chumchal 2011). The change in these tracers along an estuarine ecosystem is recorded in the isotopic signatures of primary producers, and provides a framework for interpretation of consumer isotopic data relative to conservative physical mixing (Fry 2002). The isotopic composition along the estuary does not change linearly with respect to salinity, but rather tends to be curvilinear because the concentration of the element of interest (e.g., C) generally varies between the freshwater and marine sources to the estuary. The spatial expression of this change in isotopic composition depends on the mixing dynamics specific to the estuary being studied (Fry 2002; Hoffman et al. 2010; Fry and Chumchal 2011). Measuring consumer isotopic ratios across the estuarine salinity gradient can identify the temporal and spatial scale of fish movements within the estuary (Cunjak et al. 2005; Hoffman et al. 2007).

Habitats within the marine coastal environment may have distinct isotopic compositions, facilitating movement studies. For example, mangrove leaves and primary producers within mangrove habitat are depleted in ^{13}C ($\delta^{13}C < -25$‰) (Rodelli et al. 1984) compared to primary producers and potential prey in seagrass ($\delta^{13}C > -14$‰; Nagelkerken and van der Velde 2004) and coral reef habitat ($\delta^{13}C -16$‰ to -8‰) (Nakamura et al. 2008). Nakamura et al. (2008) were able to use this difference to identify small black-tail snapper *Lutjanus fulvus* that had recently emigrated from mangrove habitat and settled on the coral reef. Fry and Ewel (2003) noted that shrimp *Metapenaeus* sp. captured in two different mangrove systems had different $\delta^{13}C$ values (-9‰ to -13‰ versus < -16‰), implying that the shrimp in the former system were feeding in nearby seagrass meadows whereas those in the latter were resident in the mangrove system. Notably, some caution in interpretation is warranted because many possible basal sources could support mangrove food webs, some with similar isotopic composition (e.g., the $\delta^{13}C$ value of mangrove leaves and mangrove-associated primary producers are similar to that of terrestrial and aquatic vascular C3 plants), and these sources can be readily transported within mangrove systems (Layman 2007).

Differences in isotopic composition between seagrass or marsh ecosystems and adjacent open water habitat have also facilitated movement studies, including studies of brown shrimp *Penaeus aztecus* (Fry 1981), larval red drum *Sciaenops ocellatus* (Herzka and Holt 2001), and juvenile weakfish *Cynoscion regalis* (Litvin and Weinstein 2004). Marine seagrasses have $\delta^{13}C$ values ranging from $-3‰$ to $-15‰$ (Fry 1981). Salt marshes dominated by C4 plants such as *Spartina alterniflora* have $\delta^{13}C$ values of ca. $-13‰$ because C4 plants have a small isotopic fractionation (ca. $6‰$; Peterson and Fry 1987). These sources are more ^{13}C enriched than plankton from adjacent open waters (ca. $-24‰$ to $-19‰$). Marshes dominated by other vegetation may also be isotopically distinct. For example, *Phragmites australis*, a vascular C3 plant, has a $\delta^{13}C$ value of $-26‰$ to $-28‰$ (Litvin and Weinstein 2004). Differences in $\delta^{34}S$ values between these ecosystems may occur, as well (Peterson and Fry 1987; Litvin and Weinstein 2004).

Recently, numerous investigators have compiled basin-scale measurements of isotopic variability in the ocean to construct isotopic landscapes, or 'isoscapes' across the coastal and offshore marine environment that might offer new opportunities to trace long distance marine migrations (Graham et al. 2010; Hobson et al. 2010; Trueman et al. 2012). Few investigators, however, have yet attempted to exploit these isotopic gradients to study fish migration using stable isotope analysis of soft tissues (but see Graham et al. 2010).

Benthic versus pelagic habitats. In both marine and freshwater shallow habitat, benthic and littoral habitats are often isotopically distinct from pelagic habitat because both benthic and epiphytic periphyton have higher $\delta^{13}C$ values than phytoplankton (France 1995). This is, presumably, due to a benthic boundary layer effect that reduces ^{13}C fractionation in benthic algae. Benthic consumers also have lower $\delta^{34}S$ values than planktonic consumers because of greater fractionations associated with sulfide production in sediments (Croisetiere et al. 2009). These same gradients may also isotopically differentiate shallow nearshore waters and deep offshore waters, if benthic primary production is incorporated into the nearshore food web but not into the offshore food web (e.g., Sierszen et al. 2006; Bertrand et al. 2011).

Within the offshore environment in lakes and oceans, the isotopic composition of the benthic food web is often enriched in ^{13}C and ^{15}N compared to the planktonic food web. The effect may arise from the carbon source, as sinking POM is generally enriched in ^{13}C and ^{15}N compared to suspended POM due to a variety of biological mechanisms (Saino and Hattori 1980; Altabet 1988; Ostrom et al. 1997), as well as from trophic enrichment in the food web (Fry and Scherr 1984). This allows isotopically distinguishing benthic versus pelagic consumers in offshore food webs in large lakes (Sierszen et al. 2006) and oceans (Pettursdottir et al. 2008).

Rivers. Isotopic differences in ^{15}N and ^{13}C among sites within watershed networks are commonly found. Differences both in N-cycling and catchment land-use characters (e.g., agricultural development, human density) can label sub-watersheds with unique

$\delta^{15}N$ values and thus help determine the origin of fish and their spatial scale of movement (Gray et al. 2004; Sepulveda et al. 2009). The ^{13}C in river systems undergoes a predictable enrichment with stream size as the influence of ^{13}C-depleted dissolved inorganic carbon derived from soils declines and that of ^{13}C-enriched atmospheric CO_2 increases (Finlay 2003). The shift can be used to trace fish movement between upstream and downstream locations (Gray et al. 2004; Rasmussen et al. 2009). Variation may occur at relatively small spatial scales, as well. For example, Finlay et al. (2002) found that consumers (scraper and collector-gatherer taxa) in stream pools are ^{13}C-enriched ($-18‰$ to $-20‰$ $\delta^{13}C$) compared to those in adjacent riffles ($< -22‰$ $\delta^{13}C$), presumably due to water velocity effects on boundary layer thickness around benthic algae.

A number of recent studies have demonstrated the utility of deuterium (^2H, or D) as a food web tracer (Doucett et al. 2007; Jardine et al. 2009; Solomon et al. 2011), particularly in tracing terrestrial versus aquatic sources to the food web. Thus, where locations or habitats vary in the respective contribution of these two sources to the local food web, ^2H may also be a useful stable isotope for tracking movement. That said, the use of δD values in aquatic food web studies is relatively new, and there are a number of residual concerns about interpreting δD values in ecological studies, including variable lipid content (Jardine et al. 2009). Among these concerns is that hydrogen from ambient water that is digested ('environmental' or 'dietary' water) can alter tissue δD values, because the δD value of water is typically different from that measured in terrestrial or aquatic primary producers (Doucett et al. 2007). Estimates for the dietary water contribution to δD values for fish are relatively scarce. Solomon et al. (2009) estimated the dietary water δD value contribution to numerous hatchery-reared salmonids (muscle tissue δD) to be an average of 12.4% ($\pm$ 2.4%). Thus, it is advisable to measure the δD value from the ambient water from the location(s) in which the fish of interest is living.

Source-specific tracers. Both point and non-point sources of anthropogenic nutrients and waste can label specific regions or sites within ocean basins, coastline, rivers, or lakes with an isotopically distinct character. For example, offshore waste disposal, sewage effluent, or seepage from septic tanks can alter the $\delta^{13}C$, $\delta^{15}N$, and $\delta^{34}S$ values of local primary producers or consumers compared to the surrounding environment (Macko and Ostrom 1994). This signal is reflected in fish, with those fish located closest to the source being most similar to its isotopic signature (Hansson et al. 1997; Steffy and Kilham 2004; Schlacher et al. 2005; Hoffman et al. 2012). The most common source-specific approach used in fish movement studies is to utilize differences among locations in the amount or form of anthropogenic nitrogen addition. Nitrate from human and animal waste is enriched in ^{15}N ($\delta^{15}N$ 10‰ to 22‰) owing to both denitrification and volatilization of ammonia (Kendall et al. 2007). In contrast, nitrate in synthetic fertilizer, which is fixed from atmospheric N, generally has much lower $\delta^{15}N$ values ($-3‰$ to 3‰) (Kendall et al. 2007), although the ^{15}N content can be elevated to a composition similar to wastewater if it is subject to denitrification in groundwater (Diebel and Vander Zanden 2009). Harrington et al. (1998) found that both nitrate and juvenile Atlantic salmon *Salmo salar* had different $\delta^{15}N$ values among small streams within the same river network, in accordance with variable agricultural land use among those catchments. Thus, they were able to use this marker to help identify the natal origin of juvenile salmon. More recently, Ramsay et al. (2012) performed a comparable study with juvenile brown trout *Salmo trutta* in the River Dee (U.K.) and found that scale $\delta^{15}N$ and $\delta^{13}C$ values could also be used to classify fish to their natal stream, and with comparable success to otolith element concentrations.

Study design

The key elements to a successful study of fish movement using stable isotopes are (1) to demonstrate that the locations or habitats of interest have an isotopic difference that is greater than the natural variation found in the fish, and (2) to use a tissue with an isotope turnover rate on a temporal scale that is appropriate to detect movement between these locations or habitats. Ideally, prior to the full study, a preliminary investigation is conducted to determine that a reliable isotopic difference between the locations or habitats of interest exists. It is important to note that stable isotope-based inferences of a fish's movement have considerable uncertainty, unless the two locations have distinct signatures or the transition timing between locations is known. The larger the difference in isotopic composition between the two habitats, the higher the sensitivity and precision of isotope approaches for detecting movements (Phillips and Gregg 2001). Analyzing tissues for multiple stable isotope ratios (e.g., $\delta^{13}C$ and $\delta^{34}S$) can improve the ability to discriminate between locations, particularly if there are more than two locations of interest or multiple potential diet sources for the fish in each location (Harrod et al. 2005). Further, a fish must reside in a location long enough to observe a change, because there is a time lag in the tissue's isotopic composition. Once the fish approaches isotopic equilibrium with its diet in its new location, it is not possible to estimate how long ago the fish arrived in that location.

The preliminary investigation should include sampling multiple sites within each location or habitat, and over multiple months or seasons to determine spatial and temporal variability at the relevant scales of interest. Sampling may primarily aim to characterize the fish species at an age or stage of interest, but the most successful studies generally also include available primary producers, primary consumers, resident fish that do not migrate, or some combination of these. It is important to develop an isotopic baseline that incorporates variability and that includes many potential prey, to successfully distinguish a shift in a fish's isotopic composition resulting from a movement versus another cause, such as a local ontogenetic diet shift or underlying isotopic shift in the ecosystem (in response to an algae bloom, for example). It is equally important to determine the ecologically relevant time-scales to successfully choose, for analysis, a tissue with an isotope turnover time that integrates over the same period as the movement of interest.

In the course of the study, it is preferable to measure the selected fish species in all locations of interest, including prior to movement and after establishment in a new location or habitat to confirm and describe the shift in isotopic composition in the tissue of interest. Generally, two approaches are used to identify migratory fish. First, fish can be identified as migratory by visually or statistically differentiating their isotopic composition from a location- or habitat-specific mean element stable isotope ratio (Herzka 2005). Formally, this can be accomplished using discriminant analysis or cluster analysis to classify fish as resident or migrant based on the distribution of stable isotope ratios (Harrod et al. 2005; Sepulveda et al. 2009; Durbec et al. 2010), or by using the data structure to determine which values do not conform to a localized mean by identifying outliers (Fry and Chumchal 2011).

Second, a theoretical model can be developed that describes the expected change in a fish's isotopic signature for a given movement scenario. Measured patterns in tissue isotopic composition, as a function of time or space, are then compared to the model to diagnose movement patterns. For example, Hoffman et al. (2008) compared weight-specific $\delta^{13}C$ values of a juvenile anadromous clupeid, American shad *Alosa sapidissima*, to an isotope turnover model that described emigration from the tidal freshwater to the ocean. Those fish with higher $\delta^{13}C$ values (at size) than the model were interpreted to have migrated to the ocean at a smaller size than stipulated by the model and, similarly, those with $\delta^{13}C$ values lower than the model were interpreted to have migrated at a larger size than stipulated by the model. It is similarly possible to compare the isotopic composition of a fish against a known spatial isotopic gradient, wherein the theoretical model describes the spatial change as a function of isotopic mixing between two geochemically distinct aquatic ecosystems, such as a coastal tributary and the ocean (Fry 2002).

Investigators have used a number of approaches to quantify spatial and temporal aspects of movement. Rasmussen et al. (2009) demonstrated that the spatial scale of movement of individual fish can be quantified by measuring the difference in the isotopic change over space (e.g., as found along freshwater-marine, inshore-offshore, upriver-downriver gradients) between a fish and its prey. First, primary consumers or prey species are measured along the spatial gradient of interest to determine the baseline isotopic shift. Then, the fish species is measured along the same spatial gradient. The difference between the slopes of the isotopic baseline and the isotopic composition of the fish, as a function of space, is used to calculate the spatial scale of movement occupied by the fish for feeding.

The temporal scale of movement can be quantified by applying stable isotope turnover models in field studies, provided that growth and metabolic isotope turnover rates are known or measured. For example, Herzka et al. (2000) was able to determine the age of settlement of red drum *Sciaenops ocellatus* larvae. Hoffman et al. (2007) quantified the duration of residency of an anadromous clupeid, American shad, within tidal freshwater habitats prior to their seaward migration.

It is also possible to quantify the amount of energy derived from the various locations between which the fish is moving using a single or multiple element stable isotope conservative mixing model. The model postulates that the fractional contribution from each habitat sums to one (Phillips and Gregg 2001). Mixing models are particularly sensitive to the difference in isotopic composition between locations. Doubling the difference between locations (e.g., from 2‰ to 4‰) will reduced the uncertainty in the contribution attributed to each location by half (Phillips and Gregg 2001). Using a two-element (C, N) stable isotope mixing model, Sierszen et al. (2012) found that the amount of energy obtained from either coastal wetlands or adjacent nearshore habitat of the Laurentian Great Lakes varied dramatically among individual adult northern pike *Esox lucius*. This indicated that movement for foraging, between open water and protected inshore habitat, was highly variable among individuals and within coastal wetlands.

Case studies

Diadromous migrations. Investigators have used stable isotope analysis to distinguish movement of adult anadromous salmonids into freshwater (Doucett et al. 1999), identify natal streams of young anadromous salmonids (Harrington et al. 1998), characterize movement patterns of juvenile anadromous clupeids during their migration to the ocean (Hoffman et al. 2007, 2008), and identify movement patterns of juveniles and sub-adult anguillids between fresh and marine water (Harrod et al. 2005).

Doucett et al. (1999) identified co-existing anadromous and non-anadromous brook trout *Salvelinus fontinalis* in the Tabusintac River (Canada) using the $\delta^{13}C$, $\delta^{15}N$, and $\delta^{34}S$ values of fish tissue. Adults were sampled from holding pools near the head of tide and their isotopic composition was compared to both age-0 brook trout (potential prey) and resident fishes caught in the river. Based on adipose fins, adults were found to have a much broader range of $\delta^{13}C$, $\delta^{15}N$, and $\delta^{34}S$ values than freshwater resident fish or benthic invertebrates (sampled as whole organisms). Larger adult brook trout generally had high $\delta^{13}C$, $\delta^{15}N$, and $\delta^{34}S$ values; therefore they were identified as anadromous brook trout that had recently migrated into the river from the Atlantic Ocean. Interestingly, when examining the isotopic composition of very small age-0 brook trout (ca. 2.0 cm fork length) that had recently emerged, only those captured in the freshwater sampling site closest to the estuary had $\delta^{13}C$, $\delta^{15}N$, and $\delta^{34}S$ values similar to anadromous brook trout. The isotopic composition of these recently emerged brook trout reflects a maternal influence because the fish's tissue is derived from its yolk sac. Thus, the spatial pattern in isotopic composition of age-0 brook trout was used to identify regions of the river that were important for anadromous brook trout production.

Harrod et al. (2005) used muscle tissue $\delta^{13}C$, $\delta^{15}N$, and C:N values of European eel *Anguilla anguilla* to study movement between salinity zones in Lough Ahalia, a tidal Atlantic lake in Ireland. The investigators captured 30 yellow-phase sub-adult European eels from each of three salinity zones: freshwater, brackish water, and marine water. They exploited the available gradients in ^{13}C and ^{15}N between fresh and marine waters, which were quantified using a resident, invasive gastropod because it was distributed across the entire salinity gradient. A discriminant analysis was used to classify eels into the three salinity zones based on $\delta^{13}C$ (lipid-corrected), $\delta^{15}N$, and C:N values. The mean classification rate was 85%, indicating that the European eels were organized into discrete groups along the freshwater-marine mixing zone. The discriminant analysis revealed movement between salinity zones, and individuals were classified into salinity zones, other than that in which the individual was caught. This occurred most frequently in freshwater and brackish water individuals, indicating European eels were moving more frequently between these salinity zones than between the marine water and either fresh or brackish water. The authors were then able to examine differences in growth and conditions among the salinity zones, because the implication of the stable isotope data was that spatial differences might arise at this scale. This analysis revealed spatial intra-population structure within this estuarine habitat.

Amphidromous migrations. To date, amphidromy has been rarely studied using stable isotope analysis.

Sorensen and Hobson (2005) used $\delta^{13}C$ and $\delta^{15}N$ values of muscle sampled from amphidromous gobiid fishes (*Lentipes concolor, Sicyopterus stimpsoni, Awaous guamensis*) to study movement from the coastal marine environment into the Hakalau Stream (Hawaii, USA). These gobies spawn in the stream. Larvae drift downstream into the ocean, where they spend three–five months before entering freshwater streams at 15–20 mm in length. Little is known, however, about habitat usage during their marine residence or their movement into streams. Over multiple years, the authors captured late-larval and juvenile stage gobies of migrating gobies, as well as resident adult gobies, by hand net along four locations within the stream; three locations were located in close proximity to the Pacific Ocean, where as the fourth was located well-upstream of the mouth. To determine habitat-specific stable isotope baselines, plausible organic matter sources, including leaf litter, stream detritus, freshwater algae, and marine POM were collected for analysis, as well as two common primary consumers (freshwater limpets *Neritina granosa* and freshwater shrimp *Atyoida bisulcata*). Comparison of putatively migrating young-of-year gobies captured near the river mouth to the resident adults provided evidence that the young-of-year gobies had not recently arrived from the ocean, but rather had been resident in the river long enough to isotopically resemble the adult fish. Moreover, the $\delta^{13}C$ and $\delta^{15}N$ values of young-of-year gobies were more similar to inshore marine POM samples than offshore marine POM samples. Together, the data indicated that amphidromous goby larvae were using freshwater plumes and stream mouth regions as rearing habitats, a previously unknown behavior. Thus, the freshwater plume region provided an important habitat linking the freshwater and marine phases.

Potamodromous migrations. Fewer studies have used stable isotope analysis to trace potamodromous migrations than to trace diadromous migrations.

Sepulveda et al. (2009) used the $\delta^{15}N$ value of Bonneville cutthroat trout *Oncorhynchus clarkii utah* muscle to identify the origin of individual Bonneville cutthroat trout captured in headwater stream spawning habitat in Bear River (Utah, USA). In the system studied, fluvial Bonneville cutthroat trout primarily inhabit mainstream rivers but migrate up to 90 km to headwater streams to spawn. Resident Bonneville cutthroat trout complete their entire life cycle within these same headwater streams. To determine the contribution of fluvial life history Bonneville cutthroat trout to the population, the investigators used differences in $\delta^{15}N$ values that arise from the diet (the fluvial form is larger and occupies a higher trophic position) and watershed land use (greater agricultural land use adjacent to mainstem rivers) between mainstem rivers and headwater streams. Fish were sampled only in headwater streams; however, fluvial fish were identified by the presence of a parasite unique to the life history form. The investigators found that the fluvial form had a $\delta^{15}N$ value about 4‰ higher than the resident form. A cluster analysis was used to test the accuracy of assignment based on size and $\delta^{15}N$ value, which helped identifying fish within a mixed assemblage.

Individuals of unknown life history were assigned to both forms, with smaller fish predominantly assigned to the resident form.

Hoffman et al. (2011) used $\delta^{13}C$ and $\delta^{15}N$ values of whole fish to trace movement of rainbow smelt *Osmerus mordax* larvae between coastal wetlands and adjacent open coastal waters (Hoffman et al. 2011). Although anadromous in marine environments, in the Laurentian Great Lakes, rainbow smelt migrate into rivers or use shorelines to spawn. Rainbow smelt larvae were sampled in the western end of Lake Superior (USA) to describe their movements by exploiting an isotopic gradient between inshore wetland habitat (^{13}C-depleted, ^{15}N-enriched) and Lake Superior (^{13}C-enriched, ^{15}N-depleted). Over a three-month period, the investigators collected larvae weekly in a large coastal wetland situated at a river mouth and behind a barrier beach, in the adjacent open coast of Lake Superior, and at the inlet connecting the two habitats. Initially, yolk-sac larvae collected at the inlet and in Lake Superior had similar stable isotope values, reflecting the similar origin of the parent stock (Lake Superior). Once the larvae had experienced a 10-fold increase in weight from the yolk-sac stage, habitat-specific isotopic signatures were apparent. Rainbow smelt captured in the wetland had lower $\delta^{13}C$ values (ca. −32‰) and higher $\delta^{15}N$ values (ca. 9‰) compared to larvae captured in the inlet (−28‰ to −26‰ $\delta^{13}C$, 5‰ to 8‰ $\delta^{15}N$) and in Lake Superior (ca. −26‰ $\delta^{13}C$, ca. 6‰ $\delta^{15}N$). Larvae had different isotopic compositions because they settled into two geochemically distinct habitats–Lake Superior and the coastal wetland. Larvae captured in the inlet had either an isotopic composition similar to larvae in Lake Superior, implying they recently entered the river, or an isotopic composition intermediate between the wetland and lake, implying that they may reside within the inlet (a hydrologic transition between the river and lake) or move regularly between Lake Superior and the wetland. Thus, the spatial pattern in isotopic composition revealed facultative use of coastal habitats.

Marine migrations and movement. A growing number of studies document movement and migrations of marine fishes wholly within the marine environment. Most of these have characterized coastal movements, utilizing inshore-offshore isotopic gradients (Estrada et al. 2003; Guelinckx et al. 2006; Tanaka et al. 2010) or differences between habitats with distinct isotopic signatures (Nakamura et al. 2008; Papastamatiou et al. 2010).

Guelinckx et al. (2006) used $\delta^{13}C$ of young-of-year Atlantic herring *Clupea harengus* and European sprat *Sprattus sprattus* muscle tissue to study movement between the North Sea (marine environment; prey with $\delta^{13}C$ ca. −23‰ to −20‰) and the Schelde River estuary (brackish environment, prey with $\delta^{13}C$ ca. −30‰ to −27‰). The investigators sampled monthly for both fish and prey items, in both the lower and upper estuary, for a year to follow movements of a single year-class. At the lower estuary station, most herring and sprat had a marine signature; however some fish captured in summer were identified as transient (intermediate $\delta^{13}C$ value), indicating seaward movement. At the upper estuary station, a large proportion of both herring and sprat captured in October through March were identified as marine, revealing migration from the North Sea to the estuary. The tracer revealed that both herring and sprat underwent almost constant immigration between the North Sea and the estuary, with pronounced movement into the upper estuary during the winter months. The investigators concluded that the isotopic evidence for simultaneous movements into and out of the estuary supports the hypothesis that migration to the estuarine nursery habitat was facultative and individual-based.

Large-scale geographic variations in isotope abundance ('isoscapes') have been used to track long distance migrations of large pelagic fishes and marine mammals, but generally using biological materials other than soft tissues. Graham et al. (2010), however, were able to use variability in the $\delta^{15}N$ value of yellowfin tuna *Thunnus albacares* and bigeye tuna *Thunnus obesus* muscle to determine movements across the equatorial Pacific Ocean. Tissue samples were collected from commercial fishing vessels from across nearly the whole of the equatorial Pacific Ocean, from latitude 20° N to 20° S. Surprisingly, the investigators found a large difference, 12‰ to 14‰, in $\delta^{15}N$ values among individuals. Further research, using compound-specific ^{15}N analysis of individual amino acids, revealed that this variability in bulk tissue $\delta^{15}N$ was caused by variability in ^{15}N at the base of the food web (i.e., not trophic position). Thus, the investigators were able to map the isotopic deviations and create a $\delta^{15}N$ value isoscape for tuna in the equatorial Pacific Ocean. Isotope turnover in juvenile yellowfin tuna is rapid; the isotope signal is integrated over about a two-month period. Thus, the study revealed that these pelagic tuna were not highly migratory but rather feeding in discrete regions, on the scale of hundreds of kilometers, for at least a few months.

Combining tagging. Combining stable isotope analysis with electronic tagging technologies such as radio or acoustic telemetry, or with other types of intrinsic tags such as otolith microchemistry analysis, can provide insight into how movement facilitates ecological connectivity among ecosystems (Hammerschalge-Peyer and Layman 2010). It also reveals why fish are undergoing movements between habitats (Cunjak et al. 2005), and yields increased certainty by providing multiple lines of evidence for migration (Hogan et al. 2007).

Cunjak et al. (2005) provide numerous examples where combining telemetry or mark-recapture data with stable isotope analysis can improve movement studies. The Mirimachi River in New Brunswick (Canada) is an important habitat for Atlantic salmon *Salmo salar*. Salmon parr are known to undertake a variety of small-scale movements in this river, including moving between large, thermally dynamic river systems and smaller, thermally stable tributaries. The authors combined Passive Integrated Transponder (PIT) tagging with fin tissue stable isotope analysis ($\delta^{13}C$, $\delta^{15}N$) to explore causes for this movement. In essence, the question was whether this movement was in search of foraging habitat or thermal refugia. The study was feasible because prior work had demonstrated that fishes and macroinvertebrates from the large and small tributary in the study area were isotopically distinct. Stable isotope analysis revealed that fish captured in the large tributary were enriched in ^{13}C with little variation among individuals, whereas those captured in the small tributary were mostly depleted in ^{13}C, but had wide variation among individuals. PIT tagging studies revealed three different forms of foraging behavior consistent with the isotope data. Those parr captured in the small tributary with the lowest $\delta^{13}C$ values were residing and foraging in the system. Those parr captured in the small tributary with intermediate $\delta^{13}C$ values were regularly moving between the two tributaries. While parr captured in the large tributary were largely residents of the large tributary, consistent with their higher

δ^{13}C values. Notably, PIT tagging revealed that some of these parr were entering the small tributary, indicating they were not using the habitat forage (they did not have an intermediate δ^{13}C value) but were using it as thermal refugia during the summer. That is, the two tagging methods together allowed the authors to identify a response to an environmental stress that neither technique alone would have revealed as effectively.

Hogan et al. (2007) combined otolith microchemistry analysis (Sr:Ca) with C and N stable isotope analysis of muscle tissue to characterize the life history of a commercially important Asian catfish, *Pangasius krempfi*, in the Mekong River (SE-Asia). Stable isotope analysis from muscle tissue of adults captured far up the Mekong River (> 700 km) in southern Laos was undertaken to determine the diet and habitat of adult *P. krempfi*. Sagittal otoliths were also obtained from spawning adults captured at the same study site; the ratio of strontium to calcium concentrations (Sr:Ca) along a transect from the core to the margin was analyzed to characterize movements throughout life between marine and freshwater. In essence, stable isotope analysis provided short-term information, whereas otolith analysis provided long-term information regarding habitat use. The δ^{13}C and δ^{15}N values of *P. krempfi* tissue were much higher than other species of resident fish and available organic matter sources to the river, suggesting these fish had recently been foraging in the marine environment. Sr:Ca measurements were highly variable between the otolith origin and margin, reaching values much higher than resident freshwater fishes, and on average were significantly higher than resident freshwater fishes. These data also suggest the fish were spending much of their life history in the marine environment, likely moving regularly between brackish and marine habitats. Together, the data demonstrate this fish is anadromous, that individuals undergo long distance spawning migrations, and that they inhabit both freshwater and marine environments.

Conclusion

While stable isotope analysis is now an established tool to trace the origins and movements of fish, recent advances in laboratory instrumentation and data analysis approaches have broadened the potential applications of this tool. Improvements in analytical throughput, as well as reduced cost, now permit the analysis of hundreds or thousands of individuals. This high analytical power allows investigators to explore variation among individuals, identify multiple migration strategies, and study movements at geographic scales that are quite small (hundreds of meters) and quite large (thousands of kilometers). The use of non-lethal methods for sampling tissues (e.g., fin clips, tissue biopsies) is increasing and these methods facilitate stable isotope studies of fish of conservation concern, as well as studies that require repeat sampling of individuals. An expanding base of stable isotope markers and the geographic scales of data sources will continue to improve our ability to interpret stable isotope data. The development of quantitative approaches has enabled investigators to better characterize movements, including using stable isotope turnover models to estimate the timing of movements, using available isotopic gradients to estimate the spatial scale of fish movement, and using stable isotope mixing models to estimate the relative amount of energy obtained from foraging in various locations. While applications have broadened, it remains fundamental that in every study, the investigators must develop ecosystem- and taxon-specific isotopic baseline signatures as a reference point for interpreting their fish data. The strength of conclusions regarding inferred movements is only as good as the isotopic context in which the data can be interpreted.

Continuing research in stable isotope turnover will undoubtedly improve this field of research. While there exist general frameworks for understanding isotope turnover and the number of isotope turnover studies is growing, it remains important to quantify isotope turnover in a broad suite of fishes, especially those fishes with complex life histories, and during different life stages (e.g., during spawning migrations, during long distance seasonal migrations). Notably, metabolic processes remain relatively unexplored with respect to stable isotope turnover and trophic fractionation. This is because, at least in part, assessing the contribution of metabolism to isotope turnover, especially in field studies, remains extremely difficult. New electronic tagging technologies, however, could be applied to this problem. For example, the contribution of metabolism could be indirectly measured in a field experiment by following individuals and repeatedly recapturing them to obtain growth estimates and measure the change in an individual's isotopic composition (based on fin clips, for example). This example demonstrates the potential of combined, complementary tagging approaches to provide new insights into the utility of stable isotope analysis for studying fish movements.

Summary

1. Element (H, C, N, S) stable isotope ratios in soft tissues are a time-lagged indicator of a fish's diet; the time period over which the diet information is integrated depends on the stable isotope turnover rate of the tissue being analyzed.
2. A difference in the isotopic composition between two habitats or ecosystems within the same fish species can result from underlying differences in ecosystem biogeochemistry or food web inputs, as well as from differences in the trophic niche of the fish or its prey.
3. Migrations and movements in fishes can be studied using stable isotope analysis if (1) the fish moves between two (or more) locations, habitats, or ecosystems in which their food source has a different isotopic composition; (2) the movement between the two locations is faster than the rate of isotope turnover in the tissues; and (3) the fish remains in the second location long enough to observe a change in the isotopic composition of its tissues.
4. The isotope turnover rate depends on somatic growth (dilution) and metabolism (replacement). The isotope turnover rate generally decreases with increasing body size. It is critical to sample a life stage and tissue that has an isotope turnover rate that is sufficiently fast to detect movements between the locations of interest. Including multiple tissues with different isotope turnover rates may provide more information on the timing of movements.
5. The ability to discriminate movements between two locations depends on the magnitude of the isotopic difference between them. The certainty of the conclusions that are drawn can be increased by choosing the stable isotope ratio

with the greatest difference between locations, using multiple stable isotope ratios (e.g., $\delta^{13}C$ and $\delta^{34}S$ values), or applying additional tracer approaches (e.g., tagging, otolith microchemistry).

6. To attribute the change in the isotopic composition of a fish's tissue to movement (versus an underlying shift in the isotopic baseline, for example), it is necessary to account for temporal and spatial isotopic variability in location-specific stable isotope ratio baselines. Such changes can be the result of changes in biogeochemistry (e.g., microbial processing), fractionation by primary producers at the base of the food web, food web inputs, trophic niche shifts, and trophic fractionation.

Acknowledgements

I thank Peter McIntyre and an anonymous reviewer for their helpful comments on the chapter. The views expressed here are those of the author and do not necessarily reflect the views or policies of the U.S. Environmental Protection Agency.

References

Alexander SA, Hobson KA, Gratto-Trevor CL and Diamond AW (1996) Conventional and isotopic determinations of shorebird diets at an inland stopover: the importance of invertebrates and *Potomageton pectinatus* tubers. Canadian Journal of Zoology 74: 1057–1068.

Altabet MA (1988) Variations in nitrogen isotopic composition between sinking and suspended particles: implications for nitrogen cycling and particle transformation in the open ocean. Deep Sea Research Part A 35: 535–554.

Arrington DA and Winemiller KO (2002) Preservation effects on stable isotope analysis of fish muscle. Transactions of the American Fisheries Society 131: 337–342.

Baker RF, Blanchfield PJ, Paterson MJ, Flett RJ and Wesson L (2004) Evaluation of nonlethal methods for the analysis of mercury in fish tissue. Transactions of the American Fisheries Society 133: 568–576.

Bertrand M, Cabana G, Marcogliese D and Magnan P (2011) Estimating the feeding range of a mobile consumer in a river-flood plain system using $\delta^{13}C$ gradients and parasites. Journal of Animal Ecology 80: 1313–1323.

Birchall J, O'Connell TC, Heaton TE and Hedges REM (2005) Hydrogen isotope ratios in animal body protein reflect trophic level. Journal of Animal Ecology 74: 877–881.

Bosley K and Wainright S (1999) Effects of preservatives and acidification on the stable isotope ratios ($^{15}N{:}^{14}N$, $^{13}C{:}^{12}C$) of two species of marine animals. Canadian Journal of Fisheries and Aquatic Sciences 56: 2181–2185.

Bosley KL, Witting DA, Chambers RC and Wainright SC (2002) Estimating turnover rates of carbon and nitrogen in recently metamorphosed winter flounder *Pseudopleuronectes americanus* with stable isotopes. Marine Ecology Progress Series 236: 233–240.

Buchheister A and Latour RJ (2010) Turnover and fractionation of carbon and nitrogen stable isotopes in tissues of a migratory coastal predator, summer flounder (*Paralichthys dentatus*). Canadian Journal of Fisheries and Aquatic Sciences 67: 445–461.

Bunn SE, Loneragan NR and Kempster MA (1995) Effects of acid washing on stable isotopes ratios of C and N in penaeid shrimp and seagrass: implications for food-web studies using multiple stable isotopes. Limnology and Oceanography 40: 622–625.

Carabel S, Godínez-Domínguez E, Verísimo P, Fernández L and Freire J (2006) An assessment of sample processing methods for stable isotope analyses of marine food webs. Journal of Experimental Marine Biology and Ecology 336: 254–261.

Caudill CC (2003) Measuring dispersal in a metapopulation using stable isotope enrichment: high rates of sex-biased dispersal between patches in a mayfly metapopulation. Oikos 101: 624–630.

Caut S, Angulo E and Courchamp F (2009) Variation in discrimination factors ($\delta^{15}N$ and $\delta^{13}C$): the effect of diet isotopic values and applications for diet reconstruction. Journal of Applied Ecology 46: 443–453.

Chesson LA, Podlesak DW, Cerling TE and Ehleringer JR (2009) Evaluating uncertainty in the calculation of nonexchangeable hydrogen fractions within organic materials. Rapid Communications in Mass Spectrometry 23: 1275–1280.

Church MR, Ebersole JL, Rensmeyer KM, Couture RB, Barrows FT and Noakes DLG (2009) Mucus: a new tissue fraction for rapid determination of fish diet switching using stable isotope analysis. Canadian Journal of Fisheries and Aquatic Sciences 66: 1–5.

Cifuentes LA, Sharp JH and Fogel ML (1988) Stable carbon and nitrogen isotope biogeochemistry in the Delaware estuary. Limnology and Oceanography 33: 1102–1115.

Cloern JE, Canuel EA and Harris D (2002) Stable carbon and nitrogen isotope composition of aquatic and terrestrial plants of the San Francisco Bay estuarine system. Limnology and Oceanography 47: 713–729.

Criss RE (1999) Principles of Stable Isotope Distribution. Oxford University Press, Inc.

Croisetière L, Hare L, Tessier A and Cabana G (2009) Sulphur stable isotopes can distinguish trophic dependence on sediments and plankton in boreal lakes. Freshwater Biology 54: 1006–1015.

Cunjak RA, Roussel JM, Gray MA, Dietrich JP, Cartwright DF, Munkittrick KR and Jardine TD (2005) Using stable isotope analysis with telemetry or mark-recapture data to identify fish movement and foraging. Oecologia 144: 636–646.

Daverat F and Martin J (2016) Microchemical and schlerochronological analyses used to infer fish migration. pp. 149–168. *In*: Morais P and Daverat F (eds.). An Introduction to Fish Migration. CRC Press, Boca Raton, FL (this book).

Deegan LA and Garritt RH (1997) Evidence for spatial variability in estuarine food webs. Marine Ecology Progress Series 147: 31–47.

Deines D (1980) The isotopic composition of reduced carbon. pp. 329–406. *In*: Fritz P and Fontes JC (eds.). Handbook of Environmental Isotope Geochemistry. Elsevier, Amsterdam, The Netherlands.

DeNiro MJ and Epstein S (1977) Mechanism of carbon isotope fractionation associated with lipid synthesis. Science 197: 261–263.

Diebel MW and Vander Zanden MJ (2009) Nitrogen stable isotope in streams: effects of agricultural sources and transformations. Ecological Applications 19: 1127–1134.

Doucett RR, Hooper W and Power G (1999) Identification of anadromous and nonanadromous adult brook trout and their progeny in the Tabusintac River, New Brunswick, by means of multiple stable isotope analysis. Transactions of the American Fisheries Society 128: 278–288.

Doucett RR, Marks JC, Blinn DW, Caron M and Hungate BA (2007) Measuring terrestrial subsidies to aquatic food webs using stable isotopes of hydrogen. Ecology 88: 1587–1592.

Durbec M, Cavalli L, Grey J, Chappaz R and Nguyen B (2010) The use of stable isotopes to trace small-scale movements by small fish species. Hydrobiology 641: 23–31.

Estep MF and Dabrowski H (1980) Tracing food webs with stable hydrogen isotopes. Science 209: 1537–1538.

Estrada JA, Rice AN, Lutcavage ME and Skomal GB (2003) Predicting trophic position in sharks of the northwest Atlantic Ocean using stable isotope analysis. Journal of the Marine Biological Association of the United Kingdom 83: 1347–1350.

Finlay JC (2003) Controls of streamwater dissolved inorganic carbon dynamics in a forested watershed. Biogeochemistry 62: 231–252.

Finlay JC, Khandwala S and Power ME (2002) Spatial scales of carbon flow in a river food web. Ecology 83: 1845–1859.

France RL (1995) Carbon-13 enrichment in benthic compared to planktonic algae: foodweb implications. Marine Ecology Progress Series 124: 307–312.

Fry B (1981) Natural stable carbon isotope tag traces Texas shrimp migrations. Fishery Bulletin 79: 337–345.

Fry B (1988) Food web structure on Georges Bank from stable C, N, and S isotopic compositions. Limnology and Oceanography 33: 1182–1190.

Fry B (2002) Conservative mixing of stable isotopes across estuarine salinity gradients: a conceptual framework for monitoring watershed influences on downstream fisheries production. Estuaries 25: 264–271.

Fry B and Arnold C (1982) Rapid $^{13}C/^{12}C$ turnover during growth of brown shrimp (*Penaeus aztecus*). Oecologia 54: 200–204.

Fry B and Sherr E (1984) $\delta^{13}C$ measurements as indicators of carbon flow in marine and freshwater ecosystems. Contributions in Marine Science 27: 13–47.

Fry B and Ewel KC (2003) Using stable isotopes in mangrove fisheries research—a review and outlook. Isotopes in Environemtal Health Studies 39: 191–196.

Fry B and Chumchal MM (2011) Sulfur stable isotope indicators of residency in estuarine fish. Limnology and Oceanography 56: 1563–1576.

Fry B, Baltz DB, Benfield MC, Fleeger JW, Gace A, Haas HL and Quiñones-Rivera ZJ (2003) Stable isotope indicators of movement and residency for brown shrimp (*Farfantepenaeus aztecus*) in coastal Louisiana marshscapes. Estuaries 26: 82–97.

Gannes LZ, O'Brien DM and Martinez del Rio C (1997) Stable isotopes in animal ecology: Assumptions, caveats, and a call for more laboratory experiments. Ecology 78: 1271–1276.

Goericke R and Fry B (1994) Variations of marine plankton $\delta^{13}C$ with latitude, temperature, and dissolved CO^2 in the world ocean. Global Biogeochemistry Cycles 8: 85–90.

Goericke R, Montoya JP and Fry B (1994) Physiology of isotopic fractionation in algae and cyanobacteria. pp. 187–221. *In*: Lajtha K and Michener RM (eds.). Stable Isotopes in Ecology and Environmental Science. Blackwell Scientific Publications, Oxford, UK.

Graham BS, Koch PL, Newsome SD, McMahon KW and Aurioles D (2010) Using isoscapes to trace the movements and foraging behavior of top predators in oceanic ecosystems. pp. 299–318. *In*: West JB et al. (eds.). Isoscapes: Understanding Movement, Pattern, and Process on Earth through Isotope Mapping. Springer Science, New York, NY, USA.

Gray MA, Cunjak RA and Munkittrick KR (2004) Site fidelity of slimy sculpin (*Cottus cognatus*): insights from stable carbon and nitrogen analysis. Canadian Journal of Fisheries and Aquatic Sciences 61: 1717–1722.

Guelinckx J, Maes J, De Brabandere L, Dehairs F and Ollevier F (2006) Migration dynamics of clupeoids in the Schelde estuary: a stable isotope approach. Estuarine, Coastal and Shelf Science 66: 612–623.

Hammerschlag-Peyer CM and Layman CA (2010) Intrapopulation variation in habitat use by two abundant coastal fish species. Marine Ecology Progress Series 415: 211–220.

Hansson S, Hobbie JE, Elmgren R, Larsson U, Fry B and Johansson S (1997) The stable nitrogen isotope ratio as a marker of food-web interactions and fish migration. Ecology 78: 2249–2257.

Harrington RR, Kennedy BP, Chamberlain CP, Blum JD and Folt CL (1998) ^{15}N enrichment in agricultural catchments: field patterns and applications to tracking Atlantic salmon (*Salmo salar*). Chemical Geology 147: 281–294.

Harrod C, Grey J, McCarthy TK and Morrissey M (2005) Stable isotope analyses provide new insights into ecological plasticity in a mixohaline population of European eel. Oecologia 144: 673–683.

Herzka SZ (2005) Assessing connectivity of estuarine fishes based on stable isotope ratio analysis. Estuarine, Coastal and Shelf Science 64: 58–69.

Herzka SZ and Holt GJ (2000) Changes in isotopic composition of red drum (*Sciaenops ocellatus*) larvae in response to dietary shifts: potential applications to settlement studies. Canadian Journal of Fisheries and Aquatic Sciences 57: 137–147.

Herzka SZ, Holt SA and Holt GJ (2001) Documenting the settlement history of individual fish larvae using stable isotope ratios: model development and validation. Journal of Experimental Marine Biology and Ecology 265: 49–74.

Hesslein RH, Capel MJ, Fox DW and Hallard K (1991) Stable isotopes of sulfur, carbon, and nitrogen as indicators of trophic level and fish migration in the lower Mackenzie River basin, Canada. Canadian Journal of Fisheries and Aquatic Sciences 48: 2258–2265.

Hesslein RH, Hallard KA and Ramlal P (1993) Replacement of sulfur, carbon, and nitrogen in tissue of growing broad whitefish (*Coregonus nasus*) in response to a change in diet traced by $\delta^{34}S$, $\delta^{13}C$, and $\delta^{15}N$. Canadian Journal of Fisheries and Aquatic Sciences 50: 2071–2076.

Hobson KA (1999) Tracing origins and migration of wildlife using stable isotopes: a review. Oecologia 120: 314–326.

Hobson KA and Clark RW (1992) Assessing avian diets using stable isotopes. I. Turnover of carbon-13. Condor 94: 181–188.

Hobson KA, Barnett-Johnson R and Cerling T (2010) Using isoscapes to track animal migration. pp. 273–298. *In*: West JB et al. (eds.). Isoscapes: Understanding Movement, Pattern, and Process on Earth through Isotope Mapping. Springer Science, New York, NY, USA.

Hoffman JC and Sutton T (2010) Lipid correction for carbon stable isotope analysis of deep-sea fishes. Deep-Sea Research Part I 57: 956–964.

Hoffman JC, Bronk DA and Olney JE (2007) Tracking nursery habitat use in the York River estuary, Virginia, by young American shad using stable isotopes. Transactions of the American Fisheries Society 136: 1285–1297.

Hoffman JC, Limburg KE, Bronk DA and Olney JE (2008) Overwintering habitats of migratory juvenile American shad in Chesapeake Bay. Environmental Biology of Fishes 81: 329–345.

Hoffman JC, Cotter AM, Peterson GS and Kelly JR (2010) Using stable isotope mixing in a Great Lakes coastal tributary to determine food web linkages in young fishes. Estuaries and Coasts 33: 1391–1405.

Hoffman JC, Cotter AM, Peterson GS, Corry TD and Kelly JR (2011) Rapid stable isotope turnover of larval fish in a Lake Superior coastal wetland: implications for diet and life history studies. Aquatic Ecosystem Health & Management 14: 403–413.

Hoffman JC, Kelly JR, Peterson GS, Cotter AM, Starry M and Sierszen ME (2012) Using $\delta^{15}N$ in fish as an indicator of watershed sources of anthropogenic nitrogen: response at multiple spatial scales. Estuaries and Coasts 35: 1453–1467.

Hogan Z, Baird IG, Radtke R and Vander Zanden MJ (2007) Long distance migration and marine habitation in the tropical Asian catfish, *Pangasius krempfi*. Journal of Fish Biology 71: 818–832.

Jardine TD, Kidd KA and Cunjak RA (2009) An evaluation of deuterium as a food source tracer in temperate streams of eastern Canada. Journal of the North American Benthological Society 28: 885–893.

Jones FRH (1968) Fish Migration. Edward Arnold Ltd., London.

Kelly B, Dempson JB and Power M (2006) The effect of preservation on fish tissue stable isotope signatures. Journal of Fish Biology 69: 1595–1611.

Kendall C, Elliott EM and Wankel SD (2007) Tracing anthropogenic inputs of nitrogen to ecosystems. pp. 375–449. *In*: Michener RH and Lajtha K (eds.). Stable Isotopes in Ecology and Environmental Science, 2nd edition. Blackwell Publishing, Oxford, UK.

Kiljunen M, Grey J, Sinisalo T, Harrod C, Immonen H and Jones RI (2006) A revised model for lipid-normalizing $\delta^{13}C$ values from aquatic organisms, with implications for isotope mixing models. Journal of Applied Ecology 43: 1213–1222.

Layman CA (2007) What can stable isotope ratios reveal about mangroves as fish habitat? Bulletin of Marine Science 80: 513–527.

Limburg KE (1998) Anomalous migration of anadromous herrings revealed with natural chemical tracers. Canadian Journal of Fisheries and Aquatic Sciences 55: 431–437.

Litvin SY and Weinstein MP (2004) Multivariate analysis of stable-isotope ratios to infer movements and utilization of estuarine organic matter by juvenile weakfish (*Cynoscion regalis*). Canadian Journal of Fisheries and Aquatic Sciences 61: 1851–1861.

Logan J, Haas H, Deegan L and Gaines E (2006) Turnover rates of nitrogen stable isotopes in the salt marsh mummichog, *Fundulus heteroclitus*, following a laboratory diet switch. Oecologia 147: 391–395.

Logan JM and Lutcavage ME (2010) Stable isotope dynamics in elasmobranch fishes. Hydrobiologia 644: 231–244.

Logan JM, Jardine TD, Miller TJ, Bunn SE, Cunjak RA and Lutcavage ME (2008) Lipid corrections in carbon and nitrogen stable isotope analyses: comparison of chemical extraction and modelling methods. Journal of Animal Ecology 77: 838–846.

MacAvoy SE, Macko SA and Garman GC (1998) Tracing marine biomass into tidal freshwater ecosystems using stable sulfur isotopes. Naturwissenschaften 85: 544–546.

MacAvoy SE, Macko SA and Garman GC (2001) Isotopic turnover in aquatic predators: quantifying the exploitation of migratory prey. Canadian Journal of Fisheries and Aquatic Sciences 58: 923–932.

Macko SA and Ostrom NE (1994) Pollution studies using stable isotopes. pp. 45–62. *In*: Lajhta K and Michener RH (eds.). Stable Isotopes in Ecology and Environmental Science. Blackwell Scientific Publications, Oxford.

Mariotti A, Germon JC, Hubert P, Kaiser P, Letolle R, Tardieux A and Tardieux P (1981) Experimental determination of nitrogen kinetic isotope fractionation: Some principles; illustration for the denitrification and nitrification processes. Plant and Soil 62: 413–430.

Martínez del Rio C, Wolf N, Carleton SA and Gannes LZ (2009) Isotopic ecology ten years after a call for more laboratory experiments. Biological Reviews 84: 91–111.

McConnaughey T and McRoy CP (1979) Food-web structure and the fractionation of carbon isotopes in the Bering Sea. Marine Biology 53: 257–262.

McCutchan JH, Lewis WM, Kendall C and McGrath CC (2003) Variation in trophic shift for stable isotope ratios of carbon, nitrogen, and sulfur. Oikos 102: 378–390.

McIntyre PB and Flecker AS (2006) Rapid turnover of tissue nitrogen of primary consumers in tropical freshwaters. Oecologia 148: 12–21.

McMahon KW, Fogel ML, Johnson BJ, Houghton LA and Thorrold SR (2011) A new method to reconstruct fish diet and movement patterns from $\delta^{13}C$ values in otolith amino acids. Canadian Journal of Fisheries and Aquatic Sciences 68: 1330–1340.

Mintenbeck K, Brey T, Jacob U, Knust R and Struck U (2008) How to account for the lipid effect on carbon stable-isotope ratio ($\delta^{13}C$): sample treatment effects and model bias. Journal of Fish Biology 72: 815–830.

Miyake Y and Wada E (1971) The isotope effect on the nitrogen in biochemical, oxidation-reduction reactions. Records of Oceanographic Works in Japan 11: 1–6.

Mook WG and Tan FC (1991) Stable carbon isotopes in rivers and estuaries. pp. 245–264. *In*: Degens ET, Kempe S and Richy JE (eds.). Biogeochemistry of Major World Rivers. Wiley & Sons, West Sussex, UK.

Mook WG, Bommerson JC and Staverman WH (1974) Carbon isotope fractionation between dissolved bicarbonate and gaseous carbon dioxide. Earth and Planetary Science Letters 22: 169–176.

Murry BA, Farrell JM, Teece MA and Smyntek PM (2006) Effect of lipid extraction on the interpretation of fish community trophic relationships determined by stable carbon and nitrogen isotopes. Canadian Journal of Fisheries and Aquatic Sciences 63: 2167–2172.

Nagelkerken I and van der Velde G (2004) Are Caribbean mangroves important feeding grounds for juvenile reef fish from adjacent seagrass beds. Marine Ecology Progress Series 274: 143–151.

Nakamura Y, Horinouchi M, Shibuno T, Tanaka Y, Miyajima T, Koike I, Kurokura H and Sano M (2008) Evidence of ontogenetic migration from mangroves to coral reefs by black-tail snapper *Lutjanus fulvus*: stable isotope approach. Marine Ecology Progress Series 355: 257–266.

O'Reilly CM, Verburg P, Hecky RE, Plisnier P-D and Cohen AS (2003) Food web dynamics in stable isotope ecology: Time integration of different trophic levels. pp. 125–134. *In*: Seuront L and Sutton P (eds.). Handbook of Scaling Methods in Aquatic Ecology: Measurement, Analysis, Simulation. CRC Press, Boca Raton, Florida.

Ostrom NE, Macko SA, Deibel D and Thompson RJ (1997) Seasonal variation in the stable carbon and nitrogen isotope biogeochemistry of a coastal cold ocean environment. Geochimica et Cosmochimica Acta 61: 2929–2942.

Papastamatiou YP, Friedlander AM, Caselle JE and Lowe CG (2010) Long-term movement patterns and trophic ecology of blacktip reef sharks (*Carcharhinus melanopterus*) at Palmyra Atoll. Journal of Experimental Marine Biology and Ecology 386: 94–102.

Peterson BJ and Fry B (1987) Stable isotopes in ecosystem studies. Annual Review of Ecology, Evolution, and Systematics 18: 293–320.

Pettursdottir H, Gislason A, Falk-Petersen S, Hop H and Svavarsson J (2008) Trophic interactions of the pelagic ecosystem over the Reykjanes Ridge as evaluated by fatty acid and stable isotope analyses. Deep-Sea Research Part II 55: 83–93.

Phillips DL and Gregg JW (2001) Uncertainty in source partitioning using stable isotopes. Oecologia 127: 171–179.

Phillips DL and Eldridge PM (2006) Estimating the timing of diet shifts using stable isotopes. Oecologia 147: 195–203.

Pinnegar JK and Polunin NVC (1999) Differential fractionation of $\delta^{13}C$ and $\delta^{15}N$ among fish tissues: implications for the study of trophic interactions. Functional Ecology 13: 225–231.

Post DM (2002) Using stable isotopes to estimate trophic position: models, methods, and assumptions. Ecology 83: 703–718.

Ramsay AL, Milner NJ, Hughes RN and McCarthy ID (2012) Fish scale $\delta^{15}N$ and $\delta^{13}C$ values provide biogeochemical tags of fish comparable in performance to element concentrations in scales and otoliths. Hydrobiologia 694: 183–196.

Rasmussen JB, Trudeau V and Morinville G (2009) Estimating the scale of fish feeding movements in rivers using $\delta^{13}C$ signature gradients. Journal of Animal Ecology 78: 674–685.

Rodelli MR, Gearing JN, Gearing PJ, Marshall N and Sasekumar A (1984) Stable isotope ratio as a tracer of mangrove carbon in Malaysian ecosystems. Oecologia 61: 326–333.

Saino T and Hattori A (1980) ^{15}N natural abundance in oceanic suspended particulate matter. Nature 283: 752–754.

Sakano H, Fujiwara E, Nohara S and Ueda H (2005) Estimation of nitrogen stable isotope turnover rate of *Oncorhynchus nerka*. Environmental Biology of Fishes 72: 13–18.

Sarakinos HC, Johnson ML and Vander Zanden MJ (2002) A synthesis of tissue-preservation effects on carbon and nitrogen stable isotope signatures. Canadian Journal of Zoology 80: 381–387.

Schaefer KM and Fuller DW (2016) Archival and pop-up satellite archival tags: designs, attachments, data analyses, and applications in studies of large-scale movements of fish. pp. 251–289. *In*: Morais P and Daverat F (eds.). An Introduction to Fish Migration. CRC Press, Boca Raton, FL (this book).

Schielke EG and Post DM (2010) Size matters: comparing stable isotope ratios of tissue plugs and whole organisms. Limnology and Oceanography: Methods 8: 348–351.

Schlacher TA, Liddell B, Gaston TF and Schlacher-Hoenlinger M (2005) Fish track waste water pollution to estuaries. Oecologia 144: 570–584.

Schlechtriem C, Focken U and Becker K (2003) Effect of different lipid extraction methods on $\delta^{13}C$ of lipid and lipid-free fractions of fish and different fish feeds. Isotopes in Environmental and Health Studies 39: 135–140.

Schoeninger MJ and DeNiro MJ (1984) Nitrogen and carbon isotopic composition of bone collagen from marine and terrestrial animals. Geochimica et Cosmochimica Acta 48: 625–639.

Sepulveda AJ, Colyer WT, Lowe WH and Vinson MR (2009) Using nitrogen stable isotopes to detect long-distance movement in threatened cutthroat trout (*Oncorhynchus clarkii utah*). Canadian Journal of Fisheries and Aquatic Sciences 66: 672–682.

Sharp Z (2007) Principles of Stable Isotope Geochemistry. Prentice Hall, Upper Saddle River, New Jersey.

Sierszen ME, Peterson GS and Scharold JV (2006) Depth-specific patterns in benthic-planktonic food web relationships in Lake Superior. Canadian Journal of Fisheries and Aquatic Sciences 63: 1496–1503.

Sierszen ME, Brazner JC, Cotter AM, Morrice JA, Peterson GS and Trebitz AS (2012) Watershed and lake influences on the energetic base of coastal wetland food webs across the Great Lakes basin. Journal of Great Lakes Research 38: 418–428.

Solomon CT, Cole JJ, Doucett RR, Pace ML, Preston ND, Smith LE and Weidel BC (2009) The influence of environmental water on the hydrogen stable isotope ratio in aquatic consumers. Oecologia 161: 313–324.

Solomon CT, Carpenter SR, Clayton MK, Cole JJ, Coloso JJ, Pace ML, Vander Zanden MJ and Weidel BC (2011) Terrestrial, benthic, and pelagic resource use in lakes: results from a three-isotope Bayesian mixing model. Ecology 92: 1115–1125.

Sorensen PW and Hobson KA (2005) Stable isotope analysis of amphidromous Hawaiian gobies suggest their larvae spend a substantial period of time in freshwater river plumes. Environmental Biology of Fishes 74: 31–42.

Steffy L and Kilham SS (2004) Elevated $\delta^{15}N$ in stream biota in areas with septic tank systems in an urban watershed. Ecological Applications 14: 637–641.

Suzuki KW, Kasai A, Nakayama K and Tanaka M (2005) Differential isotopic enrichment and half-life among tissues in Japanese temperate bass (*Lateolabrax japonicus*) juveniles: implications for analyzing migration. Canadian Journal of Fisheries and Aquatic Sciences 62: 671–678.

Tanaka H, Ohshimo S, Takagi N and Ichimaru T (2010) Investigation of the geographical origin and migration of anchovy *Engraulis japonicus* in Tachibana Bay, Japan: a stable isotope approach. Fisheries Research 102: 217–220.

Tieszen LL, Boutton TW, Tesdahl KG and Slade NA (1983) Fractionation and turnover of stable carbon isotopes in animal tissues: Implications for $\delta^{13}C$ analysis of diet. Oecologia 57: 32–37.

Tronquart NH, Mazeas L, Reuilly-Manenti L, Zahm A and Belliard J (2012) Fish fins as non-lethal surrogates for muscle tissues in freshwater food web studies using stable isotopes. Rapid Communications in Mass Spectrometry 26: 1603–1608.

Trueman CN, McGill RAR and Guyard PH (2005) The effect of growth rate on tissue-diet isotopic spacing in rapidly growing animals. An experimental study with Atlantic salmon (*Salmo salar*). Rapid Communications in Mass Spectrometry 19: 3239–3247.

Trueman CN, MacKenzie KM and Palmer MR (2012) Identifying migrations in marine fishes through stable-isotope analysis. Journal of Fish Biology 81: 826–847.

Trust BA and Fry B (1992) Stable sulphur isotopes in plants: a review. Plant, Cell & Environment 15: 1105–1110.

Vander Zanden MJ and Rasmussen JB (2001) Variation in $\delta^{15}N$ and $\delta^{13}C$ trophic fractionation: Implications for aquatic food web studies. Limnology and Oceanography 46: 2061–2066.

Vander Zanden MJ, Hulshof M, Ridgway MS and Rasmussen JB (1998) Application of stable isotope techniques to trophic studies of age-0 smallmouth bass. Transactions of the American Fisheries Society 127: 729–739.

Wada E and Hattori A (1978) Nitrogen isotope effects in the assimilation of inorganic nitrogenous compounds. Geomicrobiology Journal 1: 85–101.

Wainright SC and Fry B (1994) Seasonal variation of the stable isotopic compositions of coastal marine plankton from Woods Hole, Massachusetts and Georges Bank. Estuaries 17: 552–560.

Wassenaar LI and Hobson KA (2000) Improved method for determining the stable-hydrogen isotopic composition (δD) of complex organic materials of environmental interest. Environmental Science & Technology 34: 2354–2360.

Wassenaar LI and Hobson KA (2003) Comparative equilibration and online technique for determination of non-exchangeable hydrogen of keratins for use in animal migration studies. Isotopes in Environmental and Health Studies 39: 211–217.

Weidel BC, Carpenter SR, Kitchell JF and Vander Zanden MJ (2011) Rates and components of carbon turnover in fish muscle: insights from bioenergetics models and a whole-lake ^{13}C addition. Canadian Journal of Fisheries and Aquatic Sciences 68: 387–399.

Xu J, Yang Q, Zhang M, Zhang M, Xie P and Hansson LA (2011) Preservation effects on stable isotope ratios and consequences for the reconstruction of energetic pathways. Aquatic Ecology 45: 483–492.

CHAPTER 11

Use of Drift Nets to Infer Fish Transport and Migration Strategies in Inland Aquatic Ecosystems

*Michal Janáč** and *Martin Reichard*[a]

Principles of drift nets' use

Drift nets are stationary nets designed to capture drifting organisms, i.e., those transported via water current. For the purposes of this chapter, drift is understood as the transport itself, in which along-current movement of an organism is maintained by the current alone and not at the expense of the fish's energy reserves. This will be independent of whether (1) it results from passive dislodgement and ends with passive deposition (Pavlov 1994), or (2) plays a part in an active migration strategy (Hare et al. 2005). It is worth noting that the term 'drift nets' is also used for coastal gill nets, where the nets themselves drift with the current (FAO 2013). Drifting gill nets work on a substantially different principle and this chapter deals only with stationary drift nets.

In fact, in its broadest sense, any stationary device capturing drifting organisms could be considered a drift net. Thus, nets usually towed in order to sample ichthyoplankton (plankton, ring or bongo nets) can also serve as drift nets when set in a stationary position. Both rotary screw traps, used to sample downstream migrating juvenile salmonids (e.g., Johnson et al. 2005), and anchored stow nets, used for commercial catches in tidal zones, work on similar principles to drift nets. Drift pumps (Gale and Mohr 1978; Dahms and Qian 2004), which mechanically pump water from

Institute of Vertebrate Biology, Academy of Sciences of the Czech Republic, v.v.i., Květná 8, 603 65 Brno, Czech Republic.
[a] Email: reichard@ivb.cz
* Corresponding author: janac@ivb.cz

the current, thereby avoiding problems with drift net clogging (see below), also serve to capture drifting organisms. However, drift pumps carry a potential bias because they may also capture non-drifting organisms present in the current.

In its basic form, a drift net consists of a frame, a tapering net attached to the frame and (optionally) a collecting jar attached to the cod end of the net (Fig. 11.1). The mouth of a drift net is installed perpendicularly to the current, such that the water and the drifting organisms flow through the net mouth. Particles carried by the current are trapped by the mesh and moved towards the cod end by current pressure. After an allotted time, the organisms collected are either picked individually from the cod end or collected in a jar, or washed from the nets into a collection jar or examination basin.

In order to ensure drift nets stay in their chosen position, they need to be anchored. Anchoring points may comprise the river bottom or banks, a boat, pontoons, bridges, piers or an anchor. In shallow water (< 1 m depth), iron rods hammered into the bottom are used both to anchor the nets and to maintain the optimal position, i.e., mouth facing perpendicular to the current. In deeper water, drift nets may be attached at a stationary point by more flexible means (e.g., by rope) using a system of weights to stabilize its position.

Although towed plankton nets had been used to sample early life stages of fish since 1828 (Kelso and Rutherford 1996), stationary sampling nets, developed for sampling drifting freshwater invertebrates, were not used for another century (Needham 1928). Subsequently, the study of drift has become a major field in macroinvertebrates studies (see Waters 1972; Brittain and Eikeland 1988; Svendsen et al. 2004 for reviews). The first studies using stationary drift nets specifically design for fish collection appear around the middle of the 20th century (e.g., Brett 1948; Wolf 1951); using drift nets only slightly modified from those for macroinvertebrates (e.g., a larger mesh size). Today's nets differ little from these, though some modifications have been proposed for specific purposes. The attachment of a flow meter, either at the mouth

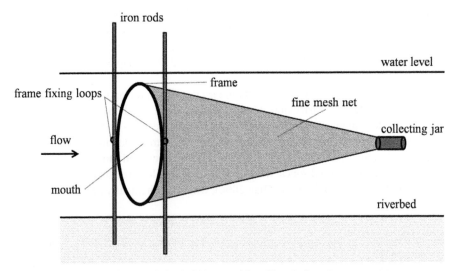

Figure 11.1. Schematic diagram of a basic drift net positioned in a shallow river.

or at the middle of the net, is probably the most common deviation from the basic scheme (Gale and Mohr 1978). In addition, many studies varied the shape of the net and/or the mesh size (see the subsection 'Net design'). Schmutz et al. (1997) proposed a sampler that could automatically collect six samples in temporal succession; allowing sampling over relatively short intervals without the presence of an operator. As fish drift mostly at night, this provides much appreciated relief in otherwise fatiguing work. The sampler works on the principle of six compartments rotating within a frame construction, with all nets except the lowest remaining closed during each interval (see Müller 1966 for macroinvertebrate drift samplers working on similar principles). However, the use of such samplers is restricted to shallow waters (Schmutz et al. 1997). Oesmann (2003) modified drift nets such that they can be opened and closed while under water, thereby allowing quantitative sampling at distinct depth strata. The method uses the principle of net collapse during retrieval to prevent contamination from other strata, as in the approach of Nester (1987). In order to study fish larval movements in an estuary, Graham and Venno (1968) ensured that their drift nets faced into the tidal current by attaching the nets to vanes attached to a line buoyed at the surface and anchored to the bottom. Similarly, a system of two buoys and two anchors allowed D'Amours et al. (2001) to set and retrieve multiple, vertically-stratified nets against the flow of a river. Finally, Hare et al. (2005) were able to continuously observe drifting larvae in an estuary by alternating two sets of vertically-stratified nets attached to a cable deployed from the stern of a ship.

Generally, drift nets are specifically designed to sample waters with unidirectional flow and are set at stationary points with easy access; hence they are mainly used in rivers. Several studies, however, have used anchored neuston (Lindsay et al. 1978; Hettler 1979), plankton (Graham and Venno 1968; Bradbury et al. 2004) or channel nets (Lewis et al. 1970; Hare et al. 2005) to sample ichthyoplankton drifting with ebb or flood tides in estuaries.

Despite well-developed procedures for sampling marine ichthyoplankton and the known importance of drift for young marine fish, the use of drift nets (in the sense described above) is rare in the marine environment. Most data on marine drifting fish has been obtained using towed nets, as the marine environment rarely provides conditions of unidirectional flow and stationary points. Indeed, only in exceptional circumstances stationary nets are installed in seawaters, namely in polar waters where nets may be attached to the ice crust and submerged below a seal hole (Sewell et al. 2008), or the installation of plankton nets in shallow coastal-waters (e.g., reef flats; Hendler et al. 1999). Crest nets are stationary nets fixed to the substrate immediately behind the crest of a reef, which sample reef-fish larvae migrating from their nurseries (e.g., mangroves or seagrass beds) to their reef settlement areas, which may also be considered as an example of marine use of drift nets (Dufour and Galzin 1993; Nolan and Danilowicz 2008).

How drift nets are used to infer fish movement

All that we can be confident about when sampling organisms captured in drift nets is that the organisms were being carried by the current at the time of capture. Any information on where, when, why and how fish larvae started drifting, or where drift would have terminated if they had not been captured in the drift nets, usually remains hidden. Only through carefully designed studies, in which drift net sampling is accompanied with other sampling methods, measurement of environmental variables and/or modelling, can we infer more detailed information on movement of fish larvae. A number of different reasons have been put forward as to why fish larvae drift; however, few studies have provided definitive proof.

For example, larvae of some species that use estuaries as nursery areas might selectively (i.e., actively) use tidal streams to enter, remain in, or leave an estuary (Hare et al. 2005). Active processes are also connected with drift of freshwater fish larvae. Studies by Robinson et al. (1998), Wolter and Sukhodolov (2008) and Schludermann et al. (2012) demonstrated that at least some of the larvae being carried in a river current do not follow the same paths as passive particles, and that these larvae are capable of (1) actively leaving the current, despite their poor swimming ability, and/or (2) actively choosing currents that would distribute them to a suitable habitat.

In contrast, drift of some organisms may be part of a completely passive process. Drifting eggs, for example, are both passively dislodged and passively distributed. Eggs of lithophilic fish (fish typically spawning on fast-flow gravel beds) (Balon 1975) may be swept by the current from gravel beds and drift for hundreds of metres before becoming adhesive enough to attach to the bottom (Hofer and Kirchhofer 1996). Pelagophilic fish, on the other hand, spawn directly into the water and the eggs can drift for hundreds of kilometres. These eggs develop into larvae during the journey, whereupon some active component may be involved in its further distribution (Jiang et al. 2010). Increased water velocity during times of elevated discharge will often sweep eggs and fish larvae away from shelters, resulting in so called 'catastrophic drift'. Following passive dislodgement, catastrophic drift is also intuitively considered to result in passive distribution, though no study to date has dealt with this question. Larvae may also 'passively' enter a current as a consequence of losing orientation in darkness (Pavlov 1994) or through some other accidental means (*sensu* 'background drift' according to Brittain and Eikeland 1988).

The degree to which active processes contribute to migration of fish larvae and juveniles in most species/age groups is presently unknown. Rather, studies have tended to combine drift net sampling with knowledge of fish life-histories in order to hypothesize on the importance of drift in different fish species/age groups.

The capture of small, early-stage larvae in drift nets set below spawning grounds indicates post-hatching migration. In cases where spawning grounds cannot serve as nurseries, such post-hatching drift is hypothesized as ensuring distribution of early-stage larvae into suitable nurseries. Fast-flow gravel beds, for example, provide optimal oxygen concentrations for eggs of lithophilic fish; however, high water velocity and low food availability make this habitat less than suitable for the larvae. In this case, drift is hypothesized as ensuring movement to slower and richer river stretches (Hofer and Kirchhofer 1996). Post-hatching drift has been reported widely for numerous species, e.g., in potamodromous percids (Priegel 1970), coregonids (Naesje et al. 1986), salmonids (Bardonnet et al. 1993) and cyprinids (Persat and Olivier 1995), as well as anadromous osmerids (Bradbury et al. 2004), lampreys (Harvey et al. 2002) and clupeids (O'Connell and Angermeier 1997). Long distance spawning migrations are common in these species and larval drift downstream is sometimes considered as compensating for the long distances travelled upstream by adult spawners.

Motivation for post-hatching drift is most apparent in amphidromous gobiids, cottids and galaxiids (Iguchi and Mizuno 1990; McDowall 2007), whose larvae hatch in freshwater. These larvae are incapable of feeding in freshwater and have very limited swimming ability; hence, it is essential that larvae of these species reach marine waters within a few days of hatching (Iguchi and Mizuno 1999).

Drift samples also commonly contain (1) newly-hatched larvae of species that attach eggs to nearshore structures, vegetation or in cavities (i.e., phytophilic, phyto-lithophilic, speleophilic or ostracophilic fish *sensu* Balon 1975) and whose spawning grounds are often close to, or identical to, their nurseries, and (2) late larvae/early juveniles (Pavlov 1994; Reichard and Jurajda 2007). Explanations for their presence vary widely and might include accidental dislodgement, high population density, or habitat shift related to ontogenetic development (Pavlov 1994; Zitek et al. 2004; Reichard and Jurajda 2007).

Drift nets, or their equivalents, are also used to track anadromous post-smoltification migrations of juvenile salmonids and the analogous movement of young acipenserids, although such movements are not completely passive (i.e., fish swim with the current) (Pavlov 1994).

Attaching a drift net to the outlet of a power-generating facility fed by a reservoir, lake or river can not only provide information on fish transfer between two water bodies but also information on young fish movement into the pelagic or benthic zones of the initial water body, depending on the vertical position of the outlet (Kelso and Leslie 1979; Baruš et al. 1986; Carter and Reader 2000). Note that this is independent of whether fish are actively following currents, as in migratory young salmonids and acipenserids (Coutant and Whitney 2000), or are passively entrained as a consequence of crossing in front of a water intake during diel vertical or horizontal migrations (Pavlov et al. 2002). Scientists often additionally record the amount of damage caused to fish by pressure changes, turbine passage or overheating when studying entrainment through power generating facilities (Cada 1991; Carter and Reader 2000).

How to prepare samples using drift nets

Net design and exposure time

Drift nets are designed to capture small organisms. The mesh size used is generally determined by the size of organism being studied. A very fine mesh (64 μm) was used by Iguchi and Mizuno (1990) when capturing larvae as small as 1 mm total length; however, most studies use a 400–500 μm mesh, as recommended by Schmutz et al. (1997). Though, a mesh as large as 1 mm can still effectively sample eggs and fish larvae as small as 6 mm standard length (Copp et al. 2002).

The choice of appropriate mesh size should be carefully considered. Too large a mesh may damage the smallest fish or let them through (Schmutz et al. 1997), while too fine a mesh may prove effective at catching coarse particulate organic matter (POM), which will lead to gradual clogging of the net. Clogging results in backflow, and thus biasing drift density estimates derived from the volume of water filtered (see below). Measuring flow velocity both at the start and end of net exposure provides an adequate estimate of the net clogging effect (with visual inspection providing a rough estimate, naturally). If clogging causes considerable backflow, one can calculate the actual volume of water filtered based in water velocity change. Note, however, that water velocity in gradually clogged nets does not change linearly with exposure time (Faulkner and Copp 2001).

The effects of backflow can be reduced by lengthening the net; however, longer nets are more difficult to handle. The net:mouth area ratio rarely exceeds 6:1 (Schmutz et al. 1997), and a ratio higher than 5:1 is usually considered to provide sufficient efficiency (Kelso and Rutherford 1996; see also Gale and Mohr 1978 for inspiration on net design). Note that reducing the net mouth area (see Elliott 1970) provides the same effect of increasing relative net length, but it will also decrease the volume of water filtered.

Iguchi and Mizuno (1990) constructed their drift net with two mesh sizes, with a fine mesh at the cod end and a coarser mesh near the mouth, allowing coarse POM to be retained in the mouth of the net (Schmutz et al. 1997), while the fine mesh at the cod end would reduce the risk of fish damage.

The negative effects of clogging on drift-net efficiency can be effectively resolved in steep rivers and above weirs or waterfalls, by replacing the net with a horizontal plastic tube (Elliott 1970). This tube feeds water into a net positioned under the tube end outside of the waterbody. Thus, net clogging does not create backflow in the tube. Water volume is calculated based on tube width and exposure time. However, the use of such mechanisms is limited by the habitats in which they can be used and, to date, they were only used for macroinvertebrate drift studies (e.g., Kubíček 1966).

A range of factors, including net clogging, compel scientists to compromise on ideal sampling effort. Exposure time, for example, varies widely between studies, ranging from 10 minutes to 12 hours, with 15–30 minutes being most common in areas containing higher amounts of coarse POM. Short exposure times (lower water volume filtered) lead to an increased probability of missing less common species; while long net exposures increase the probability that captured fish will be damaged and unidentifiable (Schmutz et al. 1997) and prevents determination of diel drift patterns at fine scales (see Elliott 1970 for accompanying problems). Moreover, high volumes of water filtered increase the amount of coarse POM retained in the net, increasing both clogging and subsequent manipulation time due to the "painstaking separation of the sample from organic and mineral debris" that follows each sampling (Faulkner and Copp 2001). Although immediate separation of the sample in the field is less comfortable than sorting preserved samples in the laboratory, it is generally preferred for a number of reasons, including reduction of preservation medium used, reduction of collateral damage to animals not of interest to the study, and, most importantly, it saves time as "living, moving and naturally pigmented animals are easier to spot amongst the debris than dead animals" (Copp et al. 2002).

Based on our experience, a shallow white basin (40×30×5 cm) originally designed for photographic development is suitable for sorting samples of live drifting organisms. Aliquots are poured into the basin in volumes sufficient for detection of small organisms, but also depending on the amount of coarse POM and water transparency. Then, fish larvae and juveniles are individually removed using tweezers. Small plastic bulb pipettes are extremely useful for removing the smallest fish larvae. The sorted fish are then sacrificed (e.g., by an overdose of anaesthetic) and preserved (4% buffered formaldehyde is widely used). Kelso and Rutherford (1996) provided a detailed discussion on preservation media for fish larvae. When captured eggs and/or larvae are too young to be identified, it is sometimes possible to hatch/raise part of a sample for later determination of more developed stages.

In general, net design and exposure time will depend on the size and density of organisms under study and on the amount of coarse POM. Final net design (shape, size, mesh size) will always represent a compromise between filtration efficiency, clogging rate, sample sorting time (Svendsen et al. 2004) and ease of net manipulation.

Temporal aspects of sample preparation

Seasonal aspects

The time of collection will be directly related to the assumed time of fish migration. For most fish species, drift occurs in the earliest life stages and is, therefore, a function of when fish spawn (Brown and Armstrong 1985). Indeed, inter-annual variability in drifting fish assemblages is usually attributed to variability in spawning success (Robinson et al. 1998; Reichard et al. 2002a), though some studies proposed that propensity to drift is density-dependent and can change between years within the same population (Economou 1991).

Strict seasonal patterns enable the planning of sampling campaigns with a precision of a few weeks, with higher precision attainable on the basis of river discharge and water temperature during the spawning season. Water temperature may speed up or slow down both spawning season and egg and larval development, and thus also timing of drift. Notably, drift seasonality in coregonids and salmonids may also be driven by changes in discharge rate (Naesje et al. 1986; Johnston et al. 1995). In general, the first sample of the seasonal cycle should precede the expected start of drift, with two consecutive negative sampling sessions usually taken as an indication that the drift season has come to an end.

Sampling frequency over the course of a season will depend on the aim of the study. Daily intervals (or nightly, as drift occurs mostly at night; see below) intuitively provide the most precise picture of seasonal drift pattern. Daily sampling can be too demanding, especially when studying partial spawners, which appear in drift net samples for long periods. Weekly sampling intervals (approximately) are usually chosen, representing a compromise between precision and effort, but sufficient to record the most important seasonal peaks in drift. Long-term studies that observe changes in drifting fish assemblages over the course of a whole year, for example, may set even longer sampling intervals of up to one month, usually with increased sampling frequency during the periods when drift is expected to peak.

Diel aspects

As larval drift occurs mostly at night, it is usually sufficient to collect samples during periods of darkness. However, further sampling may be necessary if diel movement patterns are the main objective of the study. Presence of diel drift patterns in streams is linked to water transparency, with most studies reporting no diel pattern in turbid rivers (Secchi disc transparency lower than 30 cm) (e.g., Pavlov 1994; Pavlov et al. 1995) and nocturnal drift in the vast majority of rivers with transparency higher than 30 cm (Pavlov 1994; see Reeves and Galat 2010 for review). Diel periodicity is not bound strictly to transparency, since species-specific diel patterns have been noted at the same sites (Robinson et al. 1998; Reeves and Galat 2010). Stream morphology may also play a significant role as observed by Iguchi and Mizuno (1990), which reported relaxation of otherwise strictly nocturnal goby diel drift patterns in steep, fast-flowing water courses where fish were likely unable to sustain the water current.

Whole-night, or even 24 hours, observation is tiring. Persat and Olivier (1995) and Zitek et al. (2004) suggested that a single night sample, collected during the first two hours of darkness, may be sufficient to predict drift density for the entire 24 hours. Indeed, a number of studies have reported peak drift density during the first hours of darkness (e.g., Reichard et al. 2002b). Some studies, however, have also noted a second peak just before sunrise (Brown and Armstrong 1985) or a single peak in the middle of the night (Naesje et al. 1986). In all these cases, those samples taken in the two hours after sunset appeared to be reliable predictors of overall drift density for the whole night. Note that a single sample may be insufficient where fish size is of concern, as the size of drifting fish may vary over the course of the night (Sonny et al. 2006, but consult Janáč et al. 2013b for contradictory results).

Spatial aspects of sample preparation

Longitudinal aspects

Longitudinal positioning of drift nets will depend upon the purpose of the study, as in the following examples. When studying drift of young fish from a reservoir, nets are placed at, or close by, the outlet(s) (Pavlov et al. 2002). Alternatively, multiple drift nets positioned across a stream immediately before its confluence with a main stem river, lake, reservoir or the sea will provide a reasonable measure of the amount of fish contributed (Franzin and Harbicht 1992). When the aim is to document the journey of newly hatched fish to the sea, or a lake, the best results are obtained by a longitudinal series of sampling points from the uppermost spawning site to the river's mouth (Priegel 1970).

Lateral and vertical aspects

The majority of studies agree that riverine fish drift is observed primarily in relatively shallow zones near the shore (Reichard et al. 2004). Shallow, nearshore habitats are also the most easily accessible and usually do not demand great effort or invention for net anchoring. It seems reasonable, therefore, to focus on these habitats when preparing drift net samples. One should be aware that spatial drift patterns can differ between species (Brown and Armstrong 1985; Pavlov 1994; Oesmann 2003), age groups (Gale and Mohr 1978; Reichard et al. 2004) and river morphology. Concerning the last topic, Pavlov et al. (2008) noted that "velocity gradients at river channel bends can drastically redistribute fish larvae drifting downstream".

There are two major patterns of vertical distribution when studying drift outside of the shallow zone, namely (1) prevalence of surface drift (Brown and Armstrong 1985; Oesmann 2003), or (2) homogenous distribution throughout the water column (Carter et al. 1986; Franzin and Harbicht 1992). Gale and Mohr (1978) and D'Amours et al. (2001) observed surface drift dominating only during the night, with bottom drift dominating during the day. Graham and Venno (1968), on the other hand, found larvae in surface nets during the ebb tide, while bottom catches prevailed during the flood tide. Notably, some studies suggest that lateral and/or vertical distribution of drifting fish in rivers is more dependent of flow velocity gradients along the transverse profile, rather than the gradient of distance from margin or surface (Copp et al. 2002; Lechner et al. 2013).

How to interpret and analyze data obtained from drift nets

In common with the general trend in ecological studies, statistical analysis of drift has evolved greatly since the 1960's (Svendsen et al. 2004); though the tools available for basic description of drift net samples remain the same. In general, two approaches are used to describe drift net yield: drift rate and drift density (Elliott 1970).

Drift rate represents a simple count of individuals captured in a standardized net per unit time, but it is rarely used since it is highly correlated with discharge. Intuitively, drift rate will increase with increasing discharge as a larger amount of water will carry more drifting individuals. In exceptional cases, so far only observed in invertebrates, drift rate can increase with decreasing discharge in response to stream desiccation (Elliott 1970).

Drift density, expressed as the number of individuals per volume of water filtered through the net, is generally agreed to be the most useful measure for quantifying drifting fish. The volume of filtered water (m^3) can be easily calculated as the area of submersed net mouth (m^2) multiplied by flow velocity in the mouth ($m.s^{-1}$) and exposure time (s). In reality, this calculation represents no more than an approximation, as water velocity varies both spatially (within the net mouth) and temporally. However, this approximation is generally acceptable and two measurements of water velocity per sample are considered acceptable. Measurements are usually conducted at the start and end of each sample, using portable flow meters positioned at the centre of the net mouth. Occasionally, high temporal variability in current velocity may call for intermediate measurements. In such cases, flow meters permanently attached to the nets allow continuous measurement of flow velocity and more accurate calculation of filtered water volume.

For decision makers, it is often important to know the total amount of drifting fish, namely when drift is used as an early measure of year-class strength (D'Amours et al. 2001), or when quantifying number of fish lost from a water body via entrainment into the inlet of a power-generating facility (Kelso and Leslie 1979). Estimates of total number of drifting fish are relatively rare; as they must contend with (1) spatial variability over the transverse profile of the river/outlet, which can be very high, especially in large rivers, and relatively low in reservoir outlets; (2) diel variability; (3) missing data from not sampled dates (continuous sampling is rarely conducted throughout the drift season); and (4) dependence of number of drifting individuals on river discharge.

In order to deal with high spatial heterogeneity, a pilot study should be conducted in order to reveal zones with different drift densities or, alternatively, density gradients. The subsequent full-scale monitoring study should then have drift nets situated in each of the zones observed, allowing estimation of drift density for the whole transverse profile. The total estimate is obtained by averaging the density from each zone, weighted by the proportion that each zone contributes to the total area of the transverse profile. Diel variability can be treated in a similar way, though an approximation to simple day and night densities and duration may be used. Drift density for intervening days (i.e., missing data) can be obtained by interpolation, while diel discharge values can usually be obtained from gauging stations. The following formula can then serve for calculation of total number of drifting fish over a 24 hour interval:

$$N = Q \sum_{i=1}^{n} T_i \times d_i$$

where Q is the daily average discharge (m^3 s^{-1}), T$_i$ is the duration of period i within 24 hours (s), d$_i$ is the drift density during period i (individuals m^{-3}) and, commonly n is 2, corresponding to the chosen periods (day and night). Note that, even after this process, the estimate of total number of drifting fish will still have a large degree of uncertainty, due to the number of approximations involved in the process.

By sampling drifting young fish and the source assemblages of young fish concurrently (typically encountered in nearshore areas), the propensity of particular fish species (or developmental stage/size categories) to drift can be revealed. For example, Reichard and Jurajda (2007) calculated a drift index (E) based on relative abundance (RA; % of total number of fish sampled in drift or in nurseries):

$$E = \frac{(\text{RA in drift} - \text{RA in nurseries})}{(\text{RA in drift} + \text{RA in nurseries})}.$$

Drift indices such as these will be influenced by bias inherent in the sampling gear used to sample the source assemblages, which will usually consist of species with different habitat preferences (e.g., nearshore pelagic, nearshore demersal and shelter-seeking fish larvae and juveniles).

Statistical treatment

Fish abundance data obtained from drift nets will rarely follow a normal distribution. Traditionally, drift data were analyzed using ANOVA (following data transformation) or non-parametric (e.g., Kruskal-Wallis) tests. Generalized Linear Models (GLMs), designed specifically to cope with non-normally distributed data, are now widely available and we strongly recommend that GLMs should be used to analyze abundance data from drift nets. Numbers of captured fish usually follow a Poisson distribution. However, drift density values are not integers, being standardized to volume of filtered water. In this case, the most appropriate option is using a Poisson distributed GLM on pure count data (i.e., numbers of individuals) with water volume set as an offset parameter (Zuur et al. 2009).

Data obtained from drift nets are frequently correlated, as the sampling design often involves repeated sampling over time (e.g., when assessing seasonal and/or diel patterns) using several nets distributed vertically and horizontally. Variables representing correlation structure (e.g., replicated samples from the same cross-section) often represent 'nuisance variables', which should be modelled as random factors. Several other approaches have been used to overcome the 'problem' of correlated drift net data. If a test of correlation structure undertaken prior to analysis reveals only a weak correlation, for example, data non-independence may be omitted, as outlined by D'Amours et al. (2001). Alternatively, the effect of repeated sampling can be removed by data standardization within a sampling unit (see Janáč et al. 2013a). Such approaches may raise criticism, often rightly, as they may be easily biased or be heavily dependent on subjective criteria. Overall, incorporating random factors into the model appears to be the most appropriate solution for dealing with correlated data; hence, we recommend mixed GLMs or their alternatives (e.g., GLMs with generalized estimation equations) when analyzing drift abundance data (see Zuur et al. 2009).

The effect of abiotic factors (e.g., discharge, temperature, turbidity) on drift density has traditionally been studied using correlations; but they can just as easily be studied using GLMs. In fact, as these abiotic factors are known to influence drift density, incorporating them as covariates into models dealing with drift abundance is often advisable, at least at the stage of model construction.

When size of drifting fish is of concern, two statistical approaches have commonly been applied. Some studies treat fish size as a response variable and use GLMs or traditional Kolmogorov-Smirnov tests, while others tend to categorize fish size into distinct groups and the abundance of the newly established 'pseudo-species' are compared using contingency tables or GLMs.

Multivariate methods are rarely used in the analysis of fish drift samples. However, Zitek et al. (2004) used cluster analysis to distinguish between different spawning events (repeated occurrence of the same size group in drift during a season) and the occurrence of later developmental stages; while Oesmann (2003) used canonical correspondence analysis to explain variability in drifting fish assemblages through environmental variables. Non-parametric multidimensional scaling based on ecological distance matrices (e.g., Bray-Curtis) can also be used to visualize similarities in drift assemblage between different sites or dates.

In general, it is reasonable to compare densities of drifting fish within a river (e.g., when studying seasonal or diel drift patterns or differences between sites). Inter-stream comparisons of drift density are more questionable, mainly due to low representativeness of samples taken from larger rivers where just a small proportion of the river can be sampled. On the other hand, comparisons between different streams are reasonable when comparing assemblage composition of drifting fish, fish propensity to drift or temporal and spatial drift patterns. Some studies have described drift patterns for the entire assemblage of drifting fish (species pooled rather than separated); however, this is not a good practice as spatial and temporal drift patterns and propensity to drift have proven to be species-specific, at least in some cases (Brown and Armstrong 1985; Reichard and Jurajda 2007). Therefore, we urge that analysis should be undertaken at the species or pseudo-species level, whenever possible.

Case studies using drift nets

Drift nets have been used to document stream fish drift worldwide, having been used in rivers of Amazonia (Pavlov et al. 1995), Australia (Humphries et al. 2002), South Asia (de Graaf et al. 1999), China (Jiang et al. 2010), Japan (Iguchi and Mizuno 1990), Russia (Pavlov et al. 1977), Europe (Zitek et al. 2004) and northern America (Gale and Mohr 1978), with most studies taking place in the latter three.

In a series of studies, the drift of young fish was monitored in large rivers of Europe, Asia and Amazonia (see Pavlov 1994 for a review). The numerous species captured during the studies documented the widespread occurrence of drift, confirming that it is not restricted to anadromous fishes but occurs also in many strictly freshwater fish, including not only potamodromous fish but also 'stationary species' with limited adult home ranges. Based on the results of these studies, Pavlov (1994) identified a range of mechanisms that influence how fish enter a current, how they react once in the current and how they orientate themselves once there. According to Pavlov (1994), drift primarily results from relaxation or reversing of various retention mechanisms (e.g., negative phototaxis, positive thigmotaxis, shelter seeking or rheoreaction) that have evolved in riverine fish larvae to keep them out of stronger currents.

Brown and Armstrong (1985) used drift nets to document drift of approximately 60 species in the Illinois River (USA). This thorough study was one of the first to describe basic seasonal, diel, lateral and vertical patterns in fish drift, and to demonstrate that drift is a function of when species spawn, a prevailing night-time drift pattern, and preferences for nearshore and surface drift. In combination with concurrent samples of resident larval fish, drift net samples suggested that some species that were abundant in the river may be able to resist drift.

Kennedy and Vinyard (1997) used drift nets to demonstrate drift avoidance in larvae of the small catostomid *Catostomus warnerensis*, hypothesizing that the species evolved 'drift resistance' in response to unreliability of habitats downstream. The almost complete absence of the species in drift nets, reinforced by direct observation of drift cessation in released larvae, supported the authors' hypothesis.

More recently, Schludermann et al. (2012) tested the hypothesis that there may be an active component in the drift of potamodromous common nase *Chondrostoma nasus* larvae in the River Danube. The combination of (1) a hydrodynamic model tracing transport of passive particles, and (2) drift net samples of released larvae, marked with alizarin red, revealed that larvae were not distributed in a completely passive manner.

Priegel (1970) used drift nets to document the early life-cycle stages of the potamodromous walleye *Sander vitreus*, concluding on the necessity of larvae leaving their marsh hatching grounds for lake habitats with richer food sources within three–five days. By releasing large numbers of marked (coloured dye) walleye fry and installing drift nets at several control points along the river, Priegel (1970) calculated the rate of drift spread, proving the ability of small walleye to drift 70 km within just two days.

Iguchi and Mizuno (1999) installed drift nets in several Japanese coastal streams at varying distances from the spawning grounds of the amphidromous goby *Rhinogobius brunneus*. Captures demonstrated that fish travelling further distances were in significantly worse condition than those travelling shorter distances. Thus, the significance of swift transport to the sea (larvae were incapable of feeding in freshwaters), and the related limited reproductive success observed in areas furthest from the sea, contributed greatly to knowledge of amphidromous gobiid life-cycles.

By installing drift nets at the inflow and outflow of a coal-fired power station cooling system, Carter and Reader (2000) estimated the density of fish entrained by the inlet canal from a nearby river, and concluded that all larvae die after passing through the cooling system. Concurrent drift net sampling in the river revealed that species composition and diel and seasonal patterns of drifting fish assemblages corresponded to those entrained at the inlet, confirming that the entrained fish larvae originated mostly from river drift.

Drift nets installed just below the outlet of a shallow, lowland reservoir allowed Janáč et al. (2013a) to monitor diel and seasonal changes in the assemblage of young fish leaving the reservoir through the turbine of a hydropower facility. The study showed that passage of non-native tubenose goby *Proterorhinus semilunaris* through the turbine allowed further downstream spreading of the species. Indeed, they estimated that approximately 0.5 million young tubenose gobies passed into the river, with only a 3% suffering significant damage.

Graham (1971) used drift nets to document the routes taken by Atlantic herring *Clupea harengus* larvae during ebb and flood tides within an estuary. By installing two vertically stratified sets of drift nets at both the landward and seaward ends of the estuarine channel, Graham (1971) revealed differences in the vertical distribution of larvae carried by tidal currents, and thus demonstrating "a system of larval movement that retains the larvae within the estuary". The suggested retention mechanism consists of (1) up-estuary movement via flow near the bottom, (2) upward movement through the water column upon reaching the limit of up-estuary movement, (3) down-estuary movement via surface flow, and (4) descent again into the up-estuary bottom flow.

Hare et al. (2005) used drift nets to sample larvae of several fish species (mostly *Micropogonias undulatus*, *Brevoortia tyrannus* and *Paralichthys dentatus*) migrating from continental shelf spawning sites to estuarine nurseries. With the addition of extensive measurement of physical variables, the authors tested several hypotheses regarding larval ingress mechanisms, including their relative importance. Tidally driven ingress was of particularly high importance, with higher larval concentrations present during up-estuary flood tides compared to low concentrations during down-estuary ebb tides. The authors rejected several hypotheses connecting tidally driven larval flux with purely physical processes (e.g., tidal change in water column density or vertical mixing), thus supporting the hypothesis that tidally driven larval flux results from active behaviour.

In general, knowledge of larval and juvenile migration still remains relatively limited and even purely descriptive studies are desirable. Future research on the migration of early fish life stages should include controlled experiments that test hypotheses originating from descriptive studies (Reichard et al. 2002b; Faria and Gonçalves 2010) and from individual based models (Peck et al. 2009).

Acknowledgements

The authors would like to thank K. Roche and P. Morais for their constructive comments on an earlier version of the chapter. During the writing of this text, both MJ and MR were supported by the Czech Science Foundation, Project No. P505/12/G112: ECIP.

References

Balon EK (1975) Reproductive guilds of fishes—proposal and definition. Journal of the Fisheries Research Board of Canada 32: 821–864.

Bardonnet A, Gaudin P and Thorpe JE (1993) Diel rhythm of emergence and of 1st displacement downstream in trout (*Salmo trutta*), atlantic salmon (*S. salar*) and grayling (*Thymallus thymallus*). Journal of Fish Biology 43: 755–762.

Baruš V, Gajdůšek J, Pavlov DS and Nezdolij VK (1986) Downstream fish migration from the Mostiště and Věstonice reservoirs (ČSSR) in the spring-summer period. Folia Zoologica 35: 79–93.

Bradbury IR, Campana SE, Bentzen P and Snelgrove PVR (2004) Synchronized hatch and its ecological significance in rainbow smelt *Osmerus mordax* in St. Mary's bay, Newfoundland. Limnology and Oceanography 49: 2310–2315.

Brett JR (1948) The design and operation of a trap for the capture of migrating young sockeye salmon. Transactions of the American Fisheries Society 75: 97–104.

Brittain JE and Eikeland TJ (1988) Invertebrate drift—a review. Hydrobiologia 166: 77–93.

Brown AV and Armstrong ML (1985) Propensity to drift downstream among various species of fish. Journal of Freshwater Ecology 3: 3–17.

Cada GF (1991) Effects of hydroelectric turbine passage on fish early life stages. pp. 318–326. *In*: Darling DD (ed.). Waterpower '91: A New View of Hydro Resources. American Society of Civil Engineers, New York.

Carter JG, Lamarra VA and Ryel RJ (1986) Drift of larval fishes in the upper Colorado River. Journal of Freshwater Ecology 3: 567–577.

Carter KL and Reader JP (2000) Patterns of drift and power station entrainment of 0 + fish in the River Trent, england. Fisheries Management and Ecology 7: 447–464.

Copp GH, Faulkner H, Doherty S, Watkins MS and Majecki J (2002) Diel drift behaviour of fish eggs and larvae, in particular barbel, *Barbus barbus* (L.), in an English chalk stream. Fisheries Management and Ecology 9: 95–103.

Coutant CC and Whitney RR (2000) Fish behavior in relation to passage through hydropower turbines: a review. Transactions of the American Fisheries Society 129: 351–380.

Dahms HU and Qian PY (2004) Drift-pump and drift-net-two devices for the collection of bottom-near drifting biota. Journal of Experimental Marine Biology and Ecology 301: 29–37.

D'Amours J, Thibodeau S and Fortin R (2001) Comparison of lake sturgeon (*Acipenser fulvescens*), *Stizostedion* spp., *Catostomus* spp., *Moxostoma* spp., quillback (*Carpiodes cyprinus*), and mooneye (*Hiodon tergisus*) larval drift in des prairies river, quebec. Canadian Journal of Zoology-Revue Canadienne De Zoologie 79: 1472–1489.

de Graaf GJ, Born AF, Uddin AMK and Huda S (1999) Larval fish movement in the river Lohajang, Tangail, Bangladesh. Fisheries Management and Ecology 6: 109–120.

Dufour V and Galzin R (1993) Colonization patterns of reef fish larvae to the lagoon at Moorea Island, French-Polynesia. Marine Ecology Progress Series 102: 143–152.

Economou AN (1991) Is dispersal of fish eggs, embryos and larvae an insurance against density dependence? Environmental Biology of Fishes 31: 313–321.

Elliott JM (1970) Methods of sampling invertebrate drift in running water. Annls Limnologie 6: 133–159.

FAO (Food and Agriculture Organization) (2013) Fishing Gear types. Driftnets. Technology Fact Sheets. *In*: FAO Fisheries and Aquaculture Department[online].Rome. http://www.fao.org/fishery/geartype/220/en. Accessed 5 September 2013.

Faria AM and Gonçalves EJ (2010) Ontogeny of swimming behaviour of two temperate cling fishes, *Lepadogaster lepadogaster* and *L. purpurea* (Gobiesocidae). Marine Ecology Progress Series 414: 237–248.

Faulkner H and Copp GH (2001) A model for accurate drift estimation in streams. Freshwater Biology 46: 723–733.

Franzin WG and Harbicht SM (1992) Tests of drift samplers for estimating abundance of recently hatched walleye larvae in small rivers. North American Journal of Fisheries Management 12: 396–405.

Gale WF and Mohr HW (1978) Larval fish drift in a large river with a comparison of sampling methods. Transactions of the American Fisheries Society 107: 46–55.

Graham JJ (1971) Retention of larval herring within the Sheepscot estuary of Maine. Fishery Bulletin 70: 299–305.

Graham JJ and Venno PMW (1968) Sampling larval herring from tidewaters with buoyed and anchored nets. Journal of the Fisheries Research Board of Canada 25: 1169–1179.

Hare JA, Thorrold S, Walsh H, Reiss C, Valle-Levinson A and Jones C (2005) Biophysical mechanisms of larval fish ingress into Chesapeake Bay. Marine Ecology Progress Series 303: 295–310.

Harvey BC, White JL and Nakamoto RJ (2002) Habitat relationships and larval drift of native and nonindigenous fishes in neighboring tributaries of a coastal california river. Transactions of the American Fisheries Society 131: 159–170.

Hendler G, Baldwin CC, Smith DG and Thacker CE (1999) Planktonic dispersal of juvenile brittle stars (Echinodermata: Ophiuroidea) on a caribbean reef. Bulletin of Marine Science 65: 283–288.

Hettler WF (1979) Modified neuston net for collecting live larval and juvenile fish. The Progressive Fish-Culturist 41: 32–33.

Hofer K and Kirchhofer A (1996) Drift, habitat choice and growth of the nase (*Chondrostoma nasus*, Cyprinidae) during early life stages, Vol. Birkhauser Verlag, Basel.

Humphries P, Serafini LG and King AJ (2002) River regulation and fish larvae: variation through space and time. Freshwater Biology 47: 1307–1331.

Iguchi K and Mizuno N (1990) Diel changes of larval drift among amphidromous gobies in Japan, especially *Rhinogobius brunneus*. Journal of Fish Biology 37: 255–264.

Iguchi K and Mizuno N (1999) Early starvation limits survival in amphidromous fishes. Journal of Fish Biology 54: 705–712.

Janáč M, Jurajda P, Kružíková L, Roche K and Prášek V (2013a) Reservoir to river passage of age-0 + year fishes, indication of a dispersion pathway for a non-native species. Journal of Fish Biology 82: 994–1010.

Janáč M, Šlapanský L, Valová Z and Jurajda P (2013b) Downstream drift of round goby (*Neogobius melanostomus*) and tubenose goby (*Proterorhinus semilunaris*) in their non-native area. Ecology of Freshwater Fish 22: 430–438.

Jiang W, Liu HZ, Duan ZH and Cao WX (2010) Seasonal variation in drifting eggs and larvae in the upper yangtze, china. Zoological Science 27: 402–409.

Johnson SL, Rodgers JD, Solazzi MF and Nickelson TE (2005) Effects of an increase in large wood on abundance and survival of juvenile salmonids (*Oncorhynchus* spp.) in an Oregon coastal stream. Canadian Journal of Fisheries and Aquatic Sciences 62: 412–424.

Johnston TA, Gaboury MN, Janusz RA and Janusz LR (1995) Larval fish drift in the valley river, manitoba: Influence of abiotic and biotic factors, and relationships with future year class strengths. Canadian Journal of Fisheries and Aquatic Sciences 52: 2423–2431.

Kelso JRM and Leslie JK (1979) Entrainment of larval fish by the douglas point generating-station, lake huron, in relation to seasonal succession and distribution. Journal of the Fisheries Research Board of Canada 36: 37–41.

Kelso WE and Rutherford DA (1996) Collection, preservation, and identification of fish eggs and larvae. *In*: Murphy BR and Willis DW (eds.). Fisheries Techniques. American Fisheries Society, Bethesda, MD.

Kennedy TB and Vinyard GL (1997) Drift ecology of western catostomid larvae with emphasis on warner suckers, *Catostomus warnerensis* (Teleostei). Environmental Biology of Fishes 49: 187–195.

Kubíček F (1966) Eine neue methode der quantitativen entnahme der organischen drift. Zoologické Listy 15: 284–285.

Lechner A, Keckeis H, Schludermann E, Humphries S, McCasker N and Tritthart M (2013) Hydraulic forces impact larval fish drift in the free flowing section of a large european river. Ecohydrology 7: 648–658.

Lewis RM, Hettler WF, Nelson Johnson G and Wilkens EPH (1970) A channel net for catching larval fishes. Chesapeake Science 11: 196–197, 191p191.

Lindsay JA, Radle ER and Wang JCS (1978) Supplemental sampling method for estuarine ichthyoplankton with emphasis on Atherinidae. Estuaries 1: 61–64.

McDowall RM (2007) On amphidromy, a distinct form of diadromy in aquatic organisms. Fish and Fisheries 8: 1–13.

Müller K (1966) Die tagesperiodik von fliesswasserorganismen. Zeitschrift fuer Morphologie und Oekologie der Tiere 56: 93–142.

Naesje TF, Jonsson B and Sandlund OT (1986) Drift of cisco and whitefish larvae in a norwegian river. Transactions of the American Fisheries Society 115: 89–93.

Needham PR (1928) A quantitative study of the fish food supply. NY State Conservation Department Annual Report 17: 192–206.

Nester RT (1987) Horizontal ichthyoplankton tow-net system with unobstructed net opening. North American Journal of Fisheries Management 7: 148–150.

Nolan CJ and Danilowicz BS (2008) Advantages of using crest nets to sample presettlement larvae of reef fishes in the Caribbean Sea. Fishery Bulletin 106: 213–221.

O'Connell AM and Angermeier PL (1997) Spawning location and distribution of early life stages of alewife and blueback herring in a Virginia stream. Estuaries 20: 779–791.

Oesmann S (2003) Vertical, lateral and diurnal drift patterns of fish larvae in a large lowland river, the elbe. Journal of Applied Ichthyology 19: 284–293.

Pavlov DS (1994) The downstream migration of young fishes in rivers—mechanisms and distribution. Folia Zoologica 43: 193–208.

Pavlov DS, Pakhorukov AM, Kuragina GN, Nezdolii VK, Nekrasova NP, Brodskii DA and Ersler AL (1977) Some peculiarities of downstream migrations of young fishes in the Volga and Kuban rivers. Voprosy ikhthiologii 17: 415–428.

Pavlov DS, Nezdoliy VK, Urteaga AK and Sanches OR (1995) Downstream migration of juvenile fishes in the rivers of Amazonian Peru. Voprosy ikhthiologii 35: 227–247.

Pavlov DS, Lupandin AI and Kostin VV (2002) Downstream Migration of Fish through Dams of Hydroelectric Power Plants, Vol. Oak Ridge National Laboratory, Oak Ridge, Tennessee.

Pavlov DS, Mikheev VN, Lupandin AI and Skorobogatov MA (2008) Ecological and behavioural influences on juvenile fish migrations in regulated rivers: a review of experimental and field studies. Hydrobiologia 609: 125–138.

Peck MA, Kühn W, Hinrichsen H-H and Pohlmann T (2009) Inter-annual and inter-specific differences in the drift of fish eggs and yolksac larvae in the North Sea: a biophysical modeling approach. Scientia Marina 73S1: 23–36.

Persat H and Olivier JM (1995) The first displacements in the early stages of *Chondrostoma nasus* under experimental conditions. Folia Zoologica 44: 43–50.

Priegel GR (1970) Reproduction and early life history of the walleye in the lake winnebago region. Wisconsin Department of Natural Resources Technical Bulletin 45: 1–105.

Reeves KS and Galat DL (2010) Do larval fishes exhibit diel drift patterns in a large, turbid river? Journal of Applied Ichthyology 26: 571–577.

Reichard M and Jurajda P (2007) Seasonal dynamics and age structure of drifting cyprinid fishes: an interspecific comparison. Ecology of Freshwater Fish 16: 482–492.

Reichard M, Jurajda P and Ondračková M (2002a) Interannual variability in seasonal dynamics and species composition of drifting young-of-the-year fishes in two european lowland rivers. Journal of Fish Biology 60: 87–101.

Reichard M, Jurajda P and Ondračková M (2002b) The effect of light intensity on the drift of young-of-the-year cyprinid fishes. Journal of Fish Biology 61: 1063–1066.

Reichard M, Jurajda P and Smith C (2004) Spatial distribution of drifting cyprinid fishes in a shallow lowland river. Archiv für Hydrobiologie 159: 395–407.

Robinson AT, Clarkson RW and Forrest RE (1998) Dispersal of larval fishes in a regulated river tributary. Transactions of the American Fisheries Society 127: 772–786.

Sewell MA, Van Dijken SG and Suberg L (2008) The cryopelagic meroplankton community in the shallow waters of gerlache inlet, Terra Nova Bay, Antarctica. Antarctic Science 20: 53–59.

Schludermann E, Tritthart M, Humphries P and Keckeis H (2012) Dispersal and retention of larval fish in a potential nursery habitat of a large temperate river: an experimental study. Canadian Journal of Fisheries and Aquatic Sciences 69: 1302–1315.

Schmutz S, Zitek A and Dorninger C (1997) A new automatic drift sampler for riverine fish. Archiv für Hydrobiologie 139: 449–460.

Sonny D, Jorry S, Watriez X and Philippart JC (2006) Inter-annual and diel patterns of the drift of cyprinid fishes in a small tributary of the Meuse river, belgium. Folia Zoologica 55: 75–85.

Svendsen CR, Quinn T and Kolbe D (2004) Review of Macroinvertebrate Drift in Lotic Ecosystems. Wildlife Research Program, Seattle.

Waters TF (1972) The drift of stream insects. Annual Review of Entomology 17: 253: 272.

Wolf P (1951) A trap for the capture of fish and other organisms moving downstream. Transactions of the American Fisheries Society 80: 41–45.

Wolter C and Sukhodolov A (2008) Random displacement versus habitat choice of fish larvae in rivers. River Research and Applications 24: 661–672.

Zitek A, Schmutz S and Ploner A (2004) Fish drift in a Danube sidearm-system: Ii. Seasonal and diurnal patterns. Journal of Fish Biology 65: 1339–1357.

Zuur AF, Ieno EN, Walker N, Saveliev AA and Smith GM (2009) Mixed Effects Models and Extensions in Ecology with R. Springer-Verlag, New York. 574p.

CHAPTER 12

Methodologies for Investigating Diadromous Fish Movements: Conventional, PIT, Acoustic and Radio Tagging and Tracking

Marie-Laure Bégout,[1,*] *Frédérique Bau,*[2,3] *Anthony Acou*[4] and *Marie-Laure Acolas*[2]

Tagging of fish has been carried out at least as long ago as the 17th century when, in 'The Compleat Angler' (first published in 1653), Izaak Walton reported the attachment of ribbon tags to the tail of juvenile Atlantic salmon *Salmo salar* to determine their movements (Walton and Cotton 1921; Lucas and Baras 2000). Since then, systematic tagging of fish for scientific purposes has been conducted for more than a century using natural marks or synthetic passive marks and tags, whereas the development of electronic tags arose in the 1950's with the first study of Trefethen (1956) using underwater telemetry.

Methods to study the migratory behaviour of fish can be divided in two categories: capture-dependent (based on sampling marked or unmarked fish) and capture-independent methods, such as visual or video observation, resistivity fish counters or hydroacoustics (Lucas and Baras 2000). In this chapter we focus only on capture-dependent methods using marked fish.

[1] Ifremer, Fisheries laboratory, Place Gaby Coll, 17137 L'Houmeau, France.
[2] IRSTEA, National Research Institute of Science and Technology for Environment and Agriculture, Aquatic Ecosystems and Global Changes research unit, Diadromous migratory fish team, 50 avenue de Verdun, 33612 Gazinet Cestas, France.
[3] Pôle Écohydraulique ONEMA-Irstea-INP, F-31400 Toulouse, France.
[4] UMR BOREA 7208, Muséum National d'Histoire Naturelle, Service des Stations Marines, 35800 Dinard, France.
* Corresponding author

Tagging methodologies

Prior to conducting direct observation, there is often a need to identify fish within a population. Identification is useful not only when recording immediate behaviour, but also for long term monitoring of an individual's or population's performance. Tagging and marking are used for identification purposes, both at the individual and group level (Murphy and Willis 1996). Tagging or marking fish involves treatment and handling, which may disturb and possibly stress the fish, so careful handling is most important when undertaking such procedures (Baras 1991; Murphy and Willis 1996; Thorstad et al. 2000a, 2001; Jepsen et al. 2002, 2005; Bridger and Booth 2003; Sulikowski et al. 2005; Brown et al. 2011; Bégout et al. 2012).

External tags and marks can be used for visual identification, whereas internal passive or transmitting tags and marks usually require specialized equipment for detection and identification. An advantage of internal marks, such as chemical marking of bony structures, is that they can enable large numbers of fish to be marked at an early age (Murphy and Willis 1996). In some cases an external tag, or mark, is used to signify the presence of an internal tag or mark.

External marks and tags

An external mark may be defined as a visible mark on the outside of the fish that is used to identify individual fish, or to distinguish between groups of fish, but without any additional information or specialized reporting format. External marks may be natural, based on variation in colour patterns and morphological traits, such as scale numbers, number of fin rays and distribution of melanophores (Garcia de Leaniz et al. 2004). Additionally, differences in overall body shape (morphometrics) may allow fish from different populations to be distinguished (Bergek and Björklund 2009). External marks may also be artificially applied, as when fish are marked with dye, stains or brands (Murphy and Willis 1996). External marking techniques are used by fish biologists in a range of field applications (Murphy and Willis 1996). Such marks are often simple, cheap and quick to apply, but carry limited information. External marks such as fin clipping have often been used as a means of calling attention to the presence of internal tags, but fin clipping may be stressful and affect swimming behaviour, so it should be used with caution.

External tags are visible structures that are usually attached to the fish by piercing tissues (Murphy and Willis 1996). Such tags, which may carry an individual code, batch code or visible instructions, can be easily detected without specialized equipment. External tags include ribbons, threads, wires, plates, discs, dangling tags, straps (McFarlane et al. 1990), T-bar anchor tags (Harden Jones 1979; Morgan and Walsch 1993), Carlin tags (Carlin 1955) and coloured beads (Jadot et al. 2003).

Ideally the behaviour, growth and survival of tagged and untagged fish should be similar. While this may be true for many types of tags and marks, some tags may well affect behaviour and influence growth and survival (Murphy and Willis 1996; Bridger and Booth 2003). For example, fish with external tags may be more vulnerable to predation and their growth may also be affected. By permanently penetrating the skin, the tag may provide an access route for infection. Additionally, tags may become overgrown with algae and other organisms, adding weight to the tag and increasing drag, as well as preventing tag detection or reading. The need to identify fish, individually or by group, with minimal influence on behaviour, health or survival has thus led to the development of internal tags.

Internal tags

Internal tags are defined as tags inserted or injected into tissues and carried in the body cavity, muscle or cartilage, which can be used to identify individuals or groups of fish. Most types have to be removed from the fish to be identified, but some, such as Radio Frequency Identification (RFID) tags often called Passive Integrated Transponder tags (PIT tags), can be read by an external antenna; so, once implanted, they provide a non-invasive and non-destructive means of identification without a need for recapture (Baras et al. 2000; Downing et al. 2001; Barbour et al. 2012; see later for further details).

Other types of internal tags include plastic or glass tubes, metal plates and small pieces (size 0.5–2.0 mm × 0.25 mm) of magnetized stainless steel, that may have a binary code of Arabic numbers engraved or laser etched on their surface. The latter, known as coded wire tags, are normally injected into the snout of a fish, often in combination with an external mark to aid recovery (Schurman and Thompson 1990). Such tags are extensively used for identifying large numbers of fish and, due to their small size, covering a broad range of sizes. Magnetic body cavity tags (MCTs) are steel plates inserted into the body cavity of the fish, which are detected during fish recapture by magnets placed in strategic positions.

Internal tags that are visible externally

Some tags are placed subcutaneously, but are visible by eye. One example is the Visible Implant Tag (VIT), or the newer visible implant alphanumeric (VI alpha) tag. Such tags were developed to combine the advantages of external tags with those of internal tags, and are used where minimal disturbance of the fish is important. VIT are made of plastic strips and VI alpha are made of medical-grade silicone rubber, often with the addition of fluorescent material. These tags come with printed information and are often placed in transparent tissue just behind the eye. An alternative, which can be used for batch tagging, comes with the Visible Implant Elastomer tags (VIE). These tags consist of a biocompatible two-part fluorescent silicone elastomer material that is mixed and injected into tissue, as a liquid, with a hypodermic syringe. After 24 hours at room temperature it cures into a pliable solid, providing an externally visible internal mark. The fluorescent elastomer is available in several colours. Recognition of individuals is possible through the use of different body locations and colours (Frederick 1997; Olsen and Vollestad 2001).

Transmitting tags

A large and growing array of electronic transmitter tags is available. Apart from pulsed and coded signals that identify the position of the fish, some tags carry sensors that collect additional data, such as depth, swimming direction and speed, heart or respiration rate or information about muscle contraction. Such electronic tags are larger than PIT tags, and require an internal battery to power the transmitter and microchip. The lifetime of the tag, which must be considered in telemetry studies, depends on transmitter size, power supply, range and rate of signal transmission. Telemetry studies on free swimming fish can range over periods of hours to months, and tag detection range can be up to a kilometre in some instances, but is generally less than 100 m. Microchip technology allows for specific instructions to be placed onto a tag, allowing it to be switched on or off under a given set of conditions. For example, tags may transmit only under certain conditions of water chemistry or light intensity, and such selective use can increase the longevity of the tags.

Non-programmable radio and acoustic pulsed transmitter tags transmit a simple radio or acoustic pulse at pre-set time intervals. Theoretically, large numbers of fish can be monitored simultaneously using multiple frequencies or pulse rates, but in practice it is very difficult to distinguish more than four or five pulse rates on an individual frequency. Coded tags are a special type of transmitter tag that operates by emitting a digitally encoded pulse signal at user-defined intervals on specific radio or acoustic frequencies. This allows up to 100 individual signals to be distinguished at a given frequency. This technology has the advantage that many fish can be tracked separately on a single frequency, the information being automatically recorded and downloaded to a computer. Coding can thus increase data acquisition rates and increase sample sizes in telemetry experiments (Stuehrenberg et al. 1990). Lastly, programmable microprocessor tags have been used to transmit radio or acoustic pulsed signals at intervals defined by the user. Specific on/off sequences can be set that are useful for preserving the battery life of the transmitter. New developments include the ability to include sensors that can collect information about the electromyogram, tail beat frequency and heart or respiration rate.

Acoustic tags are mostly used in seawater because sound is transmitted over long distances in salt water, whereas radio waves are attenuated very rapidly. Frequencies of 20–500 kHz are used. Signals from pulsed acoustic tags can be detected using a simple receiving system, comprising a hand-held directional hydrophone, a portable receiver and headphones. This provides only a rough indication of the position of the fish, so accurate position fixing requires triangulation using an array of fixed hydrophones (see later for further details).

Radio tags, which can only be used in water of very low salinity, are useful because radio waves are less affected by physical obstacles, turbidity, turbulence and thermal stratification than acoustic, non-electromagnetic waves (see later for further details). Radio signals also radiate through the water surface and can be detected at great distances, because there is little loss of signal strength in air. Receivers can be placed in boats, aircraft or at land-based listening stations. Radio tags operate at high frequencies (30–300 MHz), so there is little signal drift.

In experiments where simple transmitting tags are used, fish depth can be deduced from time of arrival of acoustic signals at the hydrophones (Anonymous 1968), and algorithms for the calculation of 3D coordinates from an array of four hydrophones are provided by Hardman and Woodward (1984). Signal strength from a radio source at a known location may also be used to estimate fish depth (Velle et al. 1979), although the use of pressure-sensing transmitters (Luke et al. 1973; Williams 1990) gives more reliable estimates, independent of signal attenuation.

In conclusion, and until recently, using conventional VHF radio transmitters was regarded as the method of choice for the majority of studies in freshwater systems (Sisak and Lotimer 1998, Fig. 12.1). But while conventional radio or new hybrid technologies, such as GPS-based radio telemetry, are today considered the standard in wildlife biology (Boitani and Powell 2012; Habib et al. 2014), radio telemetry is no longer the only option for research on fish in freshwater habitats (Cooke and Thorstad 2012; Cooke et al. 2013). In parallel to the innovations in radio telemetry (see later), there have been considerable technical advances in other electronic tagging and tracking technologies, including PIT tags (e.g., Barbour et al. 2012; Burnett et al. 2013; Thiem et al. 2013; see later), acoustic transmitters (e.g., Pincock and Johnston 2012; see later), and archival tags with or without transmitting capabilities (e.g., multi-sensor loggers, tri-axial accelerometers, miniaturized geopositioning systems, pop-up satellite tags—Bograd et al. 2010; Block et al. 2011; Schaefer et al. 2011; Hazen et al. 2012; see also Schaefer et al. 2016), which all now provide further effective means of tracking fish in every aquatic ecosystems. Hence, for more than a decade, there has been a steadily increasing trend particularly towards the use of acoustic telemetry for freshwater studies (Cooke et al. 2013; Fig. 12.1).

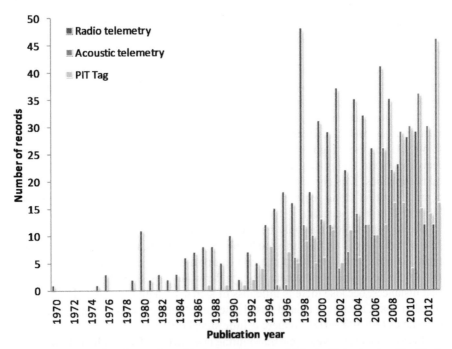

Figure 12.1. Results of a wide record search in Aquatic Sciences and Fisheries Abstracts using either [(radio telemetry) and migration], giving a total of 605 records from 1970 to 2013, or [(acoustic telemetry) and migration], giving a total of 356 records from 1985 to 2013, and [(PIT tag) and migration], giving a total of 194 records from 1987 to 2013. Data is expressed as number of records per publication year and per method.

Passive integrated transponder tags

Equipment basics and principles

The PIT tag technology is based on magnetic fields, since the PIT tag is an inert, small biocompatible glass-encapsulated integrated circuit microchip and coil antenna that receives and transmits automatically low-frequency radio signals (generally 125 to 400 kHz) carrying a unique alphanumeric identity code (ID) when triggered by an interrogator. Hence, PIT tags are powered by magnetic induction. The interrogator (or reader) consists of an antenna which generates a magnetic field inducing an alternating voltage at a specific frequency to energize the PIT tag, and receives the ID signal transmitted by the PIT tag which can be read an unlimited number of times. PIT tags allow rapid retrieval of transmitted information from large numbers of tagged individuals. The spatial resolution is less important with PIT tags than with classical radio or acoustic tags (Hyun et al. 2012), but they have an unlimited lifespan (Gibbons and Andrews 2004). The lack of a battery is the greatest advantage of the PIT tag, since it allows for the production of much smaller tags that can be used on smaller organisms. They can be easily implanted into the body cavity, or muscle, of a fish using a veterinary syringe or a scalpel (Fig. 12.2) and are well retained (Hill et al. 2006; Cucherousset et al. 2007; Hirt-Chabbert and Young 2012; Mazel et al. 2013), with no or low effect on growth and survival (Zydlewski et al. 2001; Hill et al. 2006; Cucherousset et al. 2007; Hirt-Chabbert and Young 2012), even in young life stages (Roussel et al. 2000; Acolas et al. 2007; Archdeacon et al. 2009).

Figure 12.2. Pit-tag insertion. (A) 12-mm Pit-tag injection into the dorsal muscle using a 12-gauge syringe (*Acipenser sturio*, © Irstea, R. Le Barh). (B) 23-mm Pit-tag insertion into the visceral cavity following scalpel incision (*Anguilla Anguilla*, © Irstea, F. Bau).

There are two main different technologies: full-duplex (FDX) and half-duplex (HDX). Both systems most often operate at a 134.2 kHz frequency but the signal transmission is different. A FDX system receives and transmits the signal simultaneously as opposed to a HDX system that transmits the signal, stops and then receives (Fig. 12.3). A FDX antenna receiver continuously emits a magnetic field that charges the tag and listens for the tag to transmit an identification code at a rapid rate. The HDX antenna receiver stops emitting magnetic charge while listening for a tag transition, and therefore they have a slower read rate (two times less). Because the HDX tags have the capacity to momentarily store energy, the tags transmit a stronger

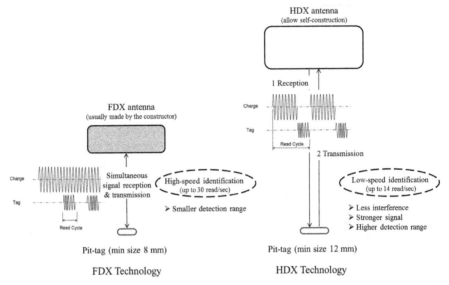

Figure 12.3. Comparison of FDX and HDX technologies: a trade-off between detection distance, tag size and reading speed requirements.

signal resulting in a greater read range. The trade-off however is a larger tag, the actual smallest FDX tags being 8 mm in length and the smallest HDX being 12 mm in length (Fig. 12.3; Chase et al. 2013; Burnett et al. 2013). However, even when using the most powerful technology in terms of detection range (i.e., HDX), the larger 23 or 32 mm PIT tags still have to be within 1 m of the antenna to be detected and decoded (Roussel et al. 2000; Zydlewski et al. 2001; Bubb et al. 2006; Tranquilli 2010; Burnett et al. 2013), the distance being much lower for 12 mm PIT tags in either HDX or FDX technology (maximum 15 to 30 cm; Cucherousset et al. 2005; Burnett et al. 2013).

The RFID technology is still improving in terms of tag miniaturization with a new generation of RFID glass microtags (Nonatec® 1 mm in diameter, 6 mm in length, and 10 mg in mass) that operate at a high frequency (13.56 MHz) and allow an electronic individual identification of small-bodied fish using a laboratory bench reader. The distance of detection being very small (around 1.5 cm), it limits their use to individuals that need to be recaptured, but it allows tagging very small fish such as zebrafish of 16–42 mm (Cousin et al. 2012), juvenile sea bass of 36 mm (400 mg) (Ferrari et al. 2014) and even glass eel of 12 mm (95 mg) (T. Podgorniak, Irstea, pers. comm.).

On the other hand, 'giant' active transponder tags (e.g., 15 × 62.5 mm and 26.5 g in air) have been developed within the frame of the NEDAP Trail-System®. This method is based on inductive coupling between an antenna loop fixed on the bottom of a river, channel or estuary, and a ferrite rod antenna within the Nedap-transponder (see Breukelaar et al. 2000). Nedap-trail stations with large antenna loops (e.g., three parallel 550 m long cables separated by 10 m or more) have been broadly deployed in the lowland reaches of the River Rhine and Meuse, where transponder detections can occur at water depths up to 15 m (Breukelaar et al. 1998, 2000; Klein Breteler et al.

2007). The frequency band used is low (33.25 kHz), and the Nedap-transponder has a functioning concept similar to PIT tag except that it can generate an interrogating signal to the antenna thanks to an internal battery, which implies that such an active tag has a predefined lifespan (2 years, ca. 2000 detections) compared to unrestricted lifespan for conventional PIT tags. Nedap-trail detection stations have been successfully used to study migration routes of up-migrating adult sea trout (Breukelaar et al. 1998; Bij de Vaate et al. 2003), out-migrating Atlantic salmon smolt (Brevé et al. 2013a), silver European eel (Winter et al. 2006, 2007; Klein Breteler et al. 2007; De Oliveira 2012) and European sturgeon (Brevé et al. 2013b), as well as large-scale migratory patterns of adult rheophilic fish (De Leeuw and Winter 2008).

Methods use to track fish migration

In the past, PIT tags were typically used for identifying unique individuals if fish were recaptured, rather than as a telemetric monitoring tool for free-swimming fish. However, PIT tagging systems have been developed whereby small fish can be implanted with PIT tags, and their movements tracked if they move past checkpoints (i.e., antennas with data loggers; Gibbons and Andrews 2004). The technique can be used in either a passive or an active mode to track fish movements (Figs. 12.4 and 12.5). Accordingly, depending on the research objectives, spatial scale, life stage, and field conditions, either mobile or stationary approach will be selected, even though combining both techniques has also become increasingly common for studies in shallow freshwater habitats (Cooke et al. 2013). Furthermore, tags and handheld readers are also available in both FDX and HDX technologies (ISO 11784/11785 standard) to easily check the tag after tagging or verify the identity of fish when recaptured. But as FDX and HDX formats are interoperable but not compatible, ISO readers inherit traits from both technologies, such as a slower FDX read rate to listen for HDX and mostly the same antenna characteristics as for FDX (i.e., high inductance loops, noise sensitivity, rigid antenna wire with air gap for underwater antennas), and therefore

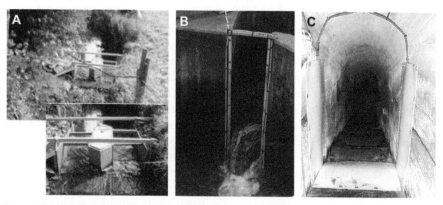

Figure 12.4. Passive tracking with fixed detection stations (antennas). Flatbed antennas in a small river, River La Roche, Normandy, France, FDX technology (© Irstea, M.L. Acolas, Inra experimentation site) (A). Swim-through antennas in a pool fish pass (B) and at a fish bypass entrance (C), HDX technology, River Gave de Pau, France (© Onema, P. Gomes).

Figure 12.5. Active tracking with portable backpack readers and hand-held antennas. (A) Active tracking in a small river, FDX technology, (Inra, © D. Huteau) and (B) search for the precise localization of a dead fish with active tracking in scuba diving, HDX technology (Irstea, © F. Bau).

ISO readers are used in either pre-defined format specification for remote active or passive monitoring.

Passive tracking using fixed detection stations can be implemented either in small, shallow natural streams (Zydlewski et al. 2001; Teixeira and Cortes 2007; Aymes and Rives 2009; Johnston et al. 2009) or within semi-natural or experimental conditions, such as in artificially confined structures (ca. < 1000 cm²) like fishways or fish traps which provide suitable narrow openings to enable PIT-tagged fish to be detected (Castro-Santos et al. 1996; Downing et al. 2001; Thiem et al. 2011; Burnett et al. 2013). Formerly, stations at hydroelectric facilities have been commonly fitted with flat-plate PIT tag interrogation systems and flatbed antennas (mainly in FDX technology) to detect fish passing over (Nunnallee et al. 1998; Lucas et al. 1999). In recent years, PIT antennas have been more frequently used as vertical loops that form swim-through windows (mainly self-built loops in HDX technology made from one or multiple turns of welding cable; see Fig. 12.4), which allow better detection in larger structures (Castro-Santos et al. 1996; Aarestrup et al. 2003; Axel et al. 2005; Travade et al. 2010; Thiem et al. 2011, 2013; Burnett et al. 2013).

In natural river environments, different constraints may also arise including site depth, width, river discharge, current velocity, conductivity, turbidity, accessibility for the operator and ambient electromagnetic noise (Bass et al. 2012; Hyun et al. 2012; Cooke et al. 2013). So, the design of PIT antennas has to be adapted to those constraints and calibrated (e.g., accordance, synchronization, multiplexing) according to the particular environment to be efficient (Castro-Santos et al. 1996; Zydlewski et al. 2006; Aymes and Rives 2009; Riley et al. 2010; Bass et al. 2012). Deep water makes the use of PIT tags ineffective to infer fish positioning by virtue of the technical difficulty of deploying PIT antennas, though technically feasible if linked *in situ* with waterproofed logger and power supply (Tranquilli 2010). Hence, the latest under-going technological developments to improve PIT-telemetry system capacity, resulting in both higher spatial and temporal resolutions in deeper and complex habitats, may go through more refinements in terms of networking of mainly long read-range HDX-antenna systems. Indeed, passive PIT telemetry arrays have allowed simultaneous detection of fine-scale movements of multiple individuals in deeper areas, and latest developments in antenna technology sturdiness in harsh stream conditions (Johnston et al. 2009) have also enabled longer-term field deployments. In small shallow rivers, passive detection can be improved by deploying small, usually, flatbed antennas that can be multiplexed to increase detection coverage. Such multiplexing leads to larger individual detection fields, that can be further extended by synchronizing one or several other multiplexed antenna systems within arrays. This technique has been used in streams to study small-scale fish movements or habitat use for periods of a few months (Teixeira and Cortes 2007; Aymes and Rives 2009; Johnston et al. 2009). In larger rivers, cross-river antenna arrays consisting of single or multiplexed larger flatbed or swim-through antennas may be deployed to form independent detection barriers to remotely monitor larger-scale fish movements and migration patterns (Zydlewski et al. 2006). So far, passive PIT telemetry has mainly been used in smaller systems. It was especially valuable in assessing fish movement through narrow passageways, such as fish passes and small streams, but is being increasingly used in wider swim-through channels (30 × 0.8 m deep; Leach 2010; Tranquilli 2010) using arrays of tag readers. The use of a single antenna may be adequate to provide descriptive data on the timing of migration, but in most cases operation of multiple arrays, even if they do not span the entire stream width, can greatly increase combined efficiency while providing movement direction (F. Kroglund, Norwegian Institute for Water Research, pers. comm.). Hence passive systems yield high-quality long-term recording of large fish samples at discrete locations in increasingly wider and deeper habitats (Zydlewski et al. 2006).

Active tracking methods have been developed in both technologies, FDX and HDX (Roussel et al. 2000; Zydlewski et al. 2001; Riley et al. 2003, 2010; Hill et al. 2006; Cucherousset et al. 2007; Linnansaari and Cunjak 2007; Hewitt et al. 2010), which also allow for enhanced prospection range at least on practicable rivers in terms of width and depth. This type of tracking allows long-distance coverage in a short period of time, however it is limited to particular cases, such as for studies on fish positions, movements and microhabitat preferences in shallow waters with low structural complexity and for relatively short discontinuous periods of time, since it needs to be made manually by an operator (Fig. 12.5). Tracking with portable backpack readers can be performed by foot within the river or on its sides, or even over ice cover (Roussel et al. 2004; Linnansaari and Cunjak 2007).

Passive and active tracking can be successfully combined to track movements of either juvenile or adult migrating fish from the upper watershed to the estuary (Zydlewski et al. 2001; Acolas et al. 2012; Barbour et al. 2012). This technology can be used in a 'mark-recapture' way: fish can be captured by different methods (electrofishing, fishing nets) and in different environments, from freshwater to seawater, individually identified for their whole life owing unlimited PIT tag lifespan and recaptured either by conventional means (electrofishing or netting) or through detection antennas disposed along the river course or by active tracking. Thanks to this technology, the same individuals can be identified from juvenile to adulthood and from freshwater to the sea.

Acoustic telemetry

Underwater telemetry actually started with Trefethen (1956) who developed the first tracking system based on acoustic signals transmission. The technology was rapidly applied (see first review on underwater telemetry by Baras 1991), mostly in marine environments, with different aims and methods detailed below, either active (Westerberg et al. 1983) or passive (Lagardère et al. 1990; Juell and Westerberg 1993), to reach now a very large number of applications (Fig. 12.1) both in basic research and in resource protection like in marine reserve (Klimley et al. 2001; Espinoza et al. 2011). Nowadays, the technique is highly complementary or even replacing radio telemetry in all types of environment (Cooke and Thorstad 2012).

Equipment basics and principles

As presented above, a transmitter, or acoustic tag, is an electronic device usually implanted or externally attached to a fish that transmits ultrasonic signals. In this case, it concerns the transmission in water of ultrasonic energy, or sound signals, at frequencies between 20–500 kHz (20,000–500,000 cycles^{-1}), generally above human hearing range. Ultrasonic frequencies are usually used for transmitting data underwater because, compared to radio frequencies (30–300 MHz—millions of cycles^{-1}, see later), acoustic frequencies are absorbed much less. However, acoustic signals, due to their lower frequencies, experience more distortion than radio and cannot transmit as much information per time unit. Physical limitations on the acoustic transmission are mostly associated with underwater noises (e.g., generated by bubbles, waves), screens due to aquatic weeds or temperature and salinity stratification. Most frequently used frequencies are between 30 and 90 kHz to maximize transmission range in fresh to salt water, recently tags were developed in 180, 307 or 416 kHz which allowed for further miniaturization (0.65 g @ 180 or 307 kHz, 0.28 g @ 416 kHz) but operating frequencies greater than 100 kHz are less effective in salt water, and are thus essentially dedicated to freshwater environments.

Tags are produced in many different shapes and sizes depending on the type of species being studied, or the type of environment in which the study is conducted. Acoustic tags consist of a piezoceramic transducer, drive/timing electronics, and a battery power source. Cylindrical or 'tube' transducers are often used, which have metalization on the inner and outer walls of the structure. In normal operation, an Alternating Current (AC) electrical signal generated by the drive/timing electronics is impressed across the two metalization layers. This voltage creates stress in the material, which in turn cause the transducer to emit an acoustic signal or 'ping', which emanates outward from the surface of the tube. Most tags transmit very short pings of ultrasound, either at very regular intervals (e.g., 1 second) or as series of several pings (6–10) that contain a digital ID (identifier) code, and sometimes physical data (e.g., temperature, pressure) that a receiver, called hydrophone in this case (see below), detects, decodes and transmits or stores in memory. Transmitters range from 5 mm up to 33 mm in diameter and are of various lengths, from 10 to 100 mm. There are two basic types of tags with unique data transmission, continuous and coded. A continuous tag transmits sequential pings at a precise or varying interval that is correlated to a physical variable, such as temperature or depth. Continuous tags are restricted to one frequency each, thus a multi-frequency acoustic receiver is needed. Continuous tags are an excellent choice for tracking a single fish from a vessel in real time, or conducting positioning studies using at least three synchronized hydrophones (see next information on tracking methods). Coded transmitter transmits a series of pings defined as a burst or code burst. Each burst contains a digital ID code and sometimes physical data (e.g., temperature, depth or acceleration). The code burst usually occurs over a few seconds followed by a delay which is usually determined by the study design. It also often depends on such factors as number of tagged animals, swimming speed or detection range. Delay is randomized to minimize the chance that the pings from any two tags will overlap or collide repeatedly. For example, a typical coded tag might have a delay that ranges between 30 and 90 seconds, which means the tag will transmit a series of pings every 30 to 90 seconds. This type of transmission scheme allows many tags to transmit on the same frequency, enabling researchers to conduct large scale population studies. Finally, within the coded tag family there are two types: tags that transmit an ID code only (referred to as a pinger), and tags that transmit both an ID code and sensor data, such as temperature, depth or acceleration (referred to as a sensor transmitter). Sensor transmitters provide a 'view' into the behaviour of animals by transmitting physiological or physical data of an animal's natural environment.

For signal reception, acoustic tags require both a hydrophone and associated receiver or data storage system. Basically, the hydrophone technology is the same as for the tag with reversal functioning: a piezoelectric transducer generates electricity when subjected to a pressure change hereby converting a sound signal into an electrical signal, which is then emitted and decoded. Hydrophones can either have omnidirectional or directional reception depending on how they are build (with a spherical body around the hydrophone for focused hydrophone). Depending on this property, they are used differently: directional hydrophone are often used for actively tracking animals, while omnidirectional hydrophones, also called passive receivers, are used in arrays moored in a fixed location to detect the acoustic signal of tags. These receivers have a single or dual frequency, with plenty of data storage, and are designed to last for greater than one to five years on a single battery.

Methods use to track fish migration

Two main tracking methods are used with acoustic systems: active or passive method. Active tracking is achieved using mobile listening receivers, and entails following tagged animals from a vessel using an acoustic receiver and directional or, sometimes, omnidirectional hydrophone. The path of the animal is approximated by way points in the course of the vessel, and accuracy is in the order of 100 m. The method is very useful for investigating large scale fish behaviour, such as mating, predation, feeding or diving over short periods of time, but has the main drawback of being time consuming (10 days maximum are spent following animals over about 100 km, Klimley et al. 2001). Usually the system is a multi-frequency Digital Signal Processing (DSP) receiver that digitally samples acoustic sound and provides output that can be heard by the human ear, and data that can be stored either in the receiver's memory or on a computer. Remarkable examples include many large species such as migrating yellowfin tuna (Holland et al. 1990; Block et al. 1997, 2011; Brill et al. 1999) or European eel (Aarestrup et al. 2009).

Passive monitoring is achieved with fixed listening receivers used for investigating positioning over particular areas and time, such as migration routes, home range, spawning and feeding areas. There are two main types of methods and associated deployment; one achieves the localization of a transmitter based on when the same pulse arrives at three stationary hydrophones aligned in a triangular array (the method is fully described and evaluated in Klimley et al. (2001)), while the other method is based on the installation of curtains of hydrophones with build-in data logger that record every ping received (the principle is described in Jackson 2011). Briefly, in the first system, the receivers are mounted with cables leading to a receiving station, or cables can be replaced by radio transmission to a base station and computer (on land or onboard a vessel), which in both cases calculate in real time the coordinates of the animal in the survey area (e.g., Bégout Anras et al. 1999; Klimley et al. 2001). Recently receivers are setup as 'curtains' or 'gates' across shorelines, or in arrays (making several triangles in an area), to quantify the amount of time animals spend in an area by recording the presence of coded transmitters (e.g., around marine reserve (Koeck et al. 2013) or along a large river, as for the European eel (Bultel et al. 2014)). Receivers are moored in many configurations, often reflecting the bathymetry and oceanographic or aquatic/weather conditions. A simple mooring design generally works well (e.g., in a lake, Roy et al. 2014). However, in rougher conditions, heavier weights and mooring lines are usually required (Jackson 2011; Bergé et al. 2012). If a bottom mooring is desired, the receiver should be positioned above the floor such that it has an unobstructed 'view' of tagged animals within their expected acoustic range. Receiver spacing is generally determined based on range testing and historical acoustic data (see procedure in Roy et al. 2014 or Bergé et al. 2012), and spacing is usually between 100 and 1000 m. Most receivers store detection data in memory along with sensor information, if the transmitter is equipped with a sensor, status and battery information. Uploading data requires retrieving the receiver and connecting it to a computer interface or some wireless devices. Translating data stored into actual positions of the animals requires most often, a database or particular software (Cote et al. 1998; Ehrenberg and Steig 2003; Espinoza et al. 2011; Roy et al. 2014). There are also statistical and GIS packages available that facilitate movement animations, home range analysis, and advanced statistical tests.

All the advances in acoustic receiver-loggers, coding and signal-processing methods have broadened the scope of acoustic telemetry, and autonomous acoustic receiver arrays are now routinely applied in inland waters, especially large rivers, lakes and reservoirs to assess migration patterns. A number of high profile acoustic marine tracking systems have been efficiently extended into estuarine and freshwater environments to record movement patterns of migratory and non-migratory fish (e.g., Melnychuk et al. 2007; McMichael et al. 2010; Bergé et al. 2012; Klimley et al. 2013; Ingraham et al. 2014; Roy et al. 2014) including small-sized individuals, thanks to the availability of relatively small acoustic transmitters (e.g., McMichael et al. 2010) compared with the Combined Acoustic and Radio Tags (CARTs) previously used (e.g., Solomon and Potter (1988) or Deary et al. (1998)).

Radio telemetry

Radio telemetry technology was first applied to terrestrial wildlife studies before being used in the late 1960s to monitor fish in freshwater ecosystems (Lonsdale and Baxter 1968; Winter et al. 1978). For nearly 40 years (Fig. 12.1), this technology has been increasingly involved in studies on freshwater fish, which have addressed a diversity of topics such as daily movements, migrations, habitat use, survival and population estimation (reviewed in Baras 1991; Lucas and Baras 2000; Spedicato et al. 2005; Adams et al. 2012), as well as physiology and bioenergetics (Cooke et al. 2008; Cocherell et al. 2011; Ueda 2012). Currently, radio technology is also commonly used to transmit data in real time from passive acoustic monitoring systems to shore or onboard based stations (Espinoza et al. 2011; McDougall et al. 2013; Kessel et al. 2014), or to communicate with satellites (pop-up archival satellite tags, Wahlberg et al. 2014).

Despite a decline in popularity (Fig. 12.1), VHF radio telemetry often remains the most appropriate or only feasible option for some types of studies in freshwater, due to advantages in antenna size (from cm to dm, see below) and reception characteristics (e.g., evaluation of dam passage especially in turbulent hydrologic conditions, microhabitat use in shallow river systems, migrations in large watersheds). The unique characteristic of radio telemetry continue to make it an extremely powerful tool for studying movements of fishes, particularly those living in fluvial habitats, and in providing information to enable effective management and conservation of aquatic resources.

Equipment basics and principles

Radio telemetry is a form of active biotelemetry, that allows the remote sensing of the movements and aspects of behavioural or physiological variables of an animal by use of radio waves. It relies upon the transmission of information between a radio transmitter, placed in or on a studied animal, and a distant receiver with a receiving antenna. Radio transmitters generate and emit electromagnetic energy in the Very High Frequency (VHF) band of the radio frequency range, usually between 30 and 300 MHz (Sisak and Lotimer 1998), but typically between 30 and 150 MHz for fisheries applications (Winter 1983, 1996). There is no one optimal radio frequency (Winter 1983; Eiler 1990; Sisak and Lotimer 1998) and the choice of operating frequency is important, conditioning the whole telemetry system since receivers and antennas are frequency specific. The actual frequency used depends on the country and government regulations, but with more and more studies on macro-scale movement patterns of fish populations, a good coordination of frequencies used has become the norm. Most usage occurs in the 140 MHz to 175 MHz range (e.g., North America, Scandinavia) while frequencies in the 30 MHz to 50 MHz range, generally assumed to have less attenuation due to water conductivity, are commonly used in other countries and tend to be gradually replaced. The frequency used and the operating environment will have major impacts on the performance and suitability of the telemetry system to research objectives. Radio telemetry systems are dependent on the surrounding environment. The distance at which transmitter signals are detected by receiving antennas mainly depends on the power radiated by the transmitter, the sensitivity of the receiving station and propagation losses.

Radio transmitters use a quartz crystal to generate the transmitting frequency and control signal modulation, and are powered by lithium or silver oxide battery (depending on tag size, lifetime or transmitting range requirements) and controlled by a stable multivibrator or microprocessor circuit (depending on field-programmable tag requirements) (see Kuechle and Kuechle 2012 for a comprehensive review). Signal modulation (i.e., pulse sequences) of transmitters can be divided into either standard or coded ID types. The transmitter itself requires a transmitting antenna that induces an electromagnetic field (radio wave) which radiates outward into the water. Most fish radio tags are fitted with a long insulated wire antenna (whip antenna) that typically trails from the transmitter, but radio tags with coated loop antenna, either to be coiled inside the fish or preferably directly pre-coiled and sealed within the transmitter package, are also available. Use of internal antenna, configurations (Cooke and Bunt 2001; Collins et al. 2002; Ovidio et al. 2002; Verbiest et al. 2012) may simplify tagging procedures (as for acoustic tags), avoid additional risks to fish such as abrasion and infection at antenna exit, entanglement with debris, vulnerability to predation (Adams et al. 1998; Mulcahy 2003; Bauer et al. 2005; Jepsen et al. 2008; Liedtke et al. 2012), and limit disruption of normal behaviour inherent to long trailing antennas (Thorstad et al. 2001; Murchie et al. 2004). However, internal antennas may negatively affect signal strength (Winter 1996; Cooke and Bunt 2001; Collins et al. 2002; Ovidio et al. 2013), reducing transmitter power output and hence decreasing significantly the signal detection range, typically about 25% relative to a transmitter with whip antenna (Kuechle and Kuechle 2012). In some cases, the loss of signal propagation may be compensated through increased transmitter power output. However, increased power output, as well as longer battery lifetime and increased pulse rate, are all variables that require larger batteries, which will increase the overall size of transmitters.

As other major components, a radio telemetry system typically includes a receiving antenna, a receiver and a data logger. The receiving antenna is used to capture the radio signal produced by the transmitter, and change it into a voltage for detection by a receiver. The receiver is used to convert the radio signal captured by the receiving antenna into an audible or electronic form that can be used to identify the tagged animal. The data logger is used to store the collected data, and may be a separate unit or may be integrated into the receiver, as in most modern fixed stations for stationary tracking arrays (Eiler 2012). Three different types of antennas are in general use for fish telemetry (i.e., dipole, loop and Yagi antennas), each with different attributes depending on the used frequency (reviewed in Kuechle and Kuechle 2012). Omnidirectional dipoles are used in large-scale river applications, while directional antennas such as loop and Yagi antennas are both particularly useful in river applications where high direction finding accuracy is required. However, if Yagi antennas are the most commonly used antennas in fish telemetry, tuned loop antennas are of interest especially for manual tracking, and particularly at lower frequencies where Yagi antennas may become too large for practical use (about 3.75 m long at 30 MHz). According to the research objectives, different types of antennas may be mixed, and antennas may also be combined into arrays to enhance gain or pattern characteristics (Larinier et al. 2005; Croze et al. 2008; Travade et al. 2010). Underwater antennas are seldom used in fisheries research relative to aerial applications. However, using common underwater antennas such as stripped coaxial cable and standard dipole types (Bunt et al. 2003; Beeman et al. 2004; Croze et al. 2008; Peters et al. 2008; Travade et al. 2010), or relatively new systems (Gingerich et al. 2012) in conjunction with aerial antennas, can be of great interest for detecting fish with more precision when located, for example, in deep-water habitats (Niemela et al. 1993; Martinelli and Shively 1997; Brown et al. 2006; Gingerich et al. 2012).

Key variables influencing the range of radio transmitters include water depth and conductivity (Velle et al. 1979). Radio telemetry is typically used in relatively shallow (< 10 m depth), low conductivity (usually < 500 μScm^{-1}) freshwater environments (Winter 1983, 1996), where radio signals are capable of travelling greater distances than acoustic signals (Thorsteinsson 2002). Most electromagnetic frequencies are rapidly attenuated by seawater (Voegeli and Pincock 1996), and are usually undetectable within a fraction of a metre from the source, explaining why normal VHF radio telemetry is useless for marine applications. Radio waves propagate omnidirectionally in the water, but only those wave vectors almost perpendicular (< 6° from the vertical) to the air-water interface propagate from water into the air (Velle et al. 1979) and can be detected by an aerial receiving antenna. Radio waves can transmit information rapidly and for long distances in air. Because radio frequencies travel well through air, receivers and antennas can be used for mobile tracking with a large range of methods (e.g., from land, boat, air). This flexibility in modes of coverage can provide greater detail on behaviour and movement of individual animals than other forms of marking or tracking (Winter 1996). The ability to detect radio transmitters in air, using highly portable aerial antennas, provides a major advantage for migration studies in large or inaccessible river systems (Koehn 1999). Furthermore, the use of radio frequencies is advantageous to acoustic frequencies when turbulent hydrologic conditions entrap air in the water column, because air bubbles shorten the distance that acoustic signals can travel through water (Monan et al. 1975; Thorstad et al. 2000b). Being more easily detected in air than acoustic transmitters, radio tags are used in turbulent locations with entrapped air, such as hydroelectric dam tailraces, and hence frequently chosen for studies on the evaluation of dam passage and fishway performance, such as the effects of bypass collectors and turbines on fish behaviour and survival (e.g., Castro-Santos and Haro 2003; Gowans et al. 2003; Hockersmith et al. 2003; Thorstad et al. 2003; Larinier et al. 2005; Ovidio et al. 2007; Croze et al. 2008; Travade et al. 2010; Calles et al. 2012; Chase et al. 2013). In addition, radio frequencies travel through ice, so this technique is the most effective in winter (including during periods of ice cover, ice break-up and flooding), for studying fish movement and migration in both lotic (Eiler 1990; Brown and Mackay 1995; Brown et al. 2001; Robertson et al. 2003; Halttunen et al. 2013) and lentic systems (Bauer and Schlott 2004). Also, radio frequencies are not adversely affected by thermoclines, algae, plant or other aquatic obstacles (Winter 1996; Jacobsen et al. 2002; Kobler et al. 2008; Jellyman 2009), although they are reflected or diffracted from rock faces.

Difficulties in using radio telemetry can arise in areas with depths greater than 15 m, areas such as estuaries and lowland rivers influenced by salt water, or in areas that have saline pools or haloclines (see earlier). Nevertheless new underwater radio antenna systems, such as the turnstile and balanced loop-vee antennas (Gingerich et al. 2012), may be another useful option in freshwater-deep habitats through their better reception characteristics than the underwater dipole antenna more commonly used for fisheries research. The balanced loop-vee antenna especially with stronger omnidirectionality and overall reception strength (–36.84 dBm) was shown to detect radio transmitters as far as 15 m through water, compared to 9 m and 12 m for dipole and turnstile antennas, respectively.

A further improvement has been the development of radio transmitters with additional capabilities, combining miniature ID with (bio)telemetry sensors to record the location of the tagged fish along with data on either its surrounding environment (e.g., depth, temperature) or its physiological status (e.g., heart rate, tail-beat frequency, swimming direction, swimming speed). Some of them (tilt sensors) have been used to indicate activity or motion (mortality) (e.g., Eiler 1990; Watry and Scarnecchia 2008). Among all biotelemetry sensors including electrocardiograms (ECGs), electromyograms (EMGs) and accelerometers, ECGs that measure the electrical activity of muscles from electrodes inserted into the axial musculature are currently the most readily used to measure fish activity (Cooke et al. 2004). However, the use of animal-borne acceleration data loggers for studying free-swimming fish is gaining popularity (e.g., Clark et al. 2010). EMG transmitters have been used to measure swimming speed and estimate energetic costs associated, for example, with migration patterns and fishway use, hydropeaking events, angling, predation/fisheries interactions (Gowans et al. 2003; Brown et al. 2006; Quintella et al. 2009; Slavík and Horký 2009; Pon et al. 2009; Cocherell et al. 2011; Makiguchi et al. 2011; Gravel and Cooke 2013; Alexandre et al. 2013; Taylor et al. 2013). Although some concerns remain regarding the reliability of the data provided by electromyogram telemetry (Geist et al. 2002), its potential for fishway research is obvious.

Methods used to track fish migration

Mobile (i.e., manual) tracking, although intermittent, provides an effective method for collecting a wide range of information, including both large and small-scale movements, macro and micro-habitat use, and activity patterns. The basic approach is to actively search for and locate the fish while travelling through the study area. Fish locations are determined either by triangulation or by following the signal back to the transmitter (i.e., zero-point tracking or successive gain reductions) (Gravel and Cooke 2008; Gillis et al. 2010), sometimes referred to as 'homing in' on the signal. Mobile tracking is dynamic and versatile; it can be performed at a variety of spatial scales and take a number of forms, including surveys on foot, and from land-based vehicles, boats, airplanes, and likely in a near future by routinely multi-UAVs (Unmanned Aerial Vehicles) localization (Jensen and Chen 2013). Although most frequently used for automated telemetry systems, coded transmitters are also useful for manual tracking because the coded signal can be detected audibly more easily during levels of high background noise, and the scan time can be reduced when searching for a large number of fish spread over a large geographical area (Eiler et al. 1991). Aerial tracking provides the greatest reception range (Winter 1983) and covers vast areas in relatively short periods. Hence, using an airplane is particularly well suited for large-scale studies where tagged fish need to be tracked over long distances (Burger et al. 1985; Hockersmith and Peterson 1997; Collins et al. 2000). In contrast, the low accuracy of aerially estimated locations can limit conclusions of studies where exact information is needed (e.g., site residency). Nonetheless, tracking fish via an aircraft is appropriate for determining raw positions (within hundreds of metres) in remote areas and rough terrain, where other tracking techniques are impractical. However, studies based only on manual monitoring are about to become less common (Cooke and Thorstad 2012), with a greater emphasis on automated systems due to increasing sample sizes that make manual tracking rather difficult. Eiler (1995) already reported much higher tracking success when using satellite-linked tracking stations compared to repeated aerial surveys for tracking adult Chinook salmon *Oncorhynchus tshawytscha* and Coho salmon *Oncorhynchus kisutch* migrating in the large isolated Taku River in Alaska (USA).

Combining station arrays with manual tracking seems to be an effective approach to collect further information on detailed behaviour patterns when fish are outside the range of fixed stations, as it was the case of studies on migration routes of Atlantic salmon *Salmo salar* (Larinier et al. 2005; Croze et al. 2008) and European eel *Anguilla anguilla* (Travade et al. 2010). These studies focused on monitoring fishes passing through or near dams, which were the monitoring stations, and this data was complemented by daily manual tracking and weekly aerial surveys on larger watershed (Croze et al. 2008).

Stationary tracking typically uses one or more tracking stations (i.e., receiver-loggers, often termed automatic listening stations, connected to one or more receiving antennas) placed at fixed locations to detect, identify, and record radio-tagged fish that move within a reception range. Stationary automated stations provide continuous coverage at specific, standardized sites and can operate autonomously under a wide range of environmental conditions and for extended periods with adequate external power supply and maintenance. Receivers available in standard or coded versions can be programmed to scan a variety of frequencies and antennas, either sequentially or simultaneously. Additional improvements of receiver-logger capabilities, including increased internal battery life, expanded data storage capacities, and remote data retrieval abilities via one-way or two-way communication uplinks (e.g., satellites, cellular networks) have allowed for continuous fish tracking and regular access to the data. The type of information provided by stationary tracking at standardized sites include information on presence-absence at site, site passage, and accurate timing and movement rates between sites, that are directly comparable for different reaches of the study area or different groups of fish. Some data loggers are also capable of recording additional information encoded within the transmitter signal, such as water depth or whether the fish is active.

Stations have been used to document residency patterns in areas of high use, and to detect fish passage at natural or man-made migration barriers, such as dams (e.g., Beeman and Maule 2001; Moser et al. 2002; Hockersmith et al. 2003; Thorstad et al. 2003; Caudill et al. 2007; Beeman et al. 2012), fishways (Castro-Santos and Haro 2003; Gowans et al. 2003; Larinier et al. 2005; Croze et al. 2008; Calles et al. 2012), or sections of river with high velocity flows (Ovidio et al. 2007; Chase et al. 2013).

Case studies for diadromous species

The technological advances in PIT, acoustic and radio telemetry systems, including those on transmitters, hydrophones, radio antennas and receivers have made it possible to track large numbers of fish (up to 500 individual fish on a single radio frequency or thousands on a single acoustic frequency), over a wide range of species (over a hundred), and small individuals, less than 100 mm (fork length) or about 8–10 g body mass based on a minimum mass of 0.2–0.25 g for the smallest currently available transmitter respecting a maximum tag burden of about 2% (Jepsen et al. 2005), with additional relevant information about the physical environment and their physiological status. But while short-term tracking of few localized individuals can be relatively simple, monitoring large numbers of migrating fish in large rivers or at sea for extended periods can be extremely challenging. Methods suitable for studying migratory patterns over substantial distances will differ considerably from those used to determine habitat preference in small streams. Accordingly, depending on the research objectives, scale, life stage, and field conditions (e.g., remote or inaccessible sites), either a mobile or stationary approach will be selected, even though combining methods has been more frequent in current studies (Cooke and Thorstad 2012) and some examples are given hereafter.

Anadromous species

Application to small rivers, salmonids downstream migration patterns

A two years study on the downstream migration patterns of young brown trout *Salmo trutta* was carried out in a small tributary, which was a spawning and nursery habitat for both anadromous and resident salmonids (Acolas et al. 2012). Young-of-the-year brown trout were captured in La Roche River, a small tributary of the L'Oir River that flows into the Baie du Mont Saint-Michel (France). Juveniles were tagged with 12 mm PIT tags and their fate in the watershed (emigration from the spawning tributary and from the Oir River) was monitored (Acolas et al. 2012). Recapture methods were combined to study the triggers of their migration: conventional methods (electrofishing and trapping) and telemetry methods (flatbed antennas in the watershed and regular active tracking in the tributary). The RFID technology used was HDX technology, and the flatbed antennas were fixed at the mouth of each tributary of the Oir River and at the Oir River itself, combined with a trapping system (Cucherousset et al. 2005). During the study, electrofishing campaigns were carried out in the nursery stream to tag the juveniles (two campaigns) and to assess their growth and survival. Active tracking was carried along the 2.2 km of the tributary every three weeks to assess fine-scale fish movements, survival and to confirm their downstream migration (crossing information between active and passive tracking in case of missing data). Thanks to these methods, data were gathered together to perform a capture-mark-recapture analysis to disentangle the effect of body size and growth on survival and downstream migration behaviour. It was identified that body size affected mainly survival probability and not migration probability, and that growth acted on downstream migration probability, with the fastest growing fish having the highest probability of emigration (Acolas et al. 2012).

Application to fishway, entrainment or bypass assessments

In areas above and below dams and through fishways, or other possible migration barriers, the use of radio station arrays involving multiple aerial and underwater antennas has greatly improved the studies on detailed individual fish movements and migrations. The use of extensive station arrays, at a large-scale hydroelectric complex on the River Garonne (France) provided valuable information to optimize the performance of a fish lift for upstream migrating Atlantic salmon and other anadromous migratory fish species, including Allis shad *Alosa alosa*. Indeed, coverage of the whole system by six elaborate radio station arrays, allowing fine-scale monitoring of fish movements, was helpful in evaluating the functionality of the fish lift, unambiguously showing its insufficient attraction efficiency, but it also showed ways to enhance fish passage and limit delays in upstream spawning migration (Croze et al. 2008). Acoustic telemetry was also applied downstream an impassable dam and, using both active tracking and two radio linked acoustic arrays, it was possible to determine the timing and spawning activity of migrating Allis shad (Acolas et al. 2004). Similarly, Larinier et al. (2005) had gained thorough knowledge to further improve the final design of a new fish passage facility for adult Atlantic salmon, at a small hydroelectric plant on the Gave de Pau (France). In all studies, combining station arrays, either radio or acoustic, with daily manual tracking was shown as an effective approach to collect further information on detailed behaviour patterns when fish were outside the range of fixed stations.

Application to surveys near dams and large scale migration

In both fields of application (i.e., surveys of small- and large-scale fish migrations), radio telemetry has been, and continues to be applied to a wide range of anadromous fish species, primarily in salmonids and also among acipenserids (Dionne et al. 2013), clupeids (e.g., Aunins et al. 2013; Grote et al. 2014) and petromyzonids (e.g., Keefer et al. 2013), in order to study animal behaviours and ecophysiological performance in their natural freshwater environments (Cooke et al. 2013). Radio telemetry remains to date the most effective and utilized technology for studies near or around dams, or other industrial and utility infrastructures (e.g., fish passages selectivity and efficiency at weirs and power stations, temporal patterns of fish passage in fishways

(Gowans et al. 2003; Hockersmith et al. 2003; Thorstad et al. 2003; Larinier et al. 2005; Croze et al. 2008; Calles et al. 2012; Chase et al. 2013). Developments in acoustic telemetry systems (new tags at higher frequency, Ehrenberg and Steig 2003) has also greatly contributed to study salmonid behaviour in the forebay of acoustically noisy hydroelectric facility, to estimate the three-dimensional position of the fish with sub-meter accuracy and track movement patterns over time. Further, the majority of these applied studies have been conducted on relatively small spatial scales, estimating survival past a single dam or through a limited river reach (i.e., migrating time between dams).

However, telemetry methods involving fixed stations and coded tags have been increasingly used for tracking large samples of animals, over long distances and large temporal scales such as watershed levels, and using radio (Robertson et al. 2003; Croze et al. 2008; Watry and Scarnecchia 2008) or acoustic systems (Welch et al. 2008, see Jackson 2011 for a complete review). Multiple stations, placed on primary travel routes and major tributaries, have been used to determine the distribution and migratory patterns of fish in large river drainages (Eiler 1995; Keefer et al. 2004; Watry and Scarnecchia 2008). For example, several station barriers have enable assessment of upstream migration rates of nearly 2,000 radio-tagged adult spring-summer Chinook salmon *Oncorhynchus tshawytscha* through 12 unimpounded river reaches (36 to 241 km long) in the Columbia River Basin (Canada and USA) from 1997 to 2002 (Keefer et al. 2004). Manual tracking is less frequently used for these applications, particularly on larger systems where there are many tagged fish. However, the development of large-scale collective telemetry systems, at the watershed-scale level, has opened new possibilities for tracking radio-tagged fish during their entire in-river migration.

Typically, radio telemetry has proven useful in understanding fish migrations between rivers and marine environments (Watry and Scarnecchia 2008; Corbett and Brenkman 2012). For example, Corbett and Brenkman (2012) reviewed the importance of radio telemetry in establishing the extent of anadromy, timing of river entry, and variability in freshwater, estuarine and marine residence times of the bull trout *Salvelinus confluentus* inhabiting coastal watersheds (Hoh and Elwha Rivers, Washington, USA). Broad-scale migratory patterns of bull trout between multiple rivers, estuaries, and the Pacific Ocean, as well as inter-basin migrations among multiple coastal watersheds, have been addressed by using this technology (Corbett and Brenkman 2012). Life history strategies and migratory patterns of bull trout were also studied with the same method in the large Secesh River watershed (Idaho, USA) (Watry and Scarnecchia 2008).

Catadromous species

Application to fishways, entrainment or bypass assessments

For a long-term study on behaviour and passage routes of downstream migrating European silver eel *Anguilla anguilla* at small hydroelectric facilities on the Gave de Pau (France), combined arrays of radio and PIT telemetry systems were used to monitor fish approach and route-specific passage (Travade et al. 2010; Larinier et al. 2012), evaluate further technical solutions and to propose rehabilitative measures to develop and create safe downstream passages. At each site, several large radio detection zones, covered by aerial loop antennas, were established to detect presence of eels in the forebay and to determine passage routes via quite large fish passage structures, such as weirs, flood gates, surface flap gates, turbines or upstream fishways. Small radio detection zones of only a few metres, covered by underwater antennas made of stripped coaxial cable, were also delimited along trash racks or around downstream migration bypasses when appropriate. In addition, some strategic passageways of limited size (including bypasses and trash troughs) were fitted with PIT antennas to detect rapidly-moving tagged fish as they transit through such narrow engineered structures.

Despite using coded radio tags in only one or two frequencies, combining both radio and 23 or 32 mm HDX PIT tagging has successfully limited risks of detection failure for silver eels transiting through such small openings at high flow velocities. To ascertain fish detection, specific bypass structures (e.g., 2 m wide × 1 m deep, bypass flow 2.2 m^3 s^{-1}, flow velocity ca. 1 m s^{-1}) were equipped with two synchronized PIT antennas, each connected to a separate reader. Antennas were constructed on-site and consisted of wire coils mechanically protected by PVC pipe, bent to the shape of passageway openings, and usually isolated from concrete walls by wood battens and/ or polyethylene plates (Fig. 12.5). To achieve required inductances while maximizing field strengths, usually two to three turns of 13-AWG (i.e., 2.5 mm^2 cross section) multi-stranded flexible copper wire were used. With adequate *in situ* calibration and regular system checking, the 11 PIT telemetry stations, deployed at strategic passageways among three of the six sites under study, have provided useful information on migration behaviour and route-specific passage of downstream migrating silver eel at small-scale hydroelectric facilities (Travade et al. 2010; Larinier et al. 2012).

Application to surveys of yearly migrant fluxes of European eel in an obstructed river system

Capture Mark Recapture (CMR) studies can be used to estimate the proportion of candidates for emigration in a given catchment in the same year (Feunteun et al. 2000; Zabel et al. 2005). The availability of such information is critical for the catadromous European eels, and particularly to the silver stage that corresponds to maturing eels on the onset of their spawning migration. Electrofishing surveys are commonly conducted in summer to characterize the status of the sedentary fraction of the eel stocks. Among sedentary eels, pre-migrant eels achieve their silvering in late summer and then wait in the catchment until migration is triggered by floods. Pre-migrant eels can be identified by measuring external signs of silvering (Acou et al. 2005; Durif et al. 2005). By enumerating and marking these pre-migrant eels, it is possible to assess the impact of river regulation on yearly migrant fluxes (Feunteun et al. 2000; McCarthy et al. 2008).

It is generally agreed that a one-to-one relation exists between pre-migrant eels and escapement of silver eels in the consecutive autumn (Acou et al. 2009). However, this assumption could lead to biased estimates, as many factors could alter migration behaviour and the final size of migrant eels. The downstream migration corresponds to specific water conditions. If these conditions are not met during autumn and winter, eel candidates to emigration are probably constrained to wait in the catchment for favourable water conditions in the following year. A minority of them can even regress to the yellow stage. Thus, with dry winter conditions, unfavourable for emigration, the quantity of emigrating silver eels does not reflect the real potential of silver eels of the catchment. Physical obstructions also play a role. The latter may lead to (i) temporary settlement or definitive stop of the migration until the next waterflood season (Durif et al. 2003), (ii) or subsequent mortalities of emigrating eels through discharge facilities and sluice gates (Legault et al. 2003) or turbines (Larinier and Travade 2002). River regulation also eliminates seasonal flow peaks, an essential cue for eel migration (Acou et al. 2008a). This is particularly true in obstructed river systems, such as the Fremur River (western France) that presents a 3×10^6 m^3 water supply reservoir created by a 14 m high dam, the Bois Joli dam, located only at 6 km from the estuary. The Bois Joli dam has been equipped with an eel lift to restore upstream migration, but no equipment has been provided for downstream migration. Apart from a minor number of silver eels that are able to find the minimum flow discharge pipe of the dam (Legault et al. 2003), most silver eels remain trapped in the reservoir and are not able to pass downstream until the dam is filled and flows pass over the crest (Acou et al. 2008a). In 2012, an acoustic telemetry survey led in the reservoir showed that among 20 silver eels marked, only one third managed to move downstream the river, despite the exceptionally favourable environmental conditions (E. Feunteun, unpubl. data).

The aim of this case study is to illustrate the type of information that can be gained from a CMR study for a threatened species. Specifically, we were interested in addressing the following questions: How many candidates for emigration are lost each year in the reservoir? For that, we assessed, during nine years (1996–2004) in the Frémur catchment, the relation between pre-migrant eels and escapement of silver eels in the consecutive year. Sampling was conducted in the low-water level period (September), i.e., after the beginning of silvering and before emigration of silver eels (Fontaine 1994). Sampled eels were measured for length and weight and released directly outside the sampled river section. Pre-migrant silver eels were identified by three criteria: colour of the back and belly, presence of a well-defined lateral line and ocular hypertrophy (Acou et al. 2005). If only two of the criteria were met, the eel was classified as yellow/silver eel which were assumed to be 'candidates for emigration' during the next season. If only one or none occurred, the eel was recorded as yellow. All eels bigger than 200 mm were marked individually using PIT tags injected with a syringe into the general cavity. Induced mortality and PIT tag rejection were tested. Fourteen percent of the tags were rejected within an hour after the injection (Feunteun et al. 2000). After this period, tag losses were very low. Therefore, the eels were kept for at least an hour after tagging before they were released in the river. Overall, 1097 eels were tagged during the study period. Among them, most were tagged at the yellow eel stage (73%). Yellow/silver and silver eels represent 10 and 17% of total marked eels respectively (Table 12.1). A Wolf trap was installed at 4.5 km from the estuary, and designed to capture every descending eel bigger than 200 mm under practically every flow condition. Daily monitoring was conducted between September 1996 and August 2004 (i.e., over eight downstream migration seasons), with count of descending eel made from September 1 to August 31 each season. Over this period, the trap was

Table 12.1. Number of eels PIT-tagged during electrofishing surveys and recaptured in a Wolf trap. Note that only 4.5% (5/111) of yellow/silver eels and 0.1% (1/800) of yellow eels, marked in the river Frémur catchment, were recaptured as silver eel in the Wolf trap the following year.

Tagging			Recapture																		
			Silver eels									Yellow/silver eels									Total
Eel stage	Electrofishing period	No.	1996	1997	1998	1999	2000	2001	2002	2003	2004	1996	1997	1998	1999	2000	2001	2002	2003	2004	
Silver eels	Sept 1996	59	12	5	2	0	0	0	0	0	0	0	0	0	0	0	0	0	0	0	19
	Sept 1997	22		5	0	0	0	0	0	0	0		0	0	0	0	0	0	0	0	5
	Sept 1998	25			7	0	0	0	0	0	0			0	0	0	0	0	0	0	7
	Sept 1999	1				1	0	0	0	0	0				0	0	0	0	0	0	1
	Sept 2000	40					17	0	0	0	0					1	0	0	0	0	18
	Sept 2001	10						0	1	0	0						0	0	0	0	1
	Sept 2002	19							4	0	0							0	0	0	4
	Sept 2003	10								7	0								0	0	7
	Sept 2004	0									0									0	0
Yellow/silver eels	Sept 1996	17	0	0	0	0	0	0	0	0	0	0	0	0	0	0	0	0	0	0	0
	Sept 1997	14		0	0	1	0	0	1	0	0		0	0	0	0	0	0	0	0	2
	Sept 1998	36			1	1	0	0	0	0	0			0	0	0	0	0	0	0	2
	Sept 1999	1				0	0	0	0	0	0				0	0	0	0	0	0	0
	Sept 2000	13					1	1	0	0	0					0	0	0	0	0	2
	Sept 2001	27						0	0	0	0						0	0	0	0	0
	Sept 2002	0								0	0								0	0	0
	Sept 2003	3								3	0									0	3
	Sept 2004	0									0										0
Yellow eels	Sept 1996	328	1	2	10	11	5	0	3	6	4	0	0	0	0	1	1	0	0	0	44
	Sept 1997	106		0	1	3	3	1	0	0	0		0	0	0	0	0	0	0	0	8
	Sept 1998	71			0	3	1	0	1	0	0			0	0	0	0	0	0	0	5
	Sept 1999	1				0	1	0	0	0	0				0	0	0	0	0	0	1
	Sept 2000	100					0	3	6	3	4					0	1	0	0	0	17
	Sept 2001	69						0	0	2	0						0	0	0	0	2
	Sept 2002	59							0	3	2							0	0	0	5
	Sept 2003	66								0	2								0	0	2
	Sept 2004	0									0										0
	Total	1097	13	12	21	20	28	5	16	24	12	0	0	0	0	2	2	0	0	0	155

inspected approximately once every two days, but every day during migration peaks. Each fish collected was measured, classified either yellow, yellow/silver or silver stage according to the same methodology presented before, and the presence of PIT tags was verified using a reader.

Among the 186 silver eels individually marked with PIT tag over the study period, 28.5% (n = 53) were recaptured in the trap the following year, and 4.3% (n = 8) probably settled in the reservoir for one or two extra years before emigration, while the fate of 67.2% (n = 125) of the silver eels marked is unknown (Table 12.1). Subsequent sampling campaigns (surveys were pursued until the summer of 2014) did not allow the recapture of these individuals, either in the river or in the trap. Natural mortality, fishing pressure and predation mortality might have reduced the tagged population. However, eel mortality is also rather low in natural populations, about 5–10% year^{-1} (Adam 1997). Moreover, both fishing pressure and predation mortality are low in the Fremur River as (i) there is no professional fishery and anglers mainly focus on cyprinids, esocids or percids, and (ii) only a few number of cormorants and herons are present in the study site. Therefore, it is likely that pre-migrant silver eels settled and finally probably died in the Bois Joli reservoir. As a consequence of eutrophication, cyanobacterial blooms regularly occur from July to September (40 µg L^{-1}, principally *Microcystis aeruginosa*) equivalent to 40 times the World Health Organization provisional guidelines value for microcystins in drinking water (Chorus and Bartram 1999). There are clear indications that such cyanobacterial concentration have severe impacts on water quality such as pH or oxygen levels (Briand et al. 2003) and induce severe damages in fish as it has been shown in silver eel livers (Acou et al. 2008b) and probably mortality (Malbrouck and Kestemont 2006). This long-term study of PIT-tagged silver eels, led between 1996 and 2004, showed that the fate of two thirds of the pre-migrant silver eels tagged above the reservoir remain unknown, the following scenario is thus envisaged. A majority (66.7%) of pre-migrant silver eels are not able to reach downstream areas but remain trapped in the reservoir, where they finally died after a while because of huge concentration of microcystins and/or bad water quality. Water reservoirs may constitute a major threat for European eels as they are widespread through the distribution range of this endangered species.

Amphidromous species

Telemetry techniques have not yet been extensively used to study amphidromous fish. This is probably due to the fact that studies on amphidromous species are more recent than for anadromous and catadromous species, for which studies are more ancient. The characterization of the amphidromous character being also more debatable than for the two others migration strategy (Myers 1949). However telemetry is defined as one of the key issues in the future to study amphidromous fish migration behaviour (Miles et al. 2014).

Smith and Kwack (2014) used both RFID techniques and radio telemetry to confirm the amphidromous character of bigmouth sleeper *Gobiomorus dormitor* (Eleotridae) and mountain mullet *Agonostomus monticola* (Mugilidae), by studying their reproductive migration patterns in freshwater. They analyze the data thanks to a capture-recapture approach that allowed them to estimate the use of a different part of the estuary (lower part to upstream part), and the survival probabilities during the spawning season.

Acoustic and radio telemetry were used to assess reproduction migration of Australian grayling *Prototroctes maraena* (Retropinnidae) coupled with drift netting for egg catching (Koster et al. 2013). The purpose was to assess the cues of this downstream migration for management and conservation purposes, since the species is under a national action plan in Australia (Backhouse et al. 2008). The study confirms that the adults migrated downstream, and linked to water flow conditions within the freshwater area just above the estuary, for reproduction in a common area. The authors explain this behaviour could benefit the eggs that would have a shorter time to reach the sea. After reproduction, all adults went back to their original capture area which demonstrates site fidelity. However the study suffers from a high tagged fish loss (especially with radio tags), once the species is highly sensitive to handling; therefore further refinement tests for transmitter insertion are needed.

As reviewed by Miles et al. (2014) for Australian diadromous fish, telemetry tools offer an opportunity to characterize migratory movements and occurrence of facultative diadromy. The differentiation between catadromy and amphidromy being allowed thanks to these techniques (i.e., Smith and Kwak 2014).

The size of the transmitters is decreasing, especially in acoustic telemetry (i.e., McMichael et al. 2010), which could allow new insights in largely unknown juvenile behaviour of amphidromous species (Miles et al. 2014). The coupling of a physiological sensor within telemetry tags (i.e., Cooke et al. 2008) is also an opportunity to understand how amphidromous fish can cope with migratory barriers in freshwater habitats, within the frame of conservation purposes.

Acknowledgements

We are grateful to the "Region Bretagne" and "Bretagne Grands Migrateurs" that funded the Fremur River study, and "Fish Pass" company (F. Charrier et V. Mazel) that conducted the sampling and data gathering.

References

Aarestrup K, Lucas MC and Hansen JA (2003) Efficiency of a natural-like bypass channel for sea trout (*Salmo trutta*) ascending a small Danish stream studied by PIT telemetry. Ecology of Freshwater Fish 12: 160–168.

Aarestrup K, Okland F, Hansen MM, Righton D, Gargan P, Castonguay M, Bernatchez L, Howey P, Sparholt H, Pedersen MI and McKinley RS (2009) Oceanic spawning migration of the European eel (*Anguilla anguilla*). Science 325: 1660.

Acolas M-L, Bégout Anras M-L, Véron V, Jourdan H, Sabatié MR and Baglinière JL (2004) Allis Shad upstream migration and reproductive behaviour: assessment using acoustic tracking methodologies. ICES Journal of Marine Science 61: 1291–1304.

Acolas ML, Roussel JM, Lebel JM and Baglinière JL (2007) Laboratory experiment on survival, growth and tag retention following PIT injection into the body cavity of juvenile brown trout (*Salmo trutta*). Ecology of Freshwater Fish 86: 280–284.

Acolas ML, Labonne J, Baglinière JL and Roussel J (2012) The role of body size versus growth on the decision to migrate: a case study with *Salmo trutta*. Naturwissenschaften 99: 11–21.

Acou A, Boury P, Laffaille P, Crivelli AJ and Feunteun E (2005) Towards a standardized characterization of the potentially migrating silver European eel (*Anguilla anguilla*, L.). Archivfür Hydrobiologie 164: 237–255.

Acou A, Laffaille P, Legault A and Feunteun E (2008a) Migration pattern of silver eel (*Anguilla anguilla*, L.) in an obstructed river system. Ecology of Freshwater Fish 17: 432–442.

Acou A, Robinet T, Mounaix B, Brient L, Gérard C, Le Rouzic B and Feunteun E (2008b) Evidence of silver eels contamination by microcystin-LR at the onset of their seaward migration: what consequences for spawning success? Journal of Fish Biology 72: 753–762.

Acou A, Gabriel G, Laffaille P and Feunteun E (2009) Differential potential production and condition indices of silver eels (*Anguilla anguilla* L.) in two small Atlantic coastal catchments of France. pp. 157–174. *In*: Casselman J and Cairns D (eds.). Eels at the Edge: Science, Status, and Conservation Concerns. American Fisheries Society Symposium, Bethesda, MD.

Adam G (1997) L'anguille européenne (*Anguilla anguilla* L. 1758): dynamique de la sous-population du lac de Grand-Lieu en relation avec les facteurs environnementaux et anthropiques. Thèse de l'Université de Toulouse. 299pp.

Adams NS, Rondorf DW, Evans SD, Kelly JE and Perry RW (1998) Effects of surgically and gastrically implanted radio transmitters on swimming performance and predator avoidance of juvenile chinook salmon (*Oncorhynchus tshawytscha*). Canadian Journal of Fisheries and Aquatic Sciences 55: 781–787.

Adams NS, Beeman JW and Eiler JH (eds.) (2012) Telemetry Techniques: A User Guide for Fisheries Research. American Fisheries Society, Bethesda, MD.

Alexandre C, Quintella BR, Silva A, Mateus C, Romão F, Branco P, Ferreira MT and Almeida PR (2013) Use of electromyogram telemetry to assess the behavior of the Iberian barbel (*Luciobarbus bocagei* Steindachner, 1864) in a pool-type. Ecological Engineering 51: 191–202.

Anonymous (1968) Aide mémoire d'acoustique sous-marine. Détection sous-marine. Marine Nationale Laboratoire Française.

Archdeacon TP, Remshardt WJ and Knecht TL (2009) Comparison of two methods for implanting passive integrated transponders in Rio Grande silvery minnow. North American Journal of Fisheries Management 29: 346–351.

Aunins AW, Brown BL, Balazik M and Garman GC (2013) Migratory movements of American shad in the James River Fall Zone, Virginia. North American Journal of Fisheries Management 33: 569–575.

Axel GA, Prentice EF and Sandford BP (2005) PIT-tag detection system for large-diameter juvenile fish bypass pipes at Columbia River Basin hydroelectric dams. North American Journal of Fisheries Management 25: 646–651.

Aymes JC and Rives J (2009) Detection efficiency of multiplexed Passive Integrated Transponder antennas is influenced by environmental conditions and fish swimming behaviour. Ecology of Freshwater Fish 18(4): 507–513.

Backhouse G, O'Connor J and Jackson J (2008) National Recovery Plan for the Australian Grayling *Prototroctes maraena*. Victorian government department of sustainability and environment, Melbourne. 16p.

Baras E (1991) A bibliography on underwater telemetry, 1956–1990. Canadian technical report of fisheries and aquatic sciences 1819: 1–55.

Baras E, Malbrouck C, Houbart M, Kestemont P and Mélard C (2000) The effect of PIT tags on growth and physiology of age-0 cultured Eurasian perch *Perca fluviatilis* of variable size. Aquaculture 185: 159–173.

Barbour AB, Adams AJ, Yess T, Behringer DC and Wolfe RK (2012) Comparison and cost-benefit analysis of pit tag antennae resighting and seine-net recapture techniques for survival analysis of an estuarine-dependent fish. Ecology of Freshwater Fish 121-122: 153–160.

Bass AL, Giannico GR and Brooks GT (2012) Performance of a full-duplex passive integrated transponder (PIT) antenna system in estuarine channels. Marine and Coastal Fisheries. Dynamics, Management, and Ecosystem Science 4: 145–155.

Bauer C and Schlott G (2004) Overwintering of farmed common carp (*Cyprinus carpio* L.) in the ponds of a central European aquaculture facility–measurement of activity by radio telemetry. Aquaculture 241: 301–317.

Bauer C, Unfer G and Loupal G (2005) Potential problems with external trailing antennas: antenna migration and ingrowth of epithelial tissue—a case study from a recaptured nase, *Chondrostoma nasus* (L.). Journal of Fish Biology 67: 885–889.

Beeman JW and Maule AG (2001) Residence times and diel passage distributions of radio-tagged juvenile spring chinook salmon and steelhead in a gatewell and fish collection channel of a Columbia river dam. North American Journal of Fisheries Management 21: 455–463.

Beeman JW, Grant C and Haner PV (2004) Comparison of three underwater antennas for use in radiotelemetry. North American Journal of Fisheries Management 24: 275–281.

Beeman JW, Hockersmith EE and Stevenson JR (2012) Design and performance of radio telemetry systems for assessing juvenile fish passage at three hydroelectric dams. pp. 281–302. *In*: Adams NS, Beeman JW and Eiler JH (eds.). Telemetry Techniques: A User Guide for Fisheries Research. American Fisheries Society, Bethesda, MD.

Bégout Anras M-L, Cooley P, Bodaly RA, Anras L and Fudge RJP (1999) Movement and habitat use by lake whitefish during spawning season in a boreal lake integrating acoustic telemetry and G.I.S. Transactions of the American Fisheries Society 128: 939–952.

Bégout M-L, Kadri S, Huntingford F and Damsgård B (2012) Techniques for studying the behaviour of farmed fish. pp. 65–86. *In*: Huntingford F, Jobling M and Kadri S (eds.). Aquaculture and Behavior. Blackwell Publishing Ltd. SPi Publishers, Pondicherry, India.

Bergé J, Capra H, Pella H, Steig T, Ovidio M, Bultel E and Lamouroux N (2012) Probability of detection and positioning error of a hydro acoustic telemetry system in a fast-flowing river: intrinsic and environmental determinants. Ecology of Freshwater Fish 125: 1–13.

Bergek S and Björklund M (2009) Genetic and morphological divergence reveals local subdivision of perch (*Perca fluviatilis* L.). Biological Journal of the Linnean Society 96: 746–758.

Bij de Vaate A, Breukelaar AW, Vriese T, De Laak G and Dijkers C (2003) Sea trout migration in the Rhine delta. Journal of Fish Biology 63: 892–908.

Block BA, Keen JE, Castillo B, Dewar H, Freund EV, Marcinek DJ, Brill RW and Farwell C (1997) Environmental preferences of yellowfin tuna (*Thunnus albacores*) at the northern extent of its range. Marine Biology 130: 119–132.

Block BA, Jonsen I, Jorgensen S, Winship A, Shaffer SA, Bograd S, Hazen E, Foley D, Breed G and Harrison A-L (2011) Tracking apex marine predator movements in a dynamic ocean. Nature 475: 86–90.

Bograd SJ, Block BA, Costa DP and Godley BJ (2010) Biologging technologies: new tools for conservation. Introduction. Endangered Species Research 10: 1–7.

Boitani L and Powell RA (eds.) (2012) Carnivore Ecology and Conservation: A Handbook of Techniques. Oxford University Press.

Boubée JAT and Williams EK (2006) Downstream passage of silver eels at a small hydroelectric facility. Fisheries Management and Ecology 13: 165–176.

Breukelaar AW, Bij de Vaate A and Fockens KTW (1998) Inland migration study of sea trout (*Salmo trutta*) into the rivers Rhine and Meuse (The Netherlands), based on inductive coupling radio telemetry. Hydrobiologia 372: 29–33.

Breukelaar AW, Fockens KTW and Bij de Vaate A (2000) Technical aspects of the NEDAP-TRAIL system used in a sea trout (*Salmo trutta* L.) migration study. pp. 7–11. *In*: Moore A and Russell IC (eds.). Advances in Fish Telemetry. CEFAS, Lowestoft, UK.

Brevé NWP, Vis H, Spierts I, de Laak GAJ, Moquette F and Breukelaar AW (2014a) Exorbitant mortality of hatchery-reared Atlantic salmon smolts *Salmo salar* L., in the Meuse river system in the Netherlands. Journal of Coastal Conservation 18: 97–109.

Brevé NWP, Vis H, Houben B, de Laak GAJ, Breukelaar AW, Acolas M-L, de Bruijn QAA and Spierts I (2014b) Exploring the possibilities of seaward migrating juvenile European sturgeon *Acipenser sturio* L., in the Dutch part of the River Rhine. Journal of Coastal Conservation 18: 131–143.

Briand JF, Jacquet S, Bernard C and Humber JF (2003) Health hazards for terrestrial vertebrates from toxic cyanobacteria in surface water ecosystems. Veterinary Research 34: 361–377.

Bridger CJ and Booth RK (2003) The effects of biotelemetry transmitter presence and attachment procedures on fish physiology and behavior. Reviews in Fisheries Science 11: 13–34.

Brill RW, Block BA, Boggs CH, Bigelow KA, Freund EV and Marcinek DJ (1999) Horizontal movements and depth distribution of large adult yellowfin tuna (*Thunnus albacores*) near the Hawaiian Islands, recorded using ultrasonic telemetry: implications for the physiological ecology of pelagic fishes. Marine Biology 133: 395–408.

Brown RS and Mackay WC (1995) Spawning ecology of cutthroat trout (*Oncorhynchus clarki*) in the Ram River, Alberta. Canadian Journal of Fisheries and Aquatic Sciences 52: 983–992.

Brown RS, Power G and Beltaos S (2001) Winter movements and habitat use of riverine brown trout, white sucker and common carp in relation to flooding and ice break-up. Journal of Fish Biology 59: 1126–1141.

Brown RS, Geist DR and Mesa MM (2006) The use of electromyogram (EMG) telemetry to assess swimming activity and energy use of adult spring chinook salmon migrating past a Columbia River dam. Transactions of the American Fisheries Society 135: 281–287.

Brown RS, Eppard MB, Murchie KJ, Nielsen JL and Cooke SJ (2011) An introduction to the practical and ethical perspectives on the need to advance and standardize the intracoelomic surgical implantation of electronic tags in fish. Reviews in Fish Biology and Fisheries 21: 1–9.

Bubb DH, Thom TJ and Lucas MC (2006) Movement patterns of the invasive signal crayfish determined by PIT telemetry. Canadian Journal of Zoology 84: 1202–1209.

Bultel E, Lasne E, Acou A, Guillaudeau J, Bertier C and Feunteun E (2014) Migration behaviour of silver eels (*Anguilla anguilla*) in a large estuary of Western Europe inferred from acoustic telemetry. Estuarine, Coastal and Shelf Science 137: 23–31.

Bunt CM, Cooke SJ and Philipp DP (2003) A simple technique for obtaining radio signals from deep water. North American Journal of Fisheries Management 23: 258–263.

Burger CV, Wilmot RL and Wangaard DB (1985) Comparison of spawning areas and times for two runs of chinook salmon (*Oncorhynchus tshawytscha*) in the Kenai River, Alaska. Canadian Journal of Fisheries and Aquatic Sciences 42: 693–700.

Burnett NJ, Stamplecoskie KM, Thiem JD and Cooke SJ (2013) Comparison of detection efficiency among three sizes of half-duplex passive integrated transponders using manual tracking and fixed antenna arrays. North American Journal of Fisheries Management 33: 7–13.

Calles O, Karlsson S, Hebrand M and Comoglio C (2012) Evaluating technical improvements for downstream migrating diadromous fish at a hydroelectric plant. Ecological Engineering 48: 30–37.

Carlin B (1955) Tagging of salmon smolts in the River Lagan. Report of the Institute of Freshwater Research Drottningholm 36: 57–74.

Castro-Santos T and Haro A (2003) Quantifying migratory delay: a new application of survival analysis methods. Canadian Journal of Fisheries and Aquatic Sciences 60: 986–996.

Castro-Santos T, Haro A and Walk S (1996) A passive integrated transponder (PIT) tag system for monitoring fishways. Ecology of Freshwater Fish 28: 253–261.

Caudill CC, Daigle WR, Keefer ML, Boggs CT, Jepson MA, Burke BJ, Zabel RW, Bjornn TC and Peery CA (2007) Slow dam passage in adult Columbia River salmonids associated with unsuccessful migration: delayed negative effects of passage obstacles or condition-dependent mortality? Canadian Journal of Fisheries and Aquatic Sciences 64: 979–995.

Chase R, Hemphill N, Beeman J, Juhnke S, Hannon J and Jenkins AM (2013) Assessment of juvenile coho salmon movement and behaviour in relation to rehabilitation efforts in the Trinity River, California, using PIT tags and radiotelemetry. Environmental Biology of Fishes 96: 303–314.

Chorus I and Bartram J (eds.) (1999) Toxic Cyanobacteria in Water: A guide to their Public Health Consequences, Monitoring and Management. Routledge, London.

Clark TD, Sandblom E, Hinch S, Patterson D, Frappell P and Farrell A (2010) Simultaneous biologging of heart rate and acceleration, and their relationships with energy expenditure in free-swimming sockeye salmon (*Oncorhynchus nerka*). Journal of Comparative Physiology B 180: 673–684.

Cocherell SA, Cocherell DE, Jones GJ, Miranda JB, Thompson LC, Cech JJ, Jr. and Klimley AP (2011) Rainbow trout *Oncorhynchus mykiss* energetic responses to pulsed flows in the American River, California, assessed by electromyogram telemetry. Environmental Biology of Fishes 90: 29–41.

Collins MR, Smith TI, Post WC and Pashuk O (2000) Habitat utilization and biological characteristics of adult Atlantic sturgeon in two South Carolina rivers. Transactions of the American Fisheries Society 129: 982–988.

Collins MR, Cooke DW, Smith TIJ, Post WC, Russ DC and Walling DC (2002) Evaluation of four methods of transmitter attachment on shortnose sturgeon, *Acipenser brevirostrum*. Journal of Applied Ichthyology 18: 491–494.

Cooke SJ and Bunt CM (2001) Assessment of internal and external antenna configurations from transmitters implanted in smallmouth bass. North American Journal of Fisheries Management 21: 236–241.

Cooke SJ and Thorstad E (2012) Is radio telemetry getting washed downstream? The changing role of radio telemetry in studies of freshwater fish relative to other tagging and telemetry technology. American Fisheries Society Symposium 76: 349–369.

Cooke SJ, Thorstad EB and Hinch SG (2004) Activity and energetics of free-swimming fish: insights from electromyogram telemetry. Fish and Fisheries 5: 21–52.

Cooke SJ, Hinch SG, Farrell AP, Patterson DA, Miller-Saunders K, Welch DW, Donaldson MR, Hanson KC, Crossin GT, Olsson I, Cooperman MS, Mathes MT, Hruska KA, Wagner GN, Thomson R, Hourston R, English KK, Larsson S, Shrimpton JM and Van Der Kraak G (2008) Developing a mechanistic understanding of fish migrations by linking telemetry with physiology, behavior, genomics and experimental biology: an interdisciplinary case study on adult Fraser River sockeye salmon. Fisheries 33: 321–338.

Cooke SJ, Midwood JD, Thiem JD, Klimley P, Lucas MC, Thorstad EB, Eiler J, Holbrook C and Ebner BC (2013) Tracking animals in freshwater with electronic tags: past, present and future. Animal Biotelemetry 1: 19.

Corbett SC and Brenkman SJ (2012) Two case studies from Washington's Olympic peninsula: radio telemetry reveals bull trout anadromy and establishes baseline information prior to large-scale dam removal. pp. 207–220. *In*: Adams NS, Beeman JW and Eiler JH (eds.). Telemetry Techniques: A User Guide for Fisheries Research. American Fisheries Society, Bethesda, MD.

Cote D, Scruton DA, Niezgoda GH, Mckinley RS, Rowsell DF, Lindstrom RT, Ollerhead LMN and Whitt CJ (1998). A coded acoustic telemetry system for high precision monitoring of fish location and movement: application to the study of nearshore nursery habitat of juvenile Atlantic cod (*Gadus morhua*). Marine Technology Society Journal 32: 54–62.

Cousin X, Daouk T, Péan S, Lyphout L, Schwartz M-E and Bégout M-L (2012) Electronic individual identification of zebrafish using radio frequency identification (RFID) microtags. The Journal of Experimental Biology 215: 2729–2734.

Croze O, Bau F and Delmouly L (2008) Efficiency of a fish lift for returning Atlantic salmon at a large-scale hydroelectric complex in France. Fisheries Management and Ecology 15: 467–476.

Cucherousset J, Roussel J-M, Keeler R, Cunjak RA and Stump R (2005) The use of two new portable 12-mm pit tag detectors to track small fish in shallow streams. North American Journal of Fisheries Management 25: 270–274.

Cucherousset J, Paillisson J-M and Roussel J-M (2007) Using PIT technology to study the fate of hatchery-reared YOY northern pike released into shallow vegetated areas. Ecology of Freshwater Fish 85: 159–164.

De Leeuw JJ and Winter HV (2008) Migration of rheophilic fish in the large lowland rivers Meuse and Rhine, the Netherlands. Fisheries Management and Ecology 15: 409–415.

De Oliveira E (2012) Determining how eels overcome a series of obstacles on the Rhine. pp. 95–97. *In*: Baran P, Basilico L, Larinier M, Rigaud C and Travade F (eds.). Management plan to save the eel: Optimising the design and management of installations. Proc. Symp. Eels and Ecological Continuity: Optimising the design and management of installations, Onema, Paris.

Deary C, Scruton DA, Niezgoda GH, McKinley S, Cote D, Clarke KD, Perry D, Lindstrom T and White D (1998) A dynamically switched combined acoustic and radio transmitting (CART) tag: an improved tool for the study of diadromous fishes. Marine Technology Society Journal 32: 63–69.

Dionne PE, Zydlewski GB, Kinnison MT, Zydlewski J and Wippelhauser GS (2013) Reconsidering residency: characterization and conservation implications of complex migratory patterns of shortnose sturgeon (*Acipenser brevirostrum*). Canadian Journal of Fisheries and Aquatic Sciences 70: 119–127.

Downing SL, Prentice EF, Frazier RW, Simonson JE and Nunnallee EP (2001) Technology developed for diverting passive integrated transponder (PIT) tagged fish at hydroelectric dams in the Columbia River Basin. Aquacultural Engineering 25: 149–164.

Durif C, Elie P, Gosset C, Rives J and Travade F (2003) Behavioural study of downstream migrating eels by radiotelemetry at a small hydroelectric power plant. pp. 343–356. *In*: Dixon DA (ed.). Biology, Management, and Protection of Catadromous Eels. American Fisheries Society, Bethesda, MD.

Durif C, Dufour S and Elie P (2005) The silvering process of *Anguilla anguilla*: a new classification from the yellow resident to the silver migrating stage. Journal of Fish Biology 66: 1025–1043.

Eiler JH (1990) Radio transmitters used to study salmon in glacial rivers. American Fisheries Society Symposium 7: 364–369.

Eiler JH (1995) A remote satellite-linked tracking system for studying Pacific salmon with radio telemetry. Transactions of the American Fisheries Society 124: 184–193.

Eiler JH (2012) Tracking aquatic animals with radio telemetry. pp. 163–204. *In*: Adams NS, Beeman JW and Eiler JH (eds.). Telemetry Techniques: A User Guide for Fisheries Research. American Fisheries Society, Bethesda, MD.

Eiler JH, Nelson BD and Bradshaw RF (1991) Radio tracking chinook salmon (*Oncorhynchus tshawytscha*) in a large turbid river. pp. 202–206. *In*: Uchiyama A and Amlaner CJ, Jr. (eds.). Biotelemetry XI. Proceedings of the 11th International Symposium on Biotelemetry. Waseda University Press, Tokyo.

Ehrenberg JE and Steig TW (2003) Improved techniques for studying the temporal and spatial behavior of fish in a fixed location. ICES Journal of Marine Science 60: 700–706.

Espinoza M, Farrugia TJ, Webber DM, Smith F and Lowe CG (2011) Testing a new acoustic telemetry technique to quantify long-term, fine-scale movements of aquatic animals. Ecology of Freshwater Fish 108: 364–371.

Ferrari S, Chatain B, Cousin X, Leguay D, Vergnet A, Vidal M-O, Vandeputte M and Bégout M-L (2014) Early individual electronic identification of sea bass using RFID microtags: a first example of early phenotyping of sex-related growth. Aquaculture 426: 165–171.

Feunteun E, Acou A, Laffaille P and Legault A (2000) European eel, *Anguilla anguilla*: population parameters and prediction of spawner escapement from continental hydrosystems. Canadian Journal of Fisheries and Aquatic Sciences 57: 1627–1635.

Fontaine YA (1994) L'argenture de l'anguille: Métamorphose, Anticipation, Adaptation. Bulletin Français de la Pêche et de la Pisciculture 335: 171–185.

Frederick JL (1997) Evaluation of fluorescent elastomer injection as a method for marking small fish. Bulletin of Marine Science 61: 399–408.

Garcia de Leaniz C, Consuegra S and Serdio A (2004) Maladaptation and phenotypic mismatch in cultured Atlantic salmon used for stocking. Journal of Fish Biology 65: 317–318.

Geist DR, Brown RS, Lepla K and Chandler J (2002) Practical application of electromyogram radiotelemetry: the suitability of applying laboratory-acquired calibration data to field data. North American Journal of Fisheries Management 22: 474–479.

Gibbons JW and Andrews KM (2004) PIT tagging: simple technology at its best. BioScience 54: 447–454.

Gillis NC, Rapp T, Hasler CT, Wachelka H and Cooke SJ (2010) Spatial ecology of adult muskellunge (*Esox masquinongy*) in the urban Ottawa reach of the historic Rideau Canal, Canada. Aquatic Living Resources 23: 225–230.

Gingerich AJ, Bellgraph BJ, Brown RS, Tavan NT, Deng ZD and Brown JR (2012) Quantifying reception strength and omnidirectionality of underwater radio telemetry antennas: advances and applications for fisheries research. Ecology of Freshwater Fish 121: 1–8.

Gowans ARD, Armstrong JD, Priede IG and McKelvey S (2003) Movements of Atlantic salmon migrating upstream through a fish-pass complex in Scotland. Ecology of Freshwater Fish 12: 177–189.

Gravel M-A and Cooke SJ (2008) Severity of barotrauma influences the physiological status, postrelease behavior, and fate of tournament-caught smallmouth bass. North American Journal of Fisheries Management 28: 607–617.

Gravel M-A and Cooke SJ (2013) Does nest predation pressure influence the energetic cost of nest guarding in a teleost fish? Environmental Biology of Fishes 96: 93–107.

Grote AB, Bailey MM and Zydlewski JD (2014) Movements and demography of spawning American shad in the Penobscot River, Maine, prior to dam removal. Transactions of the American Fisheries Society 143: 552–563.

Habib B, Shrotriya S, Sivakumar K, Sinha PR and Mathur VB (2014) Three decades of wildlife radio telemetry in India: a review. Animal Biotelemetry 2: 4.

Halttunen E, Jensen JLA, Næsje TF, Davidsen JG, Thorstad EB, Chittenden CM, Hamel S, Primicerio R, Rikardsen AH and Taylor E (2013) State-dependent migratory timing of postspawned Atlantic salmon (*Salmo salar*). Canadian Journal of Fisheries and Aquatic Sciences 70: 1063–1071.

Harden Jones FR, Arnold GP, Greer Walker M and Scholes P (1979) Selective tidal stream transport and the migration of plaice (*Pleuronectes platessa* L.) in the southern North Sea. Journal du Conseil International pour l'Exploration de la Mer 38: 331–337.

Hardman PA and Woodward B (1984) Underwater location fixing by a diver-operated acoustic telemetry system. Acustica 55: 34–42.

Hazen EL, Maxwell SM, Bailey H, Bograd SJ, Hamann M, Gaspar P, Godley BJ and Shillinger GL (2012) Ontogeny in marine tagging and tracking science: technologies and data gaps. Marine Ecology Progress Series 457: 221–240.

Hewitt DA, Janney EC, Hayes BS and Shively RS (2010) Improving inferences from fisheries capture-recapture studies through remote detection of pit tags. Fisheries 35: 217–231.

Hill MS, Zydlewski GB, Zydlewski JD and Gasvoda JM (2006) Development and evaluation of portable PIT tag detection units: PIT packs. Ecology of Freshwater Fish 77: 102–109.

Hirt-Chabbert JA and Young OA (2012) Effects of surgically implanted PIT tags on growth, survival and tag retention of yellow shortfin eels *Anguilla australis* under laboratory conditions. Journal of Fish Biology 81: 314–319.

Hockersmith EE and Peterson BW (1997) Use of the global positioning system for locating radio-tagged fish from aircraft. North American Journal of Fisheries Management 17: 457–460.

Hockersmith EE, Muir WD, Smith SG, Sandford BP, Perry RW, Adams NS and Rondorf DW (2003) Comparison of migration rate and survival between radio-tagged and PIT-tagged migrant yearling chinook salmon in the Snake and Columbia Rivers. North American Journal of Fisheries Management 23: 404–413.

Holland KN, Brill RW and Chang RKC (1990) Horizontal and vertical movement of yellowfin and bigeye tuna associated with fish aggregating devices. Fishery Bulletin US 88: 493–507.

Hyun S-Y, Keefer ML, Fryer JK, Jepson MA, Sharma R, Caudill CC, Whiteaker JM and Naughton GP (2012) Population-specific escapement of Columbia River fall Chinook salmon: Trade-offs among estimation techniques. Ecology of Freshwater Fish 129-130: 82–93.

Ingraham JM, Deng ZD, Martinez JJ, Trumbo BA, Mueller RP and Weiland MA (2014) Feasibility of tracking fish with acoustic transmitters in the Ice Harbor Dam tailrace. Scientific Reports 4: 4090.

Jackson GD (2011) The Development of the Pacific Ocean Shelf Tracking Project within the Decade Long Census of Marine Life. PLOS ONE 6(4): e18999. doi:10.1371/journal.pone.0018999.

Jacobsen L, Berg S, Broberg M, Jepsen N and Skov C (2002) Activity and food choice of piscivorous perch (*Perca fluviatilis*) in a eutrophic shallow lake: a radio-telemetry study. Freshwater Biology 47: 2370–2379.

Jadot C (2003) Comparison of two tagging techniques for *Sarpa salpa*: external attachment and intraperitoneal implantation. Oceanologica Acta 26: 497–501.

Jellyman D (2009) A review of radio and acoustic telemetry studies of freshwater fish in New Zealand. Marine & Freshwater Research 60: 321–327.

Jensen A and Chen YQ (2013) Tracking tagged fish with swarming Unmanned Aerial Vehicles using fractional order potential fields and Kalman filtering. *In*: Proceeding of the International Conference on Unmanned Aircraft Systems (ICUAS), IEEE 1144–1149.

Jepsen N, Koed A, Thorstad EB and Baras E (2002) Surgical implantation of telemetry transmitters in fish: how much have we learned? Hydrobiologia 483: 239–248.

Jepsen N, Schreck C, Clements S and Thorstad EB (2005) A brief discussion on the 2% tag/bodymass rule of thumb. pp. 255–260. *In*: Spedicato MT, Lembo G and Marmulla G (eds.). Aquatic Telemetry: Advances and Applications. FAO/COISPA, Rome.

Jepsen N, Mikkelsen JS and Koed A (2008) Effects of tag and suture type on survival and growth of brown trout with surgically implanted telemetry tags in the wild. Journal of Fish Biology 72: 594–602.

Johnston P, Bérubé F and Bergeron NE (2009) Development of a flatbed passive integrated transponder antenna grid for continuous monitoring of fishes in natural streams. Journal of Fish Biology 74: 1651–1661.

Juell JE and Westerberg H (1993) An ultrasonic telemetric system for automatic positioning of individual fish used to track atlantic salmon (*Salmo salar* L.) in sea cage. Aquacultural Engineering 12: 1–18.

Keefer ML, Peery CA, Jepson MA and Stuehrenberg LC (2004) Upstream migration rates of radio-tagged adult chinook salmon in riverine habitats of the Columbia River basin. Journal of Fish Biology 65: 1126–1141.

Keefer ML, Boggs CT, Peery CA and Caudill CC (2013) Factors affecting dam passage and upstream distribution of adult Pacific lamprey in the interior Columbia River basin. Ecology of Freshwater Fish 22: 1–10.

Kessel ST, Cooke SJ, Heupel MR, Hussey NE, Simpfendorfer CA, Vagle S and Fisk AT (2014) A review of detection range testing in aquatic passive acoustic telemetry studies. Reviews in Fish Biology and Fisheries 24: 199–218.

Klein Breteler J, Vriese T, Borcherding J, Breukelaar A, Jörgensen L, Staas S, De Laak G and Ingendahl D (2007) Assessment of population size and migration routes of silver eel in the River Rhine based on a 2-year combined mark-recapture and telemetry study. ICES Journal of Marine Science 64: 1450–1456.

Klimley AP, Le Boeuf BJ, Cantara KM, Richert JE, Davis SF and Van Sommeran S (2001) Radio-acoustic positioning as a tool for studying site-specific behavior of the white shark and other large marine species. Marine Biology 138: 429–446.

Klimley AP, MacFarlane RB, Sandstrom PT and Lindley ST (2013) A summary of the use of electronic tagging to provide insights into salmon migration and survival. Environmental Biology of Fishes 96: 419–428.

Kobler A, Klefoth T, Wolter C, Fredrich F and Arlinghaus R (2008) Contrasting pike (*Esox lucius* L.) movement and habitat choice between summer and winter in a small lake. Hydrobiologia 601: 17–27.

Koeck B, Alo's J, Caro A, Neveu R, Crec'hriou R, Saragoni G and Lenfant P (2013) Contrasting fish behavior in artificial seascapes with implications for resources conservation. PLOS ONE 8: e69303.

Koehn JD (2000) Why use radio tags to study freshwater fish? pp. 24–32. *In*: Hancock DA, Smith DC and Koehn JD (eds.). Fish Movement and Migration. Australian Society for Fish Biology, Sydney.

Koehn JD, McKenzie JA, O'Mahony DJ, Nicol SJ, O'Connor JP and O'Connor WG (2009) Movements of Murray cod (*Maccullochella peelii peelii*) in a large Australian lowland river. Ecology of Freshwater Fish 18: 594–602.

Koster WM, Dawson DR and Crook DA (2013) Downstream spawning migration by the amphidromous Australian grayling (*Prototroctes maraena*) in a coastal river in south-eastern Australia. Marine & Freshwater Research 64: 31–41.

Kuechle VB and Kuechle PJ (2012) Radio telemetry in fresh water: the basics. pp. 91–137. *In*: Adams NS, Beeman JW and Eiler JH (eds.). Telemetry Techniques: A User Guide for Fisheries Research. American Fisheries Society, Bethesda, MD.

Laffaille P, Acou A and Guillouët J (2005) The yellow European eel (*Anguilla anguilla* L.) may adopt a sedentary lifestyle in inland freshwaters. Ecology of Freshwater Fish 14: 191–196.

Lagardère JP, Ducamp JJ, Favre L, Mosneron-Dupin J and Sperandio M (1990) A method for the quantitative evaluation of fish movements in salt ponds by acoustic telemetry. Journal of Experimental Marine Biology and Ecology 141: 221–236.

Larinier M and Travade F (2002) Downstream migration: problems and facilities. Bulletin Français de la Pêcheet de la Pisciculture 364: 181–207.

Larinier M, Chanseau M, Bau F and Croze O (2005) The use of radio telemetry for optimizing fish pass design. pp. 53–60. *In*: Spedicato MT, Lembo G and Marmulla G (eds.). Aquatic Telemetry: Advances and Applications. FAO/COISPA, Rome.

Larinier M, Bau F, Baran P and Travade F (2012) Study of eel downstream migration and passage of hydroelectric installations on the Gave de Pau river. pp. 98–101. *In*: Baran P, Basilico L, Larinier M, Rigaud C and Travade F (eds.). Management plan to save the eel: Optimising the design and management of installations. Proceedings of the Symposium on "Eels and Ecological Continuity: Optimising the design and management of installations", Onema, Paris.

Leach W (2010) Comparison of FDX and HDX technologies. PIT tag Techniques and Technologies Workshop, Annual Meeting, Oregon Chapter of the American Fisheries Society, Eugene, OR.

Legault A, Acou A, Guillouët J and Feunteun E (2003) Survey of downstream migration of silver eels through discharge pipe on a reservoir dam. Bulletin Français de Pêche et de Pisciculture 368: 43–54.

Liedtke TL, Wargo-Rub M, Adams N, Beeman J and Eiler J (2012) Techniques for telemetry transmitter attachment and evaluation of transmitter effects on fish performance. pp. 45–87. *In*: Adams NS, Beeman JW and Eiler JH (eds.). Telemetry Techniques: A User Guide for Fisheries Research. American Fisheries Society, Bethesda, MD.

Linnansaari TP and Cunjak RA (2007) The performance and efficacy of a two-person operated portable pit-antenna for monitoring spatial distribution of stream fish populations. River Research and Applications 23: 559–564.

Lonsdale EM and Baxter GT (1968) Design and field tests of a radio-wave transmitter for fish tagging. The Progressive Fish-Culturist 30: 47–52.

Lucas MC and Baras E (2000) Methods for studying spatial behaviour of freshwater fishes in the natural environment. Fish and Fisheries 1: 283–316.

Lucas MC, Mercer T, Armstrong JD, McGinty S and Rycroft P (1999) Use of a flat-bed passive integrated transponder antenna array to study the migration and behaviour of lowland river fishes at a fish pass. Ecology of Freshwater Fish 44: 183–191.

Luke, McG D, Pincock DG and Stasko AB (1973) Pressure-sensing ultrasonic transmitter for tracking aquatic animals. Journal of the Fisheries Research Board of Canada 30: 1402–1404.

Makiguchi Y, Konno Y, Konishi K, Miyoshi K, Sakashita T, Nii H, Nakao K and Ueda H (2011) EMG telemetry studies on upstream migration of chum salmon in the Toyohira River, Hokkaido, Japan. Fish Physiology and Biochemistry 37: 273–284.

Malbrouck C and Kestemont P (2006) Effects of microcystin on fish. Environmental Toxicology and Chemistry 25: 72–86.

Martinelli TL and Shively RS (1997) Seasonal distribution, movements and habitat associations of northern squawfish in two lower Columbia River reservoirs. Regulated Rivers: Research & Management 13: 543–556.

Mazel V, Charrier F, Legault A and Laffaille P (2013) Long-term effects of passive integrated transponder tagging (PIT tags) on the growth of the yellow European eel (*Anguilla anguilla* (Linnaeus, 1758)). Journal of Applied Ichthyology 29: 906–908.

McCarthy K, Frankiewicz P, Cullen P, Blaszkowski M, O'Connor W and Doherty D (2008). Long-term effects of hydropower installations and associated river regulation on River Shannon eel populations: mitigation and management. Hydrobiologia 609: 109–124.

McDougall CA, Blanchfield PJ, Peake SJ and Anderson WG (2013) Movement patterns and size-class influence entrainment susceptibility of Lake Sturgeon in a small hydroelectric reservoir. Transactions of the American Fisheries Society 142: 1508–1521.

McFarlane GA, Wydowski RS and Prince ED (1990) External tags and marks, historical review of the development of external tags and marks. American Fisheries Society Symposium 7: 9–29.

McMichael GA, Eppard MB, Carlson TJ, Carter JA, Ebberts BD, Brown RS, Weiland M, Ploskey GR, Harnish RA and Deng ZD (2010) The juvenile salmon acoustic telemetry system: a new tool. Fisheries 35: 9–22.

Melnychuk MC, Welch DW, Walters CJ and Christensen V (2007) Riverine and early ocean migration and mortality patterns of juvenile steelhead trout (*Oncorhynchus mykiss*) from the Cheakamus River, British Columbia. Hydrobiologia 582: 55–65.

Miles NG, Walsh CT, Butler G, Ueda H and West RJ (2014) Australian diadromous fishes—challenges and solutions for understanding migrations in the 21st century. Marine & Freshwater Research 1: 12–24.

Monan GE, Johnson JH and Esterberg GF (1975) Electronic tags and related tracking techniques aid in study of migrating salmon and steelhead trout in the Columbia River basin. Marine Fisheries Review 37: 9–15.

Morgan MJ and Walsch SJ (1993) Evaluation of the retention of external tags by juvenile American plaice (*Hippoglossoides platessoides*) using an aquarium experiment. Fisheries Research 16: 1–7.

Moser ML, Matter AL, Stuehrenberg LC and Bjornn TC (2002) Use of an extensive radio receiver network to document Pacific lamprey (*Lampetra tridentata*) entrance efficiency at fishways in the lower Columbia River, USA. Hydrobiologia 483: 45–53.

Mulcahy DM (2003) Surgical implantation of transmitters into fish. ILAR Journal 44: 295–306.

Murchie KJ, Cooke SJ and Schreer JF (2004) Effects of radio-transmitter antenna length on swimming performance of juvenile rainbow trout. Ecology of Freshwater Fish 13: 312–316.

Murphy BR and Willis DW (1996) Fisheries Techniques, 2nd edition. American Fisheries Society, Bethesda, MD.

Myers GS (1949) Usage of anadromous, catadromous, and allied terms for migratory fishes. Copeia 1949: 89–97.

Niemela SL, Layzer JB and Gore JA (1993) An improved radiotelemetry method for determining use of microhabitats by fishes. Rivers 4: 30–35.

Nunnallee EP, Prentice EF, Jonasson BF and Patten W (1998) Evaluation of a flat-plate PIT tag interrogation system at Bonneville Dam. Aquacultural Engineering 17: 261–272.

Olsen EM and Vollestad LA (2001) An evaluation of visible implant elastomer for marking age-0 brown trout. North American Journal of Fisheries Management 21: 967–970.

Ovidio M, Baras E, Goffaux D, Giroux F and Philippart J-C (2002) Seasonal variations of activity pattern of brown trout (*Salmo trutta*) in a small stream, as determined by radio-telemetry. Hydrobiologia 470: 195–202.

Ovidio M, Capra H and Philippart J-C (2007) Field protocol for assessing small obstacles to migration of brown trout *Salmo trutta*, and European grayling *Thymallus thymallus*: a contribution to the management of free movement in rivers. Fisheries Management and Ecology 14: 41–50.

Ovidio M, Seredynski AL, Philippart J-C and Matondo BN (2013) A bit of quiet between the migrations: the resting life of the European eel during their freshwater growth phase in a small stream. Aquatic Ecology 47: 291–301.

Peters LM, Reinhardt UG and Pegg MA (2008) Factors influencing radio wave transmission and reception: use of radiotelemetry in large river systems. North American Journal of Fisheries Management 28: 301–307.

Pincock DG and Johnston SV (2012) Acoustic Telemetry Overview. pp. 305–337. *In*: Adams NS, Beeman JW and Eiler JH (eds.). Telemetry Techniques: A User Guide for Fisheries Research. American Fisheries Society, Bethesda, MD.

Pon LB, Hinch SG, Cooke SJ, Patterson DA and Farrell AP (2009) Physiological, energetic and behavioural correlates of successful fishway passage of adult sockeye salmon *Oncorhynchus nerka* in the Seton River, British Columbia. Journal of Fish Biology 74: 1323–1336.

Quintella BR, Póvoa I and Almeida PR (2009) Swimming behaviour of upriver migrating sea lamprey assessed by electromyogram telemetry. Journal of Applied Ichthyology 25: 46–54.

Riley WD, Eagle MO, Ives MJ, Rycroft P and Wilkinson A (2003) A portable passive integrated transponder multi-point decoder system for monitoring habitat use and behaviour of freshwater fish in small streams. Fisheries Management and Ecology 10: 265–268.

Riley WD, Ibbotson AT, Beaumont WRC, Rycroft P and Cook AC (2010) A portable, cost effective, pass-through system to detect downstream migrating salmonids marked with 12 mm passive integrated transponder tags. Ecology of Freshwater Fish 101: 203–206.

Robertson MJ, Clarke KD, Scruton DA and Brown JA (2003) Interhabitat and instream movements of large Atlantic salmon parr in a Newfoundland watershed in winter. Journal of Fish Biology 63: 1208–1218.

Roussel J-M, Haro A and Cunjak RA (2000) Field test of a new method for tracking small fishes in shallow rivers using passive integrated transponder (PIT) technology. Canadian Journal of Fisheries and Aquatic Sciences 57: 1326–1329.

Roussel J-M, Cunjak RA, Newbury R, Caissie D and Haro A (2004) Movements and habitat use by PIT-tagged Atlantic salmon parr in early winter: the influence of anchor ice. Freshwater Biology 49: 1026–1035.

Roy R, Beguin J, Argillier C, Tissot L, Smith F, Smedbol S and De-Oliveira E (2014) Testing the VEMCO Positioning system: spatial distribution of the probability of location and the positioning error in a reservoir. Animal Biotelemetry 2: 1–7.

Schaefer KM and Fuller DW (2016) Methodologies for investigating oceanodromous fish movements: archival and pop-up satellite archival tags. pp. 14–19. *In*: Morais P and Daverat F (eds.). An Introduction to Fish Migration. CRC Press, Boca Raton, FL, USA.

Schaefer KM, Fuller DW and Block BA (2011) Movements, behavior, and habitat utilization of yellowfin tuna (*Thunnus albacares*) in the Pacific Ocean off Baja California, Mexico, determined from archival tag data analyses, including unscented Kalman filtering. Ecology of Freshwater Fish 112: 22–37.

Schurman GC and Thompson DA (1990) Washington Department of Fisheries' mobile tagging units: construction and operation. pp. 232–236. *In*: Parker NC, Giorgi AE, Heidinger RC, Jester DB, Jr., Prince ED and Winans GA (eds.). Fish Marking Techniques American Fisheries Society Symposium 7, Bethesda, MD.

Sisak MM and Lotimer JS (1998) Frequency choice for radio telemetry: the HF vs. VHF conundrum. Hydrobiologia 371/372: 53–59.

Slavík O and Horký P (2009) When fish meet fish as determined by physiological sensors. Ecology of Freshwater Fish 18: 501–506.

Smith WE and Kwak TJ (2014) A capture-recapture model of amphidromous fish dispersal. Journal of Fish Biology 84: 897–912.

Solomon D and Potter E (1988) First results with a new estuarine fish tracking system. Journal of Fish Biology 33: 127–132.

Spedicato MT, Lembo G and Marmulla G (eds.) (2005) Aquatic Telemetry: Advances and Applications. FAO/COISPA, Rome.

Stuehrenberg L, Giorgi A and Bartlett C (1990) Pulse-coded radio tags for fish identification. American Fisheries Society Symposium 7: 370–374.

Sulikowski JA, Fairchild EA, Kennels N, Howell WH and Tsang PCW (2005) The effects of tagging and transport on stress in juvenile winter flounder, *Pseudopleuronectes americanus*, Implications for Successful Stock Enhancement. pp. 148–156.

Taylor MK, Hasler CT, Findlay CS, Lewis B, Schmidt DC, Hinch SG and Cooke SJ (2014) Hydrologic correlates of bull trout (*Salvelinus confluentus*) swimming activity in a hydropeaking river. River Research and Applications 30: 756–765.

Teixeira A and Cortes RMV (2007) PIT telemetry as a method to study the habitat requirements of fish populations: application to native and stocked trout movements. Hydrobiologia 582: 171–185.

Thiem JD, Binder TR, Dawson JW, Dumont P, Hatin D, Katopodis C, Zhu DZ and Cooke SJ (2011) Behaviour and passage success of upriver-migrating lake sturgeon *Acipenser fulvescens* in a vertical slot fishway on the Richelieu River, Quebec, Canada. Endangered Species Research 15: 1–11.

Thiem JD, Binder TR, Dumont P, Hatin D, Hatry C, Katopodis C, Stamplecoskie KM and Cooke SJ (2013) Multispecies fish passage behaviour in a vertical slot fishway on the Richelieu River, Quebec, Canada. River Research and Applications 29: 582–592.

Thorstad EB, Økland F and Finstad B (2000a) Effects of telemetry transmitters on swimming performance of adult Atlantic salmon. Journal of Fish Biology 57: 531–535.

Thorstad EB, Økland F, Rowsell D and McKinley RS (2000b) A system for automatic recording of fish tagged with coded acoustic transmitters. Fisheries Management and Ecology 7: 281–294.

Thorstad EB, Økland F and Heggberget TG (2001) Are long term negative effects from external tags underestimated? Fouling of an externally attached telemetry transmitter. Journal of Fish Biology 59: 1092–1094.

Thorstad EB, Økland F, Kroglund F and Jepsen N (2003) Upstream migration of Atlantic salmon at a power station on the River Nidelva, Southern Norway. Fisheries Management and Ecology 10: 139–146.

Thorsteinsson V (2002) Tagging methods for stock assessment and research in fisheries. Report of concerted action FAIR CT.96.1394 (CATAG). Reykjavik: Marine Research Institute Technical Report 79: 179.

Tranquilli JV (2010) Guide to Half-and Full-duplex RFID: lessons learned tracking bull trout in the Upper Willamette since 2001. PIT tag techniques and technologies workshop, Eugene, Oregon, February 23, 2010.

Travade F, Larinier M, Subra S, Gomes P and deOliveira E (2010) Behaviour and passage of European silver eels (*Anguilla anguilla*) at a small hydropower plant during their downstream migration. Knowledge and Management of Aquatic Ecosystems 398: 1–19.

Trefethen PS (1956) Sonic equipment for tracking individual fish. Special Scientific Report—Fisheries Number 179, United States Fish and Wildlife Service, Washington, D.C.

Ueda H (2012) Physiological mechanisms of imprinting and homing migration in Pacific salmon *Oncorhynchus* spp. Journal of Fish Biology 81: 543–558.

Velle JJ, Lindsay JE, Weeks RW and Long FM (1979) An investigation of the loss mechanisms encountered in propagation from a submerged fish telemetry transmitter. pp. 228–237. *In*: Long FM (ed.). Proceeding of the 2nd Internation Conference on Wildlife Biotelemetry. University of Wyoming, Laramie, WY.

Verbiest H, Breukelaar A, Ovidio M, Philippart J-C and Belpaire C (2012) Escapement success and patterns of downstream migration of female silver eel *Anguilla anguilla* in the River Meuse. Ecology of Freshwater Fish 21: 395–403.

Voegeli FA and Pincock DG (1996) Overview of underwater acoustics as it applies to telemetry. pp. 23–30. *In*: Baras E and Phillipart J-C (eds.). Underwater Biotelemetry. University of Liège, Belgium.

Wahlberg M, Westerberg H, Aarestrup K, Feunteun E, Gargan P and Righton D (2014) Evidence of marine mammal predation of the European eel (*Anguilla anguilla* L.) on its marine migration. Deep-Sea Research I 86: 32–38.

Walton and Cotton (1921) The Compleat Angler or the Contemplative Man's Recreation, with an Introduction and Bibliography by R.B. Marston. Oxford University Press, London.

Watry CB and Scarnecchia DL (2008) Adfluvial and fluvial life history variations and migratory patterns of a relict charr, *Salvelinus confluentus*, stock in west-central Idaho, USA. Ecology of Freshwater Fish 17: 231–243.

Welch DW, Rechisky EL, Melnychuk MC, Porter AD, Walters CJ, Clements S, Cleme BJ, McKinley RS and Schreck C (2008) Survival of migrating salmon smolts in large rivers with and without dams. PLOS Biology 6: e314.

Westerberg H (1983) Monitoring fish behaviour to hydrographic fine structure. Proceedings, 4th International Wildlife Biotelemetry Conference, August 22-23-24 1983, Halifax, Nova Scotia.

Williams TH (1990) Evaluation of pressure-sensitive radio transmitters used for monitoring depth selection by trout in lotic systems. pp. 390–394. *In*: Parker NC, Giorgi AE, Heidinger RC, Jester DB, Jr., Prince ED and Winans GA (eds.). Fish-Marking Techniques (American Fisheries Society Symposium 7). American Fisheries Society, Bethesda, MD.

Winter HV, Jansen HM and Bruijs MCM (2006) Assessing the impact of hydropower and fisheries on downstream migrating silver eel, *Anguilla anguilla*, by telemetry in the River Meuse. Ecology of Freshwater Fish 15: 221–228.

Winter JD (1983) Underwater biotelemetry. pp. 371–395. *In*: Nielsen LA and Johnson DL (eds.). Fisheries Techniques. American Fisheries Society, Bethesda, MD.

Winter JD (1996) Advances in underwater biotelemetry. pp. 555–590. *In*: Murphy BR and Willis DW (eds.). Fisheries Techniques 2nd edition. American Fisheries Society, Bethesda, MD.

Winter JD, Kuechle VB, Siniff DB and Tester JR (1978) Equipment and methods for radio tracking freshwater fish. University of Minnesota: Miscellaneous report 152: 20.

Winter HV, Jansen HM and Breukelaar A (2007) Silver eel mortality during downstream migration in the River Meuse, from a population perspective. ICES Journal of Marine Science 64: 1444–1449.

Zabel RW, Wagner T, Congleton JL, Smith SG and Williams JG (2005) Survival and selection of migrating salmon from capture-recapture models with individual traits. Ecological Applications 15: 1427–1439.

Zydlewski GB, Haro A, Whalen KG and McCormick SD (2001) Performance of stationary and portable passive transponder detection systems for monitoring of fish movements. Journal of Fish Biology 58: 1471–1475.

Zydlewski GB, Horton G, Dubreuil T, Letcher B, Casey S and Zydlewski J (2006) Remote monitoring of fish in small streams: a unified approach using pit tags. Fisheries 31: 492–502.

CHAPTER 13

Methodologies for Investigating Oceanodromous Fish Movements: Archival and Pop-up Satellite Archival Tags

*Kurt M. Schaefer** and *Daniel W. Fuller*

Introduction

Traditional mark-recapture studies, utilizing coded plastic tags, have long provided the foundation for determining individual fish linear displacements, and dispersion and mixing rates among fish stocks (Thorsteinsson 2002). However the method has limitations for studies of spatial dynamics of fish, including large-scale movements and migrations, as it only provides information about the release and recapture positions, and recaptures are highly dependent on spatial and temporal variability in fishing effort, fish behavior, and gear selectivity. Also, the methodology is unable to provide any information on behavior or movement of fish between the locations of release and recapture. There is also the inherent problem of these so-called conventional tags being returned with incomplete or inaccurate recapture information. Many high-seas commercial fishing vessels operate over extensive areas during a fishing trip. Since those who eventually find tagged fish, usually at the time the vessel is unloading, are unaware of when and where the fish was recaptured during the fishing trip sometimes return tags with inaccurate information. These factors along with geographic variability in reporting rates, can have a profound impact on the interpretation of the data obtained from mark-recapture studies.

Since most fish remain submerged, the use of Advanced Research and Global Observation Satellite (ARGOS) and Global Positioning Systems (GPS) tags, which provide ± 350 m or GPS precision positions, respectively, and for which data acquisition is fisheries independent, have been used successfully only with some shark and billfish species (Weng et al. 2005; Holdsworth et al. 2009) which exhibit sufficient time at the surface for communications with satellites. Archival Tags (ATs) and pop-up archival satellite tags (PSATs) with light sensors have, however, been used successfully to

Inter-American Tropical Tuna Commission, 8604 La Jolla Shores Drive, La Jolla, CA 92037-1508, USA.
* Corresponding author: kschaefer@iattc.org

estimate positions and reconstruct most probable tracks, although with much less accuracy than ARGOS and GPS tags, for many marine epipelagic species, based primarily on daily records of ambient light and sea-surface temperatures. Recent advances in light- and temperature-based geolocation methods, coupled with state-space modeling, have enabled better estimation of the geographical positions and most probable tracks of fish from ATs and PSATs (Lam et al. 2010). The large-scale movements and migrations of individual fish, along with their behavior patterns relative to fine-scale environmental influences and physiological constraints, acquired through experiments utilizing ATs and PSATs, offers the potential to understand spatial population processes (Patterson et al. 2008).

However, obtaining data from ATs, like conventional tags, is dependent on their returns, with most coming from recaptures by commercial fisheries. Normally, however far greater monetary rewards are offered for their return, than for conventional tags, and thus reporting rates are expected to be close to 100% of the recaptures. PSATs do not need to be returned, as the archived data are summarized onboard the tags and transmitted to satellite-borne ARGOS receivers, following predetermined durations, detachments, and floating to the surface. Data acquisitions from PSATs are thus considered to be fisheries independent. Compared with conventional tagging experiments, the numbers of ATs and PSATs deployed in experiments, including in multi-year tagging programs, is quite limited primarily because of the high costs of these tags.

ATs and PSATs are designed for simultaneously measuring at specified intervals, and storing in the tag's memory, depth, temperature, and light-level data while attached to fish and or aquatic animals. By ascertaining the times of dawn and dusk events from ambient light-level curves, and matching sea-surface temperatures recorded by the tags with those obtained from satellite sensors, it is possible to estimate daily positions to within about 100 nautical miles (nmi), on average, of actual GPS positions, and reconstruct the most probable tracks of tagged fish in coastal and oceanic environments (Lam et al. 2010). Current-generation ATs are capable of storing data for up to several years, providing a unique opportunity to evaluate the influence of seasonal and annual environmental variability and ontogenetic change on fish movement patterns, behavior, and habitat utilization. These tags have been used on a wide range of species, since their development in the 1990s, and proven to be a valuable research tool in biologging studies, including investigations of the large-scale movements and migrations of fish stocks.

For oceanic species such as some tunas, billfishes, sharks, salmon, and eels, which undertake fairly extensive movements or migrations over distances of several hundred nautical miles, tagging studies using ATs and PSATs can provide remarkable insights and empirical data on their spatial dynamics and behavior. But for fish which are not highly mobile, exhibiting confined home range distributions, this technology would be of limited value for investigations of their movements or migrations, because errors of around ± 100 nmi surround most probable tracks derived from State-Space Models (SSMs) utilizing light- and temperature-based geolocations (Lam et al. 2010). Furthermore, fish species which live at depths greater than about 300–400 m, exhibit extreme diving activities during dawn or dusk, or reside in murky coastal waters, there can be critical issues with inadequate light sensitivity for use in geolocation estimates, and these tags are probably not appropriate for use in investigating spatial dynamics of those species.

For more than a decade, fish tagging programs using ATs and PSATs have been generating large volumes of data on spatial dynamics, physiology and habitat utilization, providing unparalleled insights into the spatial ecology of some species (Block 2005). Analyses of data sets obtained from deployments of these tags with commercially-important species can provide essential information on movement parameters, and mortality rates, for incorporation into fish stock assessments and ultimately used for science-based management (Kurota et al. 2009; Eveson et al. 2012). Movement parameters from ATs and PSATs are far more informative than those from conventional tagging studies and for incorporating into seasonally- and spatially-explicit fisheries models to account for movements, mixing, residency and site fidelity (Taylor et al. 2011).

This chapter is intended to provide an overview of the designs of ATs and PSATs with light sensors, attachment methods, methods of analysis of data for estimating movement patterns and parameters, and examples of applications of this technology in studies of large-scale movements and migrations of teleost and cartilaginous fish, since early 2000. There is a vast amount of useful information published on these topics in scientific journals, for a wide range of fish species. Those interested in further information on these topics, including specific details relevant to studies on particular species of fish, will benefit by consulting references provided in this chapter.

TAG designs

Currently there are only a few companies manufacturing ATs and PSATs with light-based geolocation capabilities suitable for deployments with fish. Although there are some similarities among the ATs and PSATs currently being manufactured by these companies, design specifications, performance, and potential applications are unique for each tag. Numerous technological innovations implemented by these manufacturers over the past decade have made their products more reliable for utilizing in long-term experiments on large-scale movements and migrations of a variety of fish species. Some of the most important technological improvements over the past decade with ATs and PSATs include lower power components, miniaturized electronics, increased memory, sensor performance, improved component reliability, data compression techniques/ firmware, and reduced production costs. There have also been improvements, of lesser importance, with higher grade lithium batteries and satellite transmission capabilities.

For some early history, dating back about 20 years to the early 1990's, on the development and designs of ATs and PSATs, and some of the first applications with these tags for estimating large-scale movements and migrations of fish, the reader is referred to the chapters by Arnold and Dewar (2001) and Gunn and Block (2001).

Archival tags

The term AT refers to electronic tags, containing clocks and various environmental sensors, with data storage capabilities. The recovery of data files stored in the memory of ATs is generally considered fisheries dependent, since the recapture of the tagged fish and the return of the tags is normally required in order to download the archived data. Exceptions include the external attachments of ATs with positive buoyancy and timed release mechanisms with the idea that those tags could be recovered either floating or along a shoreline, or by means of a tracking signal. Another exception previously investigated in a limited number of applications is the use of automated acoustic data retrieval stations capable of downloading the data from Communicating History Acoustic Transponder (CHAT) tags attached to fish either at moored or mobile hydrophone systems (Arnold and Dewar 2001).

There are two companies that have been producing ATs with light-based geolocation capabilities since early 2000, Lotek Wireless, Inc., St. John's, Newfoundland, Canada (http://www.lotek.com/) and Wildlife Computers, Inc., Redmond, Washington, USA (http://www.wildlifecomputers.com/). Some of the technical specifications in the designs of the LTD2310, LAT2810, and LAT2910 ATs manufactured by Lotek Wireless, and the MK9 AT manufactured by Wildlife Computers, with 2000 m depth ratings and external sensor stalk configurations, are given in Table 13.1, and the tags shown in Fig. 13.1. These tag configurations are designed for internal implantation of the tag body into either the coelom or dorsal musculature of fish, with the external sensor stalk from which the ambient light-level and temperature measurements originate, exits the body of the fish through an incision. The Lotek Wireless external sensor stalk is an omni directional-light gathering fiber (and temperature sensor) in a protective sheath, which conducts light into the body of the tag and focuses it on a temperature-compensated light circuit. The Lotek Wireless LAT2800 and LAT2900 are also available with light sensors, and without external stalks, in several different configurations designed for external attachment. The Wildlife Computers external sensor stalk has a diode-type temperature-compensated light sensor (and temperature sensor) in a protective sheath, situated near the end of the stalk, for obtaining light measurements (Fig. 13.1). The MK9 is also available with a light sensor, and without the external stalk, in customer specified packaging configurations for external attachment.

The Lotek Wireless LTD2310 and Wildlife Computers MK9 tags have been used widely during the past decade. The LTD2310 has recently been discontinued, and replaced with the LAT2310, as the company has moved to a new platform configuration which is also used for its LAT2810 and LAT2910 ATs. The LTD2310 and MK9 tag bodies are similar in size and weight, and cylindrical in shape, so as to better withstand high pressure at extreme depths without fracturing. The ranges, resolutions, response times, and sensitivities of the depth, temperature, and light-level sensors of the LTD2310 and MK9 are fairly similar, but with some differences in sensor performances (Schaefer and Fuller 2006). Experiments in the equatorial eastern Pacific Ocean with hydrocasts to about 500 m, demonstrated average light sensitivities to around 400 m with both the LTD2310 and MK9. Data obtained from both these ATs while attached to bigeye tuna in the equatorial eastern Pacific have shown that the light sensors are capable of identifying dawn and dusk events to at least 300 m (Schaefer and Fuller 2009).

Table 13.1. Specifications of archival tags with external sensor stalks. The LTD2310, LAT2810L, and LAT2910 are manufactured by Lotek Wireless. The MK9 is manufactured by Wildlife Computers. Information in the table was taken from manufacturer's web sites, and provided by manufacturers.

Specification	LTD2310	LAT2810L	LAT2910	MK9
Dimensions (without stalk)	70 mm x 16 mm diameter	44 mm x 13 mm diameter	26 mm x 7.8 mm diameter	73 mm x 18 mm diameter
Weight of tag (in air)	40 g	12.8 g	2.3 g	30 g
Casing	stainless steel	polycarbonate	polycarbonate	epoxy
Memory capacity	16 MBytes	128 MBytes	56 KBytes logging	64 MBytes
Depth range and resolution	2000 m, 1 m	2000 m, 1 m	2000 m, 1 m	2000 m, 1 m
Temperature range and resolution	0 to 35°C, 0.05°C	–5° to 35°C, 0.05°C	–5° to 45°C, 0.05°C	–40° to 60°C, 0.05°C
Maximum light sensitivity	9 decades @ 470 nm	9 decades @ 470 nm	9 decades @ 470 nm	5×10^{-12} W.cm^{-2} @ 550 nm

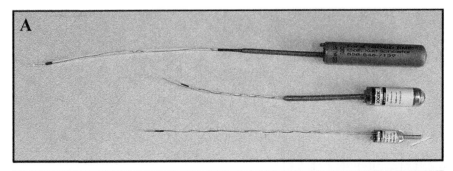

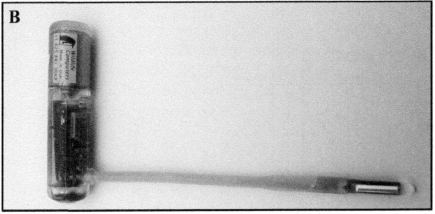

Figure 13.1. Archival tags with external sensor stalks for ambient light and temperature measurements (A) LTD2310, LAT2810L, and LAT2910PDA (top down) manufactured by Lotek Wireless, and (B) MK9 manufactured by Wildlife Computers.

In comparison, the newest generation LAT2810 and LAT2910 geolocating archival tags, which have been produced by Lotek Wireless only since early 2012, are substantially smaller and lighter (Table 13.1). Because of battery sizes and voltage requirements, these tags are not expected to have as long a life for logging data as expected from the LAT2310 or the Wildlife Computers MK9 ATs.

One substantial difference between the Wildlife Computers MK9 and the Lotek Wireless ATs is the way in which the memory of the tags is configured and utilized. The MK9 is capable of logging data for depth, internal and ambient temperatures, and light-level, at a 30-second sampling interval for 8.7 years, and with a non-volatile memory, storing that data for up to 25 years, after the battery has expired. Following a successful deployment and recovery of an MK9, the entire data set can be downloaded and then processed with software provided by the manufacturer to obtain estimates of daily geolocation, sea-surface temperature (SST) and a suite of other summaries based on the physical oceanographic parameters measured. The memory of Lotek Wireless ATs is partitioned, with a section for a day-log, based on calculations done onboard the tag, and a section for data logging. Within the day-log the estimates of geolocation are provided, along with SSTs, and other summaries of physical oceanographic data collected. There is also a substantial amount of memory allocated to the logging of raw data for depth, temperatures, and light-level at a 30-second sampling interval for multiple years for the LTD2310 and LAT2810 tags, but the memory configuration is currently limited with logging of time-series data with the LAT2910. The memory in each of these tags is also non-volatile, which means that even if the tag stops working the data previously stored is saved in the tags memory for upto 25 years.

Further advancements in integrated circuits have contributed to the miniaturization of ATs. This now makes it feasible to be manufacturing ATs for applications with fish much smaller than previously possible. With the relatively small LAT2910, the minimum size for pelagic fish potentially capable of carrying such a tag internally implanted, without probable adverse effects is possibly approaching 30 cm Fork Length (FL). Conducting experiments with captive fish to determine minimum sizes of fish for which tags are suitable for implantation, before undertaking field experiments, is strongly advised if feasible.

Pop-up satellite archival tags

Following deployment attached to a fish, PSATs record and store measurements of depth, temperature, and ambient light-level data at pre-programmed intervals, and perform onboard data processing. At a pre-programmed date the tags are designed to detach from their tethers and float to the surface to transmit processed and time-series data to the ARGOS system of polar orbiting satellites. The ARGOS data collection and Collecte Localisation Satellite (CLS), is a joint venture between the Centre National d'Etudes SPSATiales (CNES) of France and the National Aeronautics and Space Administration (NASA), and National Oceanic and Atmospheric Administration (NOAA) of the United States. Receivers on board NOAA satellites provide world coverage. The system uses UHF radio frequencies and a Doppler location system to calculate the position of the PSAT at the time of release (< 250 m, with best location class) by the shift in frequency as the satellite approaches and then moves away (Taillade 1992). The data along with a classification of their location accuracy are provided by ARGOS (http://www.argos-system.org) to the registered account for that PSAT.

Because of the relative size of PSATs, their use is intended for deployments with relatively large fish. PSATs are externally attached to fish with anchors and tethers, and at a pre-programmed date and time during a deployment, the PSAT actively corrodes the pin to which the tether is attached and floats to the surface to transmit a subset of the archival data to the ARGOS satellites. Those data which include daily geolocation estimates, depths, and temperatures, provide the opportunity of reconstructing the large-scale movements of fish, in addition to understanding of their habitat utilization throughout the deployment period. Acquiring data from PSATs is fisheries-independent, because it is not necessary to recapture the fish or recover PSAT, as with ATs, to obtain data. The full archival tag record, which is maintained in memory, can also be acquired from recovered PSATs.

Currently the two primary manufacturers of PSATs, which have been used since the late 1990s in numerous experiments with fish, are Microwave Telemetry, Inc., Columbia, Maryland, USA (http://www.microwavetelemetry.com/) and Wildlife

Computers, Redmond, Washington, USA (http://www.wildlifecomputers.com/). There are other companies currently experimenting with producing their own versions of PSATs. Some of the technical specifications in the designs of the PTT-100 and X-Tag PSATs manufactured by Microwave Telemetry, and the MK-10 PSAT and MiniPAT manufactured by Wildlife Computers, are given in Table 13.2, and the tags shown in Fig. 13.2. Because of requests from researchers desiring smaller PSATs, with lower hydrodynamic drag, to potentially improve retention rates and enable their use with smaller animals, Microwave Telemetry produced the X-Tag in 2007 and Wildlife Computers produced the MiniPAT in 2010. In the development process of these smaller PSATs numerous other improvements were also made, including sensors and relevant information returned from the tags. These smaller PSATs from both companies are considered replacements for their large tags, although they are still currently manufacturing the larger models.

Although the appearance, physical dimensions, and sensors on the PSATs manufactured by these two companies are similar (Fig. 13.2; Table 13.1), there are some fundamental differences between their products. The pre-deployment programming and data decoding and processing of the Wildlife Computers PSATs are controlled by the user, and for the Microwave Telemetry PSATs, these features are performed by the manufacturer so as to provide less chance for user errors and bias. The Wildlife Computers PSATs have enough power to sample data for at least one year and then make 10,000 transmissions, using a frequency-stability specialized ARGOS transmitter ('Cricket') to maximize the quantity and quality of those transmissions. The Microwave

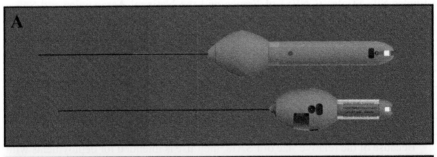

Figure 13.2. Pop-up satellite archival tags (A) MK10 and MiniPAT manufactured by Wildlife Computers (photo credit: Wildlife Computers), and (B) PTT-100 and X-Tag manufactured by Microwave Telemetry (photo credit: Microwave Telemetry).

Table 13.2. Specifications of current generation pop-up archival transmitting tags. The MK10 and MiniPAT are manufactured by Wildlife Computers. The PTT-100 and X-Tag are manufactured by Microwave Telemetry. Information in the table was taken from manufacturer's web sites, and provided by manufacturers.

Specification	MK10	MiniPAT	PTT-100	X-Tag
Dimensions	21 mm diameter x 175 mm (+ 170 mm antenna), 41 mm max. float diameter	18 mm diameter x 115 mm (+ 170 mm antenna), 38 mm max. float diameter	21 mm diameter x 166 mm (+ 171 mm antenna), 41 mm max. float diameter	16 mm diameter x 120 mm (+ 185 mm antenna), 32 mm max. float diameter
Weight of tag in air	75 g	53 g	68 g	40 g
Depth range and resolution	0–1000 m, 0.5 m	0–1700 m, 0.5 m	0–1296 m, 5.4 m	0–1296 m, 0.34 m
Temperature range and resolution	–40°C to 60°C, 0.05°C	–40°C to 60°C, 0.05°C	–4°C to 40°C, 0.23°C	–4°C to 40°C, 0.014°C
Maximum light sensitivity	5 x 10^{-12}W.cm^{-2} @ 550 nm	5 x 10^{-12}W.cm^{-2} @ 550 nm	< 4 x 10^{-5}Lux @ 555 nm	< 4 x 10^{-5}Lux @ 555 nm
Memory capacity	64 MBytes	16 MBytes	64 MBytes	64 MBytes
Tag programming and data recovery	User	User	Manufacturer	Manufacturer

Figure 13.3. Intracoelomic implantation of (A) Wildlife Computers MK9 archival tag in a yellowfin tuna (photo credit: J. Dettling II/SPC-OFP-PTTP), and (B) Lotek Wireless LTD2310 archival tag in a bigeye tuna (photo credit: D. Itano/SPC-OFP-PTTP).

Telemetry PSATs also have enough power to sample data for at least one year, have a 15,000 transmission capability, and are equipped with a Satellite-in View (SIV™) technology to maximize the efficiency of transmissions. PSATs manufactured by both companies provide daily estimates of light-based geolocation, using changes in ambient light levels so as to determine the times of sunrise and sunset. Also, while

Microwave Telemetry PSATS provide raw time-series data of temperature and depth at specified sampling rates, the Wildlife Computers PSATs provide summarized histograms for those parameters. The PSATs manufactured by both companies include similar methods to detect a mortality, such as a constant depth for a user-defined period of time, and a corrosion link on the nose cone of the PSATs to jettison the tag and initiate transmissions at the surface. The emergency mechanisms to prevent PSATs from imploding at depths below which they are rated, is the corrosion link on the nose cone of the Microwave Telemetry tags activated at 1250 m (default setting), and a mechanical guillotine (RD1800) on the Wildlife Computers tags designed for cutting a monofilament tether at 1800 m.

Although the MiniPAT and the X-Tag are substantially smaller, lighter, and with much lower drag coefficients than earlier generation PSATs, the minimum size of fish that should be considered for their use without probable adverse impacts on their behavior and survival is unknown for most species. However, as suggested before undertaking tagging experiments with ATs, if the minimum acceptable size has not already been established for the species, conducting experiments with captive fish or field experiments with dummy tags to determine minimum sizes for which PSATs may be suitable, and is advisable before undertaking costly field experiments with functional PSATs.

The data typically acquired from deployments and recoveries of ATs have been of longer average durations, and greater length and age ranges, than those obtained from PSATs, and thus more useful for investigations of large-scale movements and migrations of fish. However, in areas and times in which sufficient fishing effort on a species of interest is limited, and current experiments have not been conducted with plastic dart tags (PDTs) to establish base-line recapture rates, one should be cautious with his or her expectations in obtaining cost-effective and sufficient AT recoveries. If the fish of interest are of adequate size for deployments with PSATs, useful information on large-scale movements may be obtained. However, retention of PSATs for prolonged time periods on numerous fish species has been problematic, as will be discussed later.

Attachment methods

The importance of using the best techniques for the capture, handling, and attachment of ATs and PSATs to fish, is to achieve high post-release survivorship, tag retention, and comparable behavior to conspecifics, which cannot be overemphasized to its impact on the overall success of tagging experiments. Protocols must be carefully considered and tailored to the candidate species being tagged, since different species have different tolerance levels to capture, handling, and stress. Materials and methods that have been used with varying levels of success for attaching ATs and PSATs to a diversity of fish species are described below. Opinions vary among experts on appropriate materials and methods to be used for attachments of ATs and PSATs, even for the same species, and limited analyses of meta-data associated with such tagging experiments have been only recently published to evaluate techniques. Unpublished information gleaned from discussions with colleagues, who pioneered some of the currently acceptable PSAT attachment methods and continue to evaluate and improve their techniques, has also contributed to the contents mentioned here. It is imperative to continue to evaluate and improve the implantation techniques of ATs and attachment techniques of PSATs so as to maximize the reliability of the data that are generated or the potential benefits from tagging studies using this powerful technology (Cooke et al. 2011).

Archival tags

During the early development of ATs, there was a lack of consensus among scientists, with various levels of experience in tagging large pelagics with conventional and acoustic tags, as to whether ATs with dimensions similar to those of the LTD2310 and MK9 (Table 13.1) should be attached externally or internally to the fish (Klimley et al. 1994). The first light-based geolocating ATs were designed and manufactured with external sensor stalks, for internal implantation in tunas and other large fish (Gunn and Block 2001). Tagging experiments were conducted off Australia with southern bluefin tuna *Thunnus maccoyii*, to specifically determine the most reliable method of attaching ATs (Gunn et al. 1994). Plastic replicas, approximately the same size (55 mm x 24 mm x 12 mm) as the AT were attached in three different ways, intracoelomic implantation and two external attachment methods with tethers secured at the base of the second dorsal fin. Return rates indicated that intracoelomic implantation was superior to either of the external attachment methods for the retention of ATs. The overall recovery rate of 16.5% for 514 ATs internally implanted in the coelom of southern bluefin between 1994 and 2000 indicated this to be a suitable method for attachment for this species (Arnold and Dewar 2001; Gunn and Block 2001). Appropriate handling and procedures were also developed for the intracoelomic implantation of ATs manufactured by Northwest Marine Technology (NMT, Shaw Island, Washington, USA), in Atlantic bluefin tuna *Thunnus thynnus* through collaboration with a veterinarian and fish husbandry experts (Block et al. 1998a). During the period between 1996 and 2000, a recovery rate of 14% for 279 ATs deployed also confirmed that intracoelomic implantation of ATs was a suitable attachment method with Atlantic bluefin (Gunn and Block 2001). Appropriate handling and implantation procedures were also developed during this early period by scientists in Japan, for intracoelomic implantation of NMT ATs in a variety of fish, including Pacific bluefin tuna *Thunnus orientalis* (Tsuji et al. 1999; Kitagawa et al. 2000), chum salmon *Oncorhynchus keta* (Wada and Ueno 1999), and the ocellate puffer *Takifugu rubripes* (Nakajima and Nitta 2001).

Tagging experiments with a variety of fish species, however, have demonstrated that acoustic and archival tags attached by intracoelomic implantation can be lost through failure of the wound closure, or by transabdominal or transintestinal expulsion, and also cause inflammatory responses (Cooke et al. 2011). Freshwater fish, particularly salmonids, have been investigated more intensively regarding issues associated with intracoelomic implantation of tags (Welch et al. 2007; Teo et al. 2013). They appear to have much higher tag expulsion rates than those reported for some marine species, including bigeye tuna *Thunnus obesus* (Schaefer and Fuller 2002), Atlantic bluefin tuna (Walli et al. 2009), Pacific bluefin tuna (Boustany et al. 2010), yellowfin tuna *Thunnus albacares* (Schaefer et al. 2011), Pacific halibut *Hippoglossus stenolepis*, and other pleuronectid species (Loher and Rensmeyer 2011). External sensor stalks of ATs protruding from the coelomic cavity creates drag, precession and continuous vibrations and movement, preventing complete wound closure and can lead to tissue necrosis and potential tag loss. Based on our experience with tunas, expulsion of ATs also appears to be a function of tag body proportions, and surgical techniques used for attachment. Although there are other attachment methods for ATs, intracoelomic implantation via laparotomy is considered the most appropriate for long-term biologging applications (Bridger and Booth 2003). However, it should be recognized, as with any procedure, there are inherent risks, including potential for infection, physiological imbalance, and even post-surgical mortality (Cooke et al. 2011).

Appropriate fishing gear and fish handling methods should not be overlooked for their importance in the post-surgical survivorship of fish released with ATs. Small pelagic fish captured by hook and line, preferably using barbless circle hooks, should be lifted aboard vessels without the aid of any devices and placed directly in v-shaped tagging cradles, padded with closed-cell foam and with a wet smooth vinyl liner (Bayliff and Holland 1986). Placing fish immediately in cradles ventral side up, and covering their head and eyes with a wet synthetic chamois cloth, normally initiates a comatose-like state, lasting for up to a few minutes. Larger fish should be lifted out of the water and aboard vessel, carefully and without injury to the fish, with the aid of either seawater-filled vinyl slings (Farwell 2001), hand-held scoop nets with knotless webbing (Schaefer and Fuller 2002), or in slings using mechanical hoists, as required for large sharks and other fish on large research vessels (Musyl et al. 2011b). There are other noteworthy methods and exceptions, including a technique developed specifically for pulling large Atlantic bluefin aboard sportfishing vessels through the fish door on the transom, and onto pads with smooth vinyl lining (Block et al. 1998a). Larger fish can also be placed in appropriate sized cradles, or on large closed cell foam pads lined with smooth vinyl surfaces, and restrained, for attaching ATs as has been done with some species of billfish (Holland et al. 2006). Rapid examinations to assess any potential bleeding from the gills, eye damage, or barotrauma are also a critical part of the protocol, so as to select only fish in excellent condition for tagging with ATs. At all times during the handling and tagging process, the comfort and welfare of the fish should be assessed. The time in which fish are out of the water is dependent on the competency of the surgeon, and speed in completing the tagging process is essential to reduce trauma, and maximize survival of the fish.

Detailed descriptions of appropriate handling methods and implantation procedures, including recommended materials to be used for intracoelomic implantation of ATs, the size of the LTD2310 and MK9 (Table 13.1), in tunas are given by Block et al. (1998a) and Schaefer and Fuller (2002). Conventional tags, such as a plastic dart tag or a plastic intramuscular tag (Hallprint Pty Ltd, South Australia) in bright colors, to make them readily observed, should also be attached to fish with implanted ATs. Text printed on the conventional tag should specify the reward amount and instructions for the return of the AT located inside the fish. The early generation NMT ATs and Lotek Wireless LTD2310 models did not have a ring welded to the stainless casing, at the base of the sensor stalk, for tying the end of a suture to prevent the AT from sliding forward in the coelomic cavity of large Atlantic bluefin which occurred and occasionally prevented the collection of ambient light data (Arnold and Dewar 2001). The use of antibiotics for such intracoelomic surgeries should probably be avoided, either injected into the fish or a coating on tags to prevent potential infections as it does not appear to be beneficial (Mulcahy 2011). With respect to surgical procedures for intracoelomic implantation of ATs it is crucial to employ sterile techniques, while realizing there are limitations to such procedures when performing surgeries aboard vessels at sea. Sterile surgical techniques are essential to prevent infections which can lead to various detrimental fish health issues, including delayed post-release mortalities. Procedures should include a clean environment where surgeries are performed, the use of new surgical gloves, sterile scalpel blades, and suture materials for each surgery. ATs should be sterile by cleaning them with alcohol, drying and wrapping in a film, and then sprayed with a broad spectrum bactericide, such as Betadyne solution (a 10% povidone-iodine), immediately before implantation. Wagner et al. (2011) provided an excellent review of surgical implantation techniques for electronic tags in salmonids, also relevant to all AT fish surgeries. For other fish species, including salmon (Teo et al. 2013), cod (Righton et al. 2006), and pleuronectid species (Loher and Rensmeyer 2011), handling techniques include the use of anesthesia prior to surgeries, and subsequently holding tagged fish in enclosures to ensure complete recovery before being released.

Experiments conducted at the Inter-American Tropical Tuna Commission's laboratory in Achotines, Panama, with captive yellowfin tuna in which earlier generation ATs of size similar to that of the LAT2810 (Table 13.1), had been attached by intracoelomic implantation following procedures described by Schaefer and Fuller (2002) resulted in a high rate of expulsion from the incision site within two weeks. A method was developed that appears more suitable for the intracoelomic implantation of the LAT2810 and LAT2910 ATs in tunas, in which the external sensor stalk exits the body through a small secondary hole, so that the initial incision to accommodate the tag body can be completely closed with sutures. This surgical procedure is expected to reduce the expulsion rates of these relatively small ATs by tropical tunas. The end of the suture material (Ethicon (PDS II) size 0, cutting cp-1, 70 cm), used for closing the incision in the ventral abdominal wall through which the AT body is implanted, should be securely tied off in advance to one of the wire connectors on the tag body. An incision about 2 cm long should be made in the abdominal wall anterior of the anus and about 2 cm to the left of the centerline of the fish. Special care should be taken to cut only through the dermis and partially through the muscle, but not into the coelomic cavity. A gloved finger is inserted into the incision and forced through the muscle, with a twisting motion, into the coelomic cavity. A sterile 8 gauge stainless steel curved piercing needle, is manipulated inside the incision with the pointed end pushed through the abdominal wall to penetrate outside at about 1 cm posterior to the initial incision. When about half of the needle is protruding, the end of the external stalk is placed inside the end of the piercing needle and then the needle is pushed entirely through and removed from the stalk (Fig. 13.4A). The stalk is then pulled gently until the relief spring penetrates through the abdominal wall. The tag body, while being held by the thumb and index finger, is then inserted, anterior end first, through the initial incision, into the coelom (Fig. 13.4B). The tag stalk is then pulled gently until the tag body is directly beneath the initial incision and the spring relieve section is protruding outside the fish. With the end of the suture material tied securely to a connection wire on the tag body, two sutures are used to tightly close the incision and hold the AT in place (Fig. 13.4C).

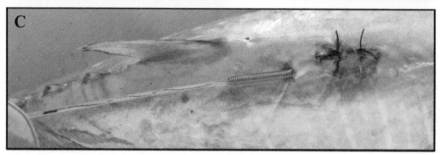

Figure 13.4. Procedures illustrating the intracoelomic implantation of an LAT2810 in a yellowfin tuna. (A) The external stalk being directed through a second incision created with the use of a curved hollow piercing needle, (B) the tag body being inserted into the coelomic cavity through the primary incision, and (C) the primary incision closed with two sutures, with external stalk exiting from secondary incision.

Some early concerns about handling of large pelagics, specifically billfishes, for attachments of ATs, and considerations of viable alternatives to intracoelomic implantation resulted in a laboratory experiment using scaled-down (1/25) model ATs with small (1–2 kg) captive yellowfin tuna to evaluate the feasibility of long-term intramuscular implantation (Brill et al. 1997). The results of these experiments were that 10 of 15 fish survived until the conclusion of the experiment at 10 months, and seven of the 10 had retained the model ATs. A similar procedure of dorsal intramuscular implantation of NMT ATs was used in some tagging experiments in Hawaii with bigeye tuna, which resulted in return rates similar to those of bigeye tagged with only conventional PDTs and with no reported AT shedding (Musyl et al. 2003). Considering the relatively small size of the LAT2910 AT, it seemed prudent to evaluate the feasibility of implanting it at the base of the second dorsal fin of tunas and other fish. The specific impetus was to establish a rapid and reliable technique to attach ATs the size of the LAT2910, or smaller, to skipjack tuna *Katsuwonus pelamis*. A plastic dart, such as that used on a plastic dart tag, was attached to a stainless shaft extending from the anterior of the LAT2910 (Fig. 13.1) for placement between the pterygiophores of the second dorsal fin to attempt to attach the AT in place, similar to that of a conventional plastic dart tag. Results from a recent tagging experiment in 2012, using this intramuscular implantation technique with yellowfin deployments ($n=97$, mean FL=111.3 cm, range: 69–160 cm) were disappointing, as a 52.6% shedding rate was observed for the LAT2910PDA tags based on 19 fish recaptured with conventional plastic dart tags (K. Schaefer and D. Fuller, unpubl. data).

It stands to reason that if proper handling and surgical techniques are used, the recapture rates of fish tagged with ATs, and AT return rates, should be comparable or greater than those for fish released during the same tagging experiments with only PDTs. This is because post-release tagging mortality should not be greater for the fish with implanted ATs than the fish with just PDTs. Tag reporting rates can be greater, however, for ATs due to publicity campaigns and greater monetary rewards for their return (Schaefer and Fuller 2009; Schaefer et al. 2011).

Pop-up satellite archival tags

Nelson (1978) proposed externally attaching several single-point PSATs to sharks as a means for reconstructing their large-scale movements. Several years later, an international group of tuna scientists met to discuss limitations of data from tuna tagging studies, using conventional plastic dart tags for understanding the movements and behavior of tunas, and concluded that a technological solution may be the development of PSATs (Hunter et al. 1986). The first PSATs were attached externally to large Atlantic bluefin in 1997 to investigate their movements and behavior (Block et al. 1998b). Following the success of those experiments, experiments by Lutcavage et al. (1999), also with PSATs externally attached to large Atlantic bluefin, confirmed and validated the concept. The technology evolved rapidly, and within a short time numerous experiments were underway with several large pelagic species, including tunas, marlins, and sharks (Arnold and Dewar 2001; Gunn and Block 2001).

The external attachment of PSATs to a variety of large fish species has provided useful information on their horizontal and vertical movements, habitat utilization, and post-release mortality, which most likely was not obtainable with other types of tags. The size of the MK-10 and PTT-100 PSATs (Table 13.2) has, however, limited their suitability for external attachment to large fish, due to concerns associated with the inability of smaller fish to swim and behave normally with the increased drag and flotation of the PSATs, and also to survive the tagging process, including the stress associated with capture and handling. With considerable reduction in the size of the MiniPAT and X-Tag (Table 13.2), the size of fish to which PSATs could be attached externally, became considerably reduced. The minimum size has not been evaluated for which PSATs should not have a prolonged adverse impact on their behavior and survival for most species of fish. This is a critical consideration for the application of PSATs, although it would be difficult to determine without considerable experimentation, including laboratory and field experiments using PSATs or dummy

PSATs on a range of fish sizes. Nevertheless, irregular behavior has been reported to last for three to 60 days for individual large pelagic fish, following their release with PSATs attached (Hoolihan et al. 2011). PSATs have also been reported to affect the swimming performance of the European eel *Anguilla anguilla* (Methling et al. 2011). The probable cause of the altered behavior is the physiological stress associated with the capture, handling, and tagging process (Wells et al. 1986; Skomal 2007), and acclimatization to swimming with the newly-attached PSAT.

A persistent problem with the use of PSATs for obtaining information on the long-term (> 1 year) movements and migrations of fish, has been the short-term (< 3 months) median retention times. Premature detachments of PSATs has occurred for most species for which experiments have been undertaken (Gunn et al. 2003; Musyl et al. 2011a; Hammershclag et al. 2011; Patterson and Hartmann 2011). By early 2000, a large number of PSATs had been deployed on a variety of marine animals, and issues surrounding the performance and reliability of PSATs were recognized as a serious problem by several research groups, so a workshop was convened to address a list of concerns, including attachment methods (Holland and Braun 2003). It was concluded that there were numerous factors that could contribute to poor performance, short retention times, and failure of PSATs, including inadequate holding power of anchors, tissue rejection, PSAT hardware failures, and predation by sharks or other large predators on PSATs while attached to the fish. Progress has been made with some species of fish to increase median retention times of PSATs by evaluating and improving handling and tagging techniques, including using different anchors and tethers for attaching PSATs. Manufacturers have also made modifications to their hardware on newer generations of PSATs, including increasing the breaking strength of the pins in the nosecone to which tethers are attached. Issues pertaining to early detachments and overall performance of PSATs remain controversial among researchers and manufacturers.

Methods used for attaching PSATs to fish have been described for tunas (Block et al. 1998b; Patterson et al. 2008), billfish (Domeier 2006; Prince and Goodyear 2006; Nielson et al. 2009), sharks (Weng et al. 2007a; Domeier et al. 2008; Musyl et al. 2011b), Pacific halibut (Seitz et al. 2003; Loher and Setiz 2006), eels (Jellyman and Tsukamoto 2002; Aarestrup et al. 2009; Manabe et al. 2011), sturgeon (Erickson 2007), Atlantic salmon kelts (Chittenden et al. 2013), and other fish. For some of these groups of fish, PSATs have been attached to individuals captured by hook and line with recreational or commercial fishing gear, and either brought aboard a vessel and placed in a cradle or on a tagging pad, or restrained alongside a vessel in the water. Proper tag placement can more easily be accomplished with fish restrained aboard vessels, but this may cause high levels of stress or infections due to removal of their protective mucous coating. Large tunas, billfishes, and sharks have been tagged alongside the vessel with tags attached to the ends of long poles by taggers standing on the deck of the vessel, but accurate placement of the tags is a problem with this method, and it is infeasible to measure the fish. The use of a 'snooter' for restraining billfish alongside sportfishing vessels has proven to be useful for improved placement of PSAT anchors, and resuscitation while moving a vessel forward at 1–2 knots for an adequate time period before release has been recommended (Prince and Goodyear 2006). Other innovative methods have been used for attachments of PSATs with anchors to free-swimming individuals of some species. Free-swimming Atlantic bluefin tuna, swordfish *Xiphias gladius*, and great white sharks *Carcharodon carcharias* have all been tagged with PSATs, using long tagging poles (Chaprales et al. 1998; Weng et al. 2007a; Nielsen et al. 2009). Divers have also tagged whale sharks *Rhincodon typus* (Wilson et al. 2006) and other fish underwater with PSATs using spearguns.

There are several types of PSAT anchors available that have been designed for implanting in the dorsal musculature near the base of the dorsal fin, passing between pterygiophores. Three custom-made anchors have been used fairly extensively by different research groups for attaching PSATs in this manner to tunas, billfishes, sharks, and other species. The titanium dart shown in Fig. 13.6, which measures $59 \times 13 \times 1.5$ mm, with the posterior 10 mm section canted at a 17° angle, has been used widely with a variety of fish (Block et al. 1998b; Weng et al. 2007a). The medical-grade nylon toggle anchor shown in Fig. 13.6, measuring 62×23 mm with blades retracted, and 80×23 mm with blades fully expanded, has been used in numerous tagging experiments with billfishes (Prince et al. 2010; Hoolihan et al. 2011a). This PSAT anchor was originally designed by Musyl et al. (unpubl. results) by adding stainless steel, spear gun flopper blades to the nylon intramuscular anchor originally used by Prince and Goodyear (2006) (Musyl et al. 2011a). The nylon umbrella-style anchor shown in Fig. 13.6, measuring 31×24 mm, has been used widely with a variety of fish (Domeier et al. 2005). The materials used for tethers connecting the anchors and PSATs also differ, depending on the objectives of the experiment (i.e., determining long-term movements or estimating post-release mortality), and also among research groups. Lengths of tethers should be designed as to minimum lengths for the species and size of fish to which PSATs are being attached. For tethers with heat-shrink tubing covering the monofilament, to stiffen the tether and reduce movement at the insertion site, instructions can be printed regarding the reward for the return of the PSAT and/ or tether. A MiniPAT rigged with a primary tether consisting of 135 kg monofilament connected to a large titanium dart, using stainless steel crimps, and covered with heat-shrink tubing is shown in Fig. 13.7A. A secondary tether, rigged in a similar manner, but with a different stainless steel dart, intended to minimize the range of motion and stress on the primary anchor site is also shown. The secondary tether has proven effective at increasing the retention times of PSATs attached to tunas (Patterson et al. 2008). A MiniPAT rigged with a tether made of abrasion-resistant cable, covered with heat-shrink tubing, and connected to a 'Wilton' dart, cast in urethane, is shown in Fig. 13.7B. The tag and tether are designed so that the tag's attachment pin is the weak link to facilitate identifying the causes of attachment failures (M. Holland, pers. comm.). A MK10 with the RD1800 (depth severance device designed to cut through the tether material if depth exceeds ~ 1800 m), rigged with a 300-lb. monofilament tether, connected to the white nylon flopper-blade anchor, using stainless steel crimps, and mounted on a tagging pole is shown in Fig. 13.7C. Fluorocarbon line, being more abrasion-resistant, may be preferable for rigging tethers, rather than monofilament, which hydrates and gets brittle over time. It is important to confirm whether PSATs, with tethers and anchors attached, are positively buoyant before deployments, as that information can be useful for interpretation of data obtained from early detachments of PSATs and post-release mortality (Musyl et al. 2011a,b; J. Hoolihan, pers. comm.).

Six species that have been tagged with PSATs using intramuscular implantation of anchors with tethers are shown in Fig. 13.7.

Aside from externally attaching PSATs to fish with anchors, another method that has been used for eels (Jellyman and Tsukamoto 2002; Aarestrup et al. 2009) and salmonids (Chittenden et al. 2013) is a bridle system, consisting of cushioned nylon plates securely fastened together on both sides of the dorsal region to which the tether is attached. This method has proven to work effectively in a limited number of experiments for those species, with some PSATs retained up to several months, but there have also been early detachments. PSATs have also been securely attached with a fluorocarbon tether crimped to a harness, made of braided stainless cable inside soft plastic tubing, inserted through a hole drilled at the base of the dorsal fins of several species of sharks (Moyes et al. 2006; Musyl et al. 2011b).

An evaluation of the performance of PSATs, based on statistical analyses of meta-data from 731 PSAT deployments for 19 species, mostly tunas, marlins, and sharks, and a total of 1433 PSAT deployments on 24 marine animals from 53 articles provides useful insights on retention times and early detachments of PSATs (Musyl et al. 2011a). Results indicated that the three types of anchors illustrated in Fig. 13.5 appear to have similar performances with respect to retention times, but 80% of 491 PSATs affixed to teleosts and sharks with those anchors detached before the programmed pop-off date. The longest median retention times for Microwave Telemetry PSATs, reported in this study, were those deployed on great white sharks (207 days), followed by oceanic whitetip sharks *Carcharhinus longimanus* (164 days), and Atlantic bluefin tuna (102 days). Median retention times for MT PSATs deployed on striped marlin *Kajikia audax* (98 days) were considerably longer than those for black *Istiompax indica* (46 days), and blue marlins *Makaira mazara* (54 days). Analyses of the shark data separately, indicated 65% of all PSATs detached before the programmed pop-up date, and sharks that were immobilized on the deck of vessels for attaching PSATs had significantly shorter retention times than those tagged in the water. PSAT retention times increased significantly for teleosts and sharks which inhabited greater depths and cooler waters. The results suggested that pressure and/or temperature experienced at greater depths, biofouling on PSATs increasing drag, and possible infections and tissue necrosis at the insertion site of the anchors were major factors to explain variable PSAT retention times (Musyl et al. 2011a). A separate review of shark satellite tagging studies by Hammerschlag et al. (2011) reported that PSAT premature releases averaged 66% across 27 studies reporting such data, even though most were programmed for relatively short deployment periods of 30, 60, or 90 days. Other researchers have hypothesized that early detachments of PSATs affixed to fish may have been caused by increased drag from biofouling (Gunn et al. 2003; Kerstetter et al. 2004; Wilson et al. 2006), tissue necrosis at the site of the implanted anchor and tether (Jellyman and Tsukamoto 2002; De Metrio et al. 2004; Wilson et al. 2005), and predation by sharks or other predators on either the tagged animals or the PSATs (Klimley et al. 1994; Kerstetter et al. 2004; Polovina et al. 2007). PSAT tag retention is obviously a complex issue and probably the result of several factors. In many instances it has been difficult to ascertain causes, except in those rare instances when fish are returned with PSATs or tethers still attached.

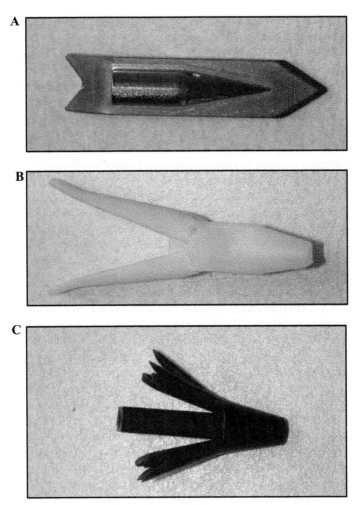

Figure 13.5. Three types of anchors which have been used for attaching PSATs with tethers to a variety of fish. (A) titanium dart, (B) medical grade nylon toggle anchor, and (C) nylon umbrella-style anchor.

Data analyses

The most important feature of ATs and PSATs is their ability to measure and record ambient light and temperature data, and other environmental data, which can be analyzed using various methods to reconstruct the most probable tracks of individual fish. Analyses of the spatial and environmental data collected by ATs and PSATs enables us to better understand the large-scale movements and migrations of individual fish, along with their behaviors, relative to their habitat. This type of information is indispensable for gaining a better understanding of the spatial dynamics of fish stocks. Elucidating the spatial dynamics of fish stocks, including movement patterns, dispersion, mixing, residency, site fidelity, and homing, and ultimately including those

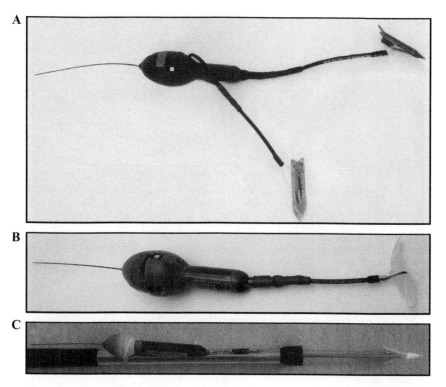

Figure 13.6. PSATs rigged with different types of tethers and anchors for different applications. (A) MiniPAT rigged with primary tether and a titanium dart, along with a secondary restraining loop and stainless steel dart (photo credit: B. Block/Stanford University). (B) MiniPAT rigged with a tether and 'Wilton' dart (photo credit: Wildlife Computers) and (C) MK10 rigged with tether and medical grade nylon toggle anchor, and mounted on a tagging pole with an applicator (photo credit: J. Hoolihan/NMFS).

factors in assessment models, requires data sets from large numbers of individuals, and the inclusion of SSMs in the data analysis (Patterson et al. 2008).

The ambient light-level data measured and recorded by archival and pop-up satellite archival tags can provide the time of dawn and dusk events and the ability to calculate daily geolocations of fish during the time they were carrying those tags (Smith and Goodman 1986; Wilson et al. 1992; Hill 1994). The manufacturer's basic tag processing software calculates longitude from the time of local noon, and latitude from the local day length using what is referred to as the threshold method. The raw (i.e., unfiltered and uncorrected) light-based geolocations obtained for longitude can be reasonably accurate and reliable, whereas those for latitude are highly variable and unreliable especially around the time of the equinoxes due to the nearly constant day length at all latitudes (Hill 1994; Hill and Braun 2001; Ekstrom 2004). The accuracy of the light-based geolocation estimates have been reported from ATs tethered to moored buoys, at higher latitudes, as ± 0.2–0.9° in longitude and ± 0.6–4.4° in latitude (Welch and Eveson 1999, 2001; Musyl et al. 2001). Comparisons of the calculated light-based geolocation within a day of recapture of bigeye tuna with ATs attached, to the reported recapture position indicated an accuracy of ± 0.5° of longitude and ± 1.5–2.0°

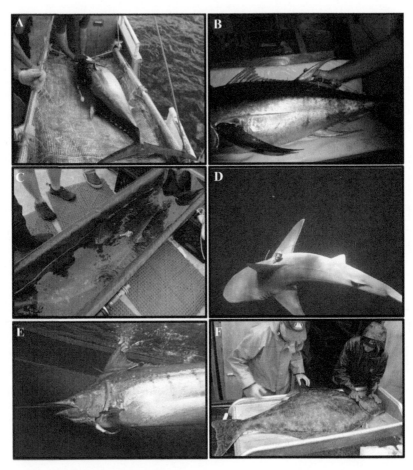

Figure 13.7. (A) Southern bluefin tuna *Thunnus maccoyii* with a Wildlife Computers PAT-4 attached using umbrella-style anchor on primary tether, and secondary restraining loop around the PSAT shaft, is a modified Hallprint PIMA tag (photo credit: Kristi Wright /CSIRO). (B) Yellowfin tuna *Thunnus albacares* with a MiniPAT attached using a 'Wilton' dart on primary tether and secondary restraining loop around the PSAT shaft is a modified Hallprint PIMA tag (photo credit: K. Holland/UH, HIMB). (C) Great white shark *Carcharodon carcharias* with a MiniPAT attached using a titanium dart on primary tether, and secondary restraining loop around the PSAT shaft is a modified Floy FH-69 tag (photo credit: Chuck Winkler/Monterrey Bay Aquarium). (D) Sandbar shark *Carcharhinus plumbeus* with a MiniPAT attached using a 'Wilton' dart (photo credit: M. Royer/UH, HIMB). (E) Indo-Pacific blue marlin *Makaira mazara* with MK10 attached using a nylon umbrella-style anchor (photo credit:B. Boyce/ www.boyceimage.com). (F) Pacific halibut *Hippoglossus stenolepis* with a MK10 attached using a titanium dart (photo credit: R. Ames/Pacific States Mar. Fish. Comm.).

latitude (Schaefer and Fuller 2002). Problems with light-based geolocation estimates and associated sources of error include depth distributions and diving behavior, meteorological and oceanographic conditions, light attenuation, and proximity to the equator (Metcalfe 2001). The large errors in the light-based geolocation estimates can be improved somewhat through utilizing the narrow-band blue light and the template-fit method (Ekstrom 2004, 2007).

Daily SSTs recorded by ATs and PSATs matched to SSTs from remote sensing have been shown to significantly improve estimates of latitude (Teo et al. 2004; Domeier et al. 2005; Wilson et al. 2007). Teo et al. (2004) developed an algorithm for estimation of daily positions of latitude by comparing SSTs measured by the tag with that measured by satellite sensors, along the longitudinal meridian estimated from the threshold technique, using light-level data. From double-tagging experiments, using Wildlife Computers PSAT and Smart Position Or Temperature transmitting (SPOT) tags on salmon sharks (*Lamna ditropis*) and blue sharks (*Prionace glauca*), Teo et al. (2004) estimated Root Mean Square (RMS) error of light-based longitude was within 0.89°, and the RMS error of SST-based latitude was within 1.54°. Domeier et al. (2005) also developed an algorithm, called 'PSAT Tracker', which automatically matches SST data from ATs and PSATs with SSTs measured by satellite sensors. They demonstrated the usefulness of their algorithm through the analyses of data collected from ATs and PSATs attached to Pacific bluefin tuna tagged and released in the eastern Pacific. Tsontos et al. (2006) improved on this approach by integrating the Fishtracker SST-matching geocorrection algorithm within EASy-GIS (http://www.runeasy.com), a time dynamic mapping system for oceanographic applications used in marine biogeographical studies.

Incorporating light-based geolocation estimates within a movement model has proved extremely useful for reconstructing tracks of fish tagged with ATs and PSATs. Sibert and Fournier (2001) first introduced a state-space statistical model, utilizing the Kalman Filter (KF) (Kalman 1960), to estimate a 'most probable' track along with geolocation errors for longitude and latitude and movement parameters applicable to population-level models. Their KF model parameterizes movement as a biased random walk, with the movement partitioned into directed and dispersive movements. The method uses the raw geolocations to estimate the random walk parameters in the periods during which the data are reliable, but relies mainly on the random walk predictions in periods during which the data are unreliable. The model estimates the geolocation errors as the longitude and average latitude standard deviations. The utility of the KF model was first demonstrated by application to AT data sets recovered from bigeye tuna released in the vicinity of the Hawaiian Islands, so as to describe their horizontal movement patterns and parameters (Sibert et al. 2003). SSMs have also been developed, utilizing the Bayesian approach (Jonsen et al. 2003; Jonsen et al. 2005), for analyses of animal movements data, including reconstruction of fish or other animal tracks from raw geolocation estimates and ancillary data.

Since daily geolocations and tracks can be improved by including SSTs measured by tag sensors and matched to SST fields derived from satellite sensors, Nielsen et al. (2006) extended the KF model by integrating SST data. The KFsst model was evaluated by attaching PSATs directly on drifter buoys with thermistors and GPS, and then applied to tracks derived from blue sharks in the central Pacific Ocean. The inclusion of SST in the KFsst model produced substantially more probable tracks than those estimated from the light-level data alone. Royer et al. (2005) also developed a Kalman filter model with SST measurements along with bathymetry integrated for use with AT and PSAT data sets obtained from marine animals. The unscented KF model with SST measurements integrated (UKFsst) (Lam et al. 2008), is similar to the KFsst model. The UKFsst model is a better model for handling non-linearities, and also has the advantage that every model parameter is handled within a statistical framework. The UKFsst model can also utilize remotely-sensed SST data of various spatial resolutions, and it automatically estimates the amount of smoothing required for the SST field. This approach is preferable to ad hoc SST matching algorithms (Lam et al. 2008).

Nielsen and Sibert (2007) developed the state-space model TrackIt, which begins the geolocation estimation process, using the records of ambient light data recorded by tags, rather than those provided from other algorithms. TrackIt assumes an underlying movement model and estimates the track that best matches both the assumed model and the relationship estimated empirically within TrackIt for the light time series data. Lam et al. (2010) incorporated SST data into the light-based geolocation model TrackIt, making it a single coherent model to estimate positions, founded on a set of much more reliable light-based geolocation estimates. TrackIt is a generic and statistically-sound modeling framework that can be extended to utilize new data streams such as geomagnetic data or ocean chemistry data. Double-tagging experiments with Wildlife Computers PSATs and SPOTs, attached to mako sharks *Isurus oxyrinchus*, demonstrated that TrackIt with SST data provides improved movement estimation to within 100 nmi, on average, in both the coastal environment and the open ocean (Lam et al. 2010). TrackIt, however, does not contain land masks, but Lam et al. (2010) indicate that by applying other filtering techniques, such as hidden Markov models (Pedersen et al. 2008; Thygesen et al. 2009) should enable that feature to be incorporated. TrackIt with SST, along with Trackit, UKFsst, KFsst, and KF models are all freely available as plug-in packages for the open-source statistical software R (https://code.google.com/p/geolocation/).

Tremblay et al. (2009) developed an alternative model to that of an SSM for reconstructing animal tracks that is simpler and more computationally efficient. The approach uses a particle filter method, consisting of bootstrapping random walks biased by forward particles. The model uses light-based geolocations and SST or other environmental data recorded by tags, and can easily incorporate land boundaries. The model was tested and validated from double-tagging experiments with elephant seals, using ARGOS and GPS tags, and results showed that for geolocation data, 50% of errors were less than 56.6 nm (< 0.94°), and 90% were less than 107.9 nmi (< 1.80°) (Tremblay et al. 2009).

SSMs are a preferred method to estimate animal movement behavior because of their statistical robustness and predictive ability, but also because they lend themselves to combined inference from a set of biologically plausible models, which is more reliable than predicting from a single best model (Patterson et al. 2008). An interesting example of this approach is given in Tancell et al. (2012) who compared and integrated the results of four models, kernel, first-passage time, SSM, and minimum displacement rate applied to data from the wandering albatross *Diomedea exulans* fitted with ARGOS transmitters. A gridded overlap approach applied to all tracks revealed core areas of habitat utilization not apparent from results of any single analysis and spatial overlap between methods based on different assumptions and among individuals (Tancell et al. 2012).

Wildlife Computers developed an integrated user-friendly software package called Wildlife Computers Data Analysis Programs (WC-DAP), which includes utilities to decode, summarize, visualize, analyze, and export Wildlife Computers AT and PSAT data. Visualization is done via 'Instrument Helper', within WC-DAP, which provides the flexibility to display time-series data for one hour to multiple years, for each recorded data channel, individually or all at once. Wildlife Computers Global Position Estimator (GPE) program integrates into WC-DAP to produce daily geolocation estimates from a tag's light-level data, and allows the user the flexibility to manage light-level geolocation operations. The integrated GPE software also provides the option, without exporting the data, for processing light-based geolocation estimates with the UKFsst state-space model. However, in order to use the TrackIt or TrackIt with SST models, the user must first use the GPE to automatically sub-sample and export raw light-level data recorded on the tag. At the same time an R-script is created for processing the sub-sampled light-levels in the R framework. For additional visualization flexibility and the ability to analyze dive patterns, AT or PSAT data can be exported from WC-DAP in an IGOR Pro format, a graphing and data analysis tool. Wildlife Computers has created a macro which operates in both the free trial and licensed versions of IGOR Pro, which is specific to the graphical display and analyses of AT data, giving the user high levels of control.

Lotek Wireless has developed LAT Viewer Studio software package, which is a Windows-based application for importing, processing, and managing data from Lotek ATs. LAT Viewer Studio operates independently from the Lotek software program TagTalk2000, which is used for setup and download of ATs. LAT Viewer studio provides several options for tabular, graphical, and statistical summaries of imported data sets. However, to accomplish some of the more complex queries and analyses of AT data, the user must be familiar with SQL script language. The program also provides the user with some flexibility in processing methods for geolocation estimates from light-level data, including Lotek's proprietary Template-Fit method, and also incorporating SST data with both template fit and threshold methods. However, in order to use the UKFsst, Trackit, or Trackit with SST models, data must first be exported, and compiled in a format compatible within R. This requires that the user have knowledge of operating these models within the R framework. These Wildlife Computers and Lotek Wireless data management and analysis programs provide attractive features for quick visualization of large volumes of time-series data, some individual data analyses and summaries, using various analysis routines, and plotting tracks in Google Earth maps.

Multi-species long-term tagging programs, which utilize ATs and PSATs, can produce huge volumes of data from various tag manufacturers. The programs available from tag manufacturers are designed primarily for initial processing of data sets from individual ATs and PSATS, but not for the management of large numbers of data sets, nor from tags produced by other companies. The management of these large AT and PSAT data bases for storing and accessing data is critical. Previously, unless institutions and organizations had resources for dedicated data management (Block et al. 2002; Hartog et al. 2009), this was an issue for many groups, until Lam and Tsontos (2011) developed Tagbase, a robust and user-friendly data base solution for ATs and PSATs. Tagbase provides a data management system for rapid assimilation of tag data from multiple tag types within a generalized platform and requires minimal user intervention. The system also provides a set of integrated tools for visualizing and summarizing data, and operates within an open-source development model (Lam and Tsontos 2011). Tagbase is implemented in Access and SQL Server running on MS Windows operating systems, providing secure storage, data base queries, and some processing suitable for small to extremely large AT and PSAT data bases. Queries can be written to extract information from the data base in required formats for use in other programs, including Microsoft Excel, R and ArcView (Environmental Systems Research Institute, ESRI).

Applications

There are many good studies describing successful applications of ATs and PSATs for determining the large-scale movements and migrations of fish. We have chosen four species to showcase the enormous potential of this technology, because of the relatively large numbers of tag deployments with each of these species, subsequent data sets acquired, analyses performed, and relevance of the results to fishery stock assessments and management considerations.

Pacific bluefin tuna

Pacific bluefin tuna have an extensive distributional range throughout the North Pacific, extending into the western South Pacific. They are a migratory species, with spawning restricted to an area from the northern Philippines to southeastern Japan and the Sea of Japan during April through August. They undertake extensive migrations throughout much of the western and central Pacific, with variable portions of the population making transPacific migrations to and from the eastern Pacific Ocean (EPO) (Bayliff 1994).

Considering the expansive distribution and migratory habits, along with concerns about fishery impacts and population status, Pacific bluefin have been studied fairly extensively with deployments of ATs in both the western and eastern Pacific. Investigations of movements of juveniles tagged and released in the western Pacific have shown primary residence within the Kuroshio current, and Sea of Japan, but also some long-distance migrations, including across the Pacific to off northern Baja California, Mexico (Kitagawa et al. 2000; Inagake et al. 2001; Itoh et al. 2003). Seasonal movements of Pacific bluefin released in the EPO with archival tags were reported to extend along the coast of North America, residing in southern Baja California during winter and spring, and moving to northern California during the summer and fall, with movements correlated with SSTs and sardine catch distributions (Domeier et al. 2005; Kitagawa et al. 2007).

Boustany et al. (2010) reported the results of a series of tagging experiments in which 253 Pacific bluefin were captured, tagged, and released with ATs off the coast of California (USA) and Baja California (Mexico), between August 2002 and August 2005. A total of 143 AT data sets were obtained from fish recaptures with mean days at liberty 359 ± 248 (SD) days. From those AT data sets a total of 38,012 geolocation positions were estimated, using the methods reported in Teo et al. (2004). Results indicate a similar pattern of movement along the coast of North America to those previously reported, with fish found farthest south off southern Baja California in the spring, and furthest north off central and northern California in the fall. Latitudinal movement patterns and residence along the coast of North America were clearly shown to be correlated with peaks in coastal upwelling-induced primary productivity. The geographic range in movements of these bluefin released with ATs was extensive, with fish distributed from the coast of North America to the Sea of Japan (Fig. 13.8). Seventeen fish exhibited movements ≥ 300 nm offshore. Seven fish exhibited trans-Pacific movement, and two of those fish moved back to the eastern Pacific, with one traveling back to the western Pacific where it was recaptured. All the other fish remained in the area near the North American coast. The authors emphasize that the transPacific migration rates from the east to the west are expected to be much higher for older fish returning to their spawning grounds.

This work was part of the Tagging of Pacific Predators (TOPP), a field program of the Census of Marine Life which deployed 4,306 electronic tags on 23 species of marine animals in the North Pacific Ocean resulting in a tracking data set of unprecedented scale and species diversity (Block et al. 2011).

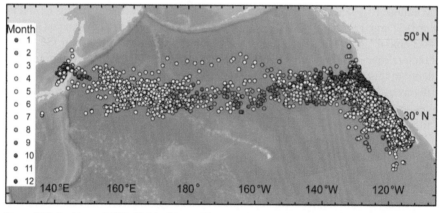

Figure 13.8. Positions of all Pacific bluefin tuna *Thunnus orientalis* tracked with archival tags, color-coded by month. Tracking data were obtained from 143 fish and 38,012 positions were obtained. Pacific bluefin tuna primarily occupy the eastern Pacific. Movement to the western Pacific (*n* = 7 fish) occurs in the winter while migrations back to the eastern Pacific occurred in the summer. Background is North Pacific bathymetry. Figure is from Boustany et al. (2010); reprinted with permission.

Yellowfin tuna in the eastern Pacific Ocean

Yellowfin tuna are distributed across the world's oceans in tropical and subtropical seas (Collette and Nauen 1983). They show regional fidelity to the eastern Pacific, with evidence of sub-stocks from conventional tagging and other biological data, but further research is required to elucidate the extents and interactions of the sub-stocks (Schaefer 2008). Yellowfin are not a highly migratory species as they exhibit spawning throughout subtropical and tropical regions where SSTs are in excess of about 24°C (Schaefer 2001).

Recent stock assessments of yellowfin in the eastern Pacific Ocean (EPO) indicate that the population appears to be fully exploited, but results of the assessment are somewhat uncertain due to uncertainty in several parameters, including movements and stock structure throughout the Pacific Ocean (Maunder and Aires-da-Silva 2011). Studies are being undertaken in the EPO with ATs to obtain information on movement patterns and parameters to help clear up some of the uncertainty regarding spatial dynamics of yellowfin.

Schaefer et al. (2011) reported the results of a series of tagging experiments in which 144 yellowfin were captured, tagged, and released with ATs off northern Baja California (Mexico), between October 2003 and October 2006, and 354 yellowfin were captured, tagged, and released with ATs off southern Baja California (Mexico) between October 2002 and December 2008. The movements off Baja California were described from analyses of 31,357 days of data, downloaded from 126 archival tags recovered from yellowfin (57 to 162 cm in length and 1.2 to 5.2 years of age) at liberty from 90 to 1161 days ($\bar{x}$ = 273.2 days), collected from 2002 to 2010. The UKFsst model (Lam et al. 2008) was used to process the AT data sets in order to obtain improved estimates of geographic positions and Most Probable Tracks (MPTs) and parameters. The median parameter estimates from the UKFsst model for errors in longitude (σ_x) and latitude (σ_y) were 0.32° and 1.36°, respectively, for directed movements (u and v) were 0.27 nmi d^{-1} and 0.77 nmi d^{-1}, respectively, and for dispersive movement (D) was 144.3 nmi^2 d^{-1}. The MPTs for 120 (95%) of the yellowfin remained within 733.3 nmi of their release locations, indicating restricted horizontal utilization distributions, and fidelity to this area of high biological productivity (Fig. 13.9A). There are observed differences in the movement patterns and parameters for fish released in different areas, and also evidence of fidelity to release locations off southern and northern Baja California (Fig. 13.9B-E). This work was also part of the TOPP program of the Census of Marine Life (Block et al. 2011).

Bigeye tuna in the eastern Pacific Ocean

Bigeye tuna inhabit tropical and subtropical seas across the world's oceans (Collette and Nauen 1983). Tagging studies with conventional and ATs indicate regional fidelity to the EPO, and evidence of probable northern and southern sub-stocks, but further research is required to elucidate the extents and interactions of these sub-stocks (Schaefer 2008). Bigeye does not appear to be a highly-migratory species as they spawn widely throughout subtropical and tropical regions where SSTs are in excess of about 24°C (Schaefer 2001).

Recent stock assessments of bigeye in the EPO (Aires-da-Silva and Maunder 2011) indicate that the population appears to be fully exploited, but results of the assessment are somewhat unpredictable due to the uncertainity in several parameters, including movements and stock structure throughout the Pacific Ocean. Studies are being undertaken in the eastern, central, and western Pacific Ocean utilizing conventional and ATs to obtain information on the spatial dynamics and stock structure of bigeye in the Pacific Ocean.

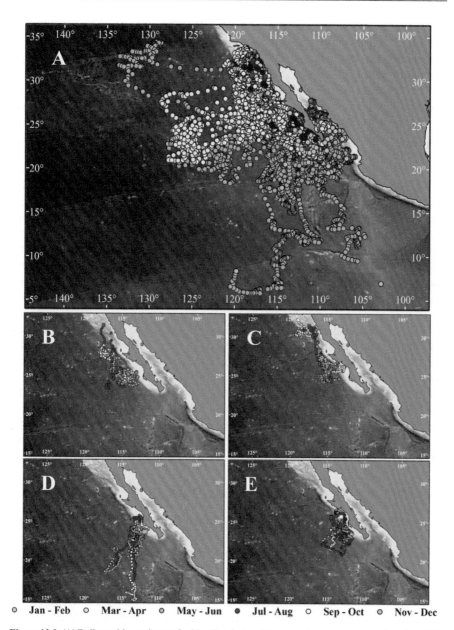

○   **Jan - Feb**   ○   **Mar - Apr**   ○   **May - Jun**   ●   **Jul - Aug**   ○   **Sep - Oct**   ○   **Nov - Dec**

Figure 13.9. (A) Daily position estimates for 38 yellowfin tuna *Thunnus albacares* released off northern Baja California, Mexico (yellow circles, *n* = 11,072) and 88 yellowfin tuna released off southern Baja California (Mexico) (orange circles, *n* = 20,285) during 2002–2008. Release positions indicated by the black circles, recapture positions for northern Baja California releases indicated by red circles, and recapture positions for southern Baja California releases indicated by green circles. Most probable tracks for yellowfin tuna exhibiting site fidelity and homing, (B) released off northern Baja (60 cm FL), at liberty for 372 d, (C) released off northern Baja (61 cm FL), at liberty for 407 days, (D) released off southern Baja (122 cm FL), at liberty for 781 days, and (E) released off southern Baja (90 cm FL) at liberty for 1161 days. Release and recapture positions indicated by the squares and triangles.

Schaefer and Fuller (2009) reported the results of a series of tagging experiments in which 323 bigeye were captured, tagged, and released with ATs in the equatorial EPO, during March to May of 2000 and 2002 through 2005, of which 163 of the tags were returned. The UKFsst model (Lam et al. 2008) was used to process 98 archival tag data sets from bigeye at liberty from 31 to 2,291 days ($\bar{x}$ = 258.4, SE = 35.5), in order to obtain improved estimates of geographic positions and MPTs and parameters. The median parameter estimates from the UKFsst model for errors in longitude (σ_x) and latitude (σ_y) were 0.27° and 2.34°, for directed movements (u and v) were 1.25 nmi d^{-1} and 0.35 nmi d^{-1}, and for dispersive movement (D) was 465 nmi^2 d^{-1}. The 95 and 50% utilization distributions, based on 11,585 positions for the combined 98 bigeye archival tag data sets, were 716,158 nmi^2 and 32,757 nmi^2, respectively, and were centered between about 3° N and 5° S and 90° W and 105° W. Based on the AT data, there are observed differences in the spatial patterns of movements by year of deployment and fish age or size. These data indicate that bigeye exhibit restricted movements, with regional fidelity to this area of high biological productivity (Fig. 13.10A), with one exception. The exception was a bigeye at liberty for 4.1 years. Because of various computational problems encountered for processing of this very large data set with UKFsst, the most probable track was derived using the TrackIt model with SST incorporated (Lam et al. 2010). During the first two years at liberty, the fish remained within 993 nm of its release location, a restricted area similar to that occupied by the other 97 fish, but centered slightly to the east (Fig. 13.10B-D). However, during the third (Fig. 13.10E) and fourth (Fig. 13.10F) years at liberty the fish undertook two very similar cyclical movements to the central Pacific. The first movement began with a departure in early July 2007, arriving at about 151° W in early November 2007, and returning to the EPO, at about 84° W, in early May 2008. The second movement began with a departure in early July 2008, arriving at about 162° W in early December 2008, and returning to the EPO, at about 84°W, in early May 2009. The fish was recaptured just 672 nmi from where it was released.

Great white sharks in the Northeastern Pacific Ocean

The great white shark has a circumglobal distribution extending through temperate to tropical waters and coastal to pelagic habitats (Compagno 1984). There is a concentration of great white sharks in the Northeastern Pacific (NEP), but little is known about their breeding, parturition, and early life history phases. Young of the year great white sharks have been observed in the southern California Bight and off Baja California, and the nursery habitat in the Northeastern Pacific is hypothesized to include the coast of North America south of Point Conception (Klimley 1985). Great white sharks are listed for international protection under the Convention on International Trade in Endangered Species of Wild Fauna and Flora (CITES) and the International Union for Conservation of Nature (IUCN) (Dulvy et al. 2008).

Since understanding the spatial dynamics and population structure of white sharks is critical for conducting appropriate population assessments, and implementing effective management strategies, there has been a considerable number of tagging experiments in the NEP utilizing PSATs in recent years, with deployments from northern Baja California (Mexico) to central California (USA). Data from those PSATs

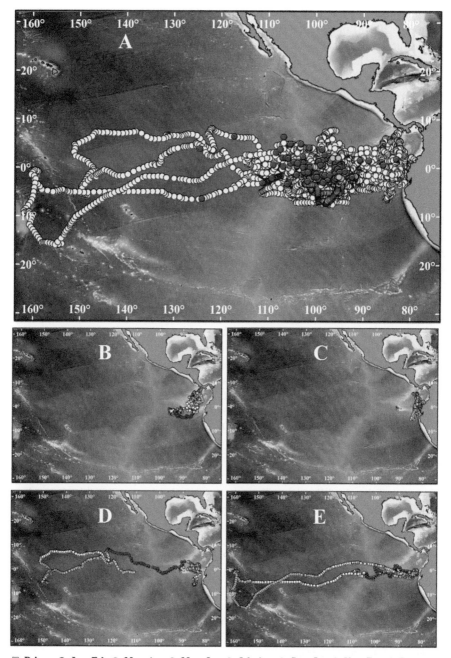

☐ **Release** ○ **Jan - Feb** ○ **Mar - Apr** ○ **May - Jun** ● **Jul - Aug** ○ **Sep - Oct** ○ **Nov - Dec** △ **Recapture**

Figure 13.10. (A) Daily position estimates from the UKFsst and Trackit models (yellow circles, $n = 16,721$) for 96 bigeye tuna *Thunnus obesus* with archival tags, at liberty for 31 to 1508 days released (green circles) in the equatorial eastern Pacific during 2000–2005, recovery positions indicated by the red circles. Most probable track, color coded by bi-monthly periods, for bigeye tuna released April 15, 2005 (67cm FL) for the (B) first, (C) second, (D) third, and (E) fourth year at liberty.

have demonstrated that great white sharks tagged in California (Boustany et al. 2002; Weng et al. 2007a) and at Guadalupe Island off northern Baja California (Domeier and Nasby-Lucas 2008) share common oceanic habitats, but also exhibit unique homing and site fidelity to coastal habitats. Longer tracks from those studies have revealed that adult great white sharks made some long-distance seasonal migrations from coastal waters offshore into the eastern and central Pacific, as far west as Hawaii and remained for up to about four months. Juvenile great white sharks tagged with PSATs have demonstrated restricted movements, remaining in waters of southern California (USA) and Baja California (Mexico) (Weng et al. 2007b).

Jorgensen et al. (2009) used a combination of PSAT tagging, passive acoustic telemetry, and genetic analysis, to determine the spatial dynamics and demographic scope of great white sharks in the NEP. They deployed 97 PSATs on great white sharks in coastal waters of central California. Data from 68 of the 97 PSATs were retrieved, 54 satellite-transmitted data sets and 14 recovered PSATs with archival records. PSAT deployments rendered 6978 longitude and 6144 latitude estimates, and 60 ARGOS endpoint positions. Latitude and longitude estimates and tracks from the PSAT data sets were determined using the methods of Teo et al. (2004). The data revealed site fidelity and consistent homing in a seasonal migratory cycle with fixed destinations, schedules, and routes (Fig. 13.11). The geolocation estimates and acoustic detection locations ($n = 74,354$) across the Northeastern and central Pacific demonstrated that great white sharks were consistently focused on three core areas: (1) the North America shelf waters, (2) the slope and offshore waters of the Hawaiian archipelago, and (3) the offshore white shark 'Café' (Jorgensen et al. 2009). Their findings from using this integrated approach combining PSAT tagging, passive acoustic telemetry, and genetics to investigate the NEP white shark population, culminates in their demonstration of a demographically independent management unit that shows fidelity to discrete and predictable locations (Jorgensen et al. 2009).

Acknowledgements

We are grateful to the following colleagues and friends whom unselfishly shared their knowledge, insights, graphics, and photographs, which contributed to enhancing the contents of this chapter: B. Block, A. Boustany, W. Boyce, M. Domeier, K. Evans, K. Holland, M. Holland, J. Hoolihan, D. Itano, S. Jorgenson, C. Lam, T. Lohr, M. Musyl, P. Oflaherty, J. O'Sullivan, T. Patterson, E. Prince, D. Righton, A. Seitz, K. Weng, and C. Winkler. Thanks are also extended to Lotek Wireless, Microwave Telemetry, and Wildlife Computers for generously providing information and photographs of their tags. We gratefully acknowledge the constructive comments provided, on a previous draft of the entire manuscript, by W. Bayliff, the section on tag designs by L. Jordan, the section on tag attachments by M. Musyl, and the section on data analyses by C. Lam.

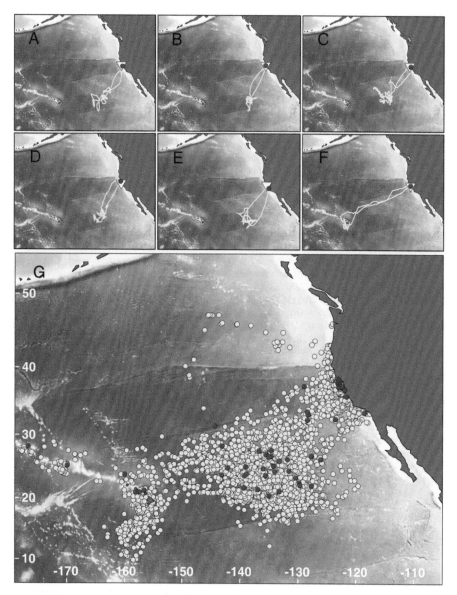

Figure 13.11. Site fidelity and homing of great white sharks *Carcharodon carcharias* tagged along the central California coast during 2000–2007 revealed by PAT records. (A–F) Site fidelity demonstrated by six individual tracks (yellow lines; based on five point moving average of geolocations). Triangles indicate tag deployment locations and red circles indicate satellite tag pop-up endpoints (Argos transmissions) for white sharks returning back to central California shelf waters after offshore migrations. (G) Site fidelity of all satellite tagged white sharks (*n* = 68) to three core areas in the NEP included the North American continental shelf waters, the waters surrounding the Hawaiian Island Archipelago and the white shark 'Café'. Yellow circles represent position estimates from light- and SST-based geolocations (Teo et al. 2004), and red circles indicate satellite tag endpoint positions (Argos transmissions), respectively. Figure is from Jorgenson et al. (2009); reprinted with permission.

References

Aarestrup K, Okland F, Hansen MM, Righton D, Gargan P, Castonguay M, Bernatchez L, Howey P, Sparholt H, Pedersen MI and McKinley RS (2009) Oceanic spawning migration of the European eel (*Anguilla anguilla*). Science 325: 1660.

Aires-da-Silva A and Maunder MN (2011) Status of bigeye tuna in the eastern Pacific Ocean in 2009 and outlook for the future. Inter-American Tropical Tuna Commission, Stock Assessment Report 11: 17–156.

Arnold G and Dewar H (2001) Electronic tags in marine fisheries research: a 30-year perspective. pp. 7–64. *In*: Sibert JR and Nielsen JL (eds.). Electronic Tagging and Tracking in Marine Fisheries. Kluwer, Dordrecht, The Netherlands.

Bayliff WH (1994) A review of the biology and fisheries for northern bluefin tuna, *Thunnus thynnus*, in the Pacific Ocean. FAO Fisheries Technical Papers 336: 244–295.

Bayliff WH and Holland KN (1986) Materials and methods for tagging tunas and billfishes, recovering the tags, and handling the recapture data. FAO Fisheries Technical Papers 279: 1–36.

Block BA (2005) Physiological ecology in the 21st century: advancements in biologging science. Integrative and Comparative Biology 45: 305–320.

Block BA, Dewar H and Williams T (1998a) Archival tagging of Atlantic bluefin tuna (*Thunnus thynnus thynnus*). Marine Technology Society Journal 32: 37–46.

Block BA, Dewar H, Farwell C and Prince ED (1998b) A new satellite technology for tracking the movements of Atlantic bluefin tuna. Proceedings of the National Academy of Sciences of the United States of America 95: 9384–9389.

Block BA, Costa DP, Boehlert GW and Kochevar RE (2002) Revealing pelagic habitat use: the Tagging of Pacific Pelagics program. Oceanologica Acta 25: 255–266.

Block BA, Jonsen ID, Jorgensen SJ, Winship AJ, Shaffer SA, Bograd SJ, Hazen EL, Foley DG, Breed GA, Harrison AL, Ganong JE, Swithenbank A, Castleton M, Dewar H, Mate BR, Shillinger GL, Schaefer KM, Benson SR, Weise MJ, Henry RW and Costa DP (2011) Tracking apex marine predator movements in a dynamic ocean. Nature 475: 86–90.

Boustany AM, Davis SF, Pyle P, Anderson SD, Le Boeuf BJ and Block BA (2002) Expanded niche for white sharks. Nature 415: 35–36.

Boustany AM, Matteson R, Castleton M, Farwell C and Block BA (2010) Movements of Pacific bluefin tuna (*Thunnus orientalis*) in the eastern North Pacific revealed with archival tags. Progress in Oceanography 86: 94–104.

Bridger CJ and Booth RK (2003) The effects of biotelemetry transmitter presence and attachment procedures on fish physiology and behavior. Reviews in Fisheries Science 11: 13–34.

Brill RW, Cousins K and Kleiber P (1997) Test of the feasibility and effects of long-term intra-muscular implantation of archival tags in pelagic fishes using scale model tags and captive juvenile yellowfin tuna (*Thunnus albacares*). NMFS Administrative Report H-97-11.

Chaprales W, Lutcavage ME, Brill RW, Chase B and Skomal GB (1998) Harpoon method for attaching ultrasonic and popup satellite tags to giant bluefin tuna and large pelagic fishes. Marine Technology Society Journal 32: 104–105.

Chittenden CM, Adlandsvik B, Pedersen OP, Righton D and Rikardsen AH (2013) Testing a model to track fish migrations in polar regions using pop-up satellite archival tags. Fisheries Oceanography 22: 1–13.

Collette BB and Nauen CE (1983) Scombrids of the World. FAO species catalogue. Vol. 2. FAO Fisheries Synopsis 125: 1–137.

Compagno L (1984) Sharks of the world. FAO species catalogue. Vol. 4. Part 1 FAO Fisheries Synopsis 125: 1–249.

Cooke SJ, Woodley CM, Eppard MB, Brown RS and Nielsen JL (2011) Advancing the surgical implantation of electronic tags in fish: a gap analysis and research agenda based on a review of trends in intracoelomic tagging effects studies. Reviews in Fish Biology and Fisheries 21: 127–151.

De Metrio GD et al. (2004) Joint Turkish-Italian research in the eastern Mediterranean: bluefin tuna tagging with pop-up satellite tags. Collective Volume of Scientific Papers ICCAT 56: 1163–1168.

Domeier ML (2006) An analysis of Pacific striped marlin (*Tetrapturus audax*) horizontal movement patterns using pop-up satellite archival tags. Bulletin of Marine Science 79: 811–825.

Domeier ML and Nasby-Lucas N (2008) Migration patterns of white sharks *Carcharodon carcharias* tagged at Guadalupe Island, Mexico, and identification of an eastern Pacific shared offshore foraging area. Marine Ecology Progress Series 370: 221–237.

Domeier ML, Kiefer D, Nasby-Lucas N, Wagschal A and O'Brien F (2005) Tracking Pacific bluefin tuna (*Thunnus thynnus orientalis*) in the northeastern Pacific with an automated algorithm that estimates latitude by matching sea-surface-temperature data from satellites with temperature data from tags on fish. Fisheries Bulletin 103: 292–306.

Dulvy NK, Baum JK, Clarke S, Compagno LJV, Cortés E, Domingo A, Fordham S, Fowler S, Francis MP, Gibson C, Martínez J, Musick JA, Soldo A, Stevens JD and Valenti S (2008) You can swim but you can't hide: the global status and conservation of oceanic pelagic sharks and rays. Aquatic Conservation: Marine and Freshwater Ecosystems 18: 459–482.

Ekstrom PA (2004) An advance in geolocation by light. pp. 210–226. *In*: Naito Y (ed.). Memoirs of the National Institute of Polar Research, Special Issue. National Institute of Polar Research, Tokyo.

Ekstrom PA (2007) Error measures for template-fit geolocation based on light. Deep-Sea Res. II 54: 392–403.

Erickson DL (2007) Oceanic distribution and behavior of green sturgeon. American Fisheries Society Symposium 56: 196–211.

Eveson JP, Basson M and Hobday AJ (2012) Using electronic tag data to improve mortality and movement estimates in a tag-based spatial fisheries assessment model. Canadian Journal of Fisheries and Aquatic Sciences 69: 869–883.

Farwell CJ (2001) Tunas in captivity. pp. 391–412. *In*: Block BA and Stevens ED (eds.). Tuna Physiology, Ecology, and Evolution. Academic Press, San Diego, California.

Gunn J and Block B (2001) Advances in acoustic, archival, and satellite tagging of tunas. pp. 167–224. *In*: Block BA and Stevens ED (eds.). Tuna Physiology, Ecology, and Evolution. Academic Press, San Diego, California.

Gunn JS, Polacheck T, Davis TL, Sherlock M and Betlehem A (1994) The application of archival tags to study the movement, behaviour and physiology of southern bluefin tuna, with comments on the transfer of the technology to ground Fisheries Research. Proceedings of ICES mini-symposium on fish migration. ICES Vol Palaegade 2-4, DK-1261.

Gunn JS, Patterson TA and Pepperell JG (2003) Short-term movement and behavior of black marlin *Makaira indica* in the Coral Sea as determined through a pop-up satellite archival tagging experiment. Marine & Freshwater Research 54: 515–525.

Hammerschlag N, Gallagher AJ and Lazarre DM (2011) A review of shark satellite tagging studies. Journal of Experimental Marine Biology and Ecology 398: 1–8.

Hartog JR, Patterson TA, Hartmann K, Jumppanen P, Cooper S and Bradford R (2009) Developing integrated database systems for the management of electronic tagging data. pp. 367–380. *In*: Nielsen JL, Arrizabalaga H, Fragoso N, Hobday A, Lutcavage M and Sibert J (eds.). Tagging and Tracking of Marine Animals with Electronic Devices. Springer, Dordrecht, The Netherlands.

Hill RD (1994) Theory of geolocation by light levels. pp. 227–236. *In*: Le Boeuf BJ and Laws RM (eds.). Elephant Seals: Population Ecology, Behavior, and Physiology. University of California Press, Berkley.

Hill RD and Braun MJ (2001) Geolocation by light levels—The next step: Latitude. pp. 315–330. *In*: Sibert J and Nielsen J (eds.). Electronic Tagging and Tracking in Marine Fisheries. Kluwer Academic Publs, Dordrecht, The Netherlands.

Holdsworth JC, Sippel TJ and Block BA (2009) Near real time satellite tracking of striped marlin (*Kajikia audax*) movements in the Pacific Ocean. Marine Biology l 156: 505–514.

Holland KN and Braun MJ (2003) Proceedings of 'tying one on'—a workshop on tag attachment techniques for large marine animals. Honolulu, HI, 4–5 December 2002. School of Ocean and Earth Science and Technology (SOEST) publ. no. 03-02. Joint Institute for Marine and Atmospheric Research Contribution (JIMAR) 03–349.

Holland KN, Itano DG and Domeier M (2006) First successful surgical implantation of electronic tags in marlin. Bulletin of Marine Science 79: 871–874.

Hoolihan JP, Luo J, Goodyear CP, Orbesen ES and Prince ED (2011a) Vertical habitat use of sailfish (*Istiophorus platypterus*) in the Atlantic and eastern Pacific, derived from pop-up satellite archival tag data. Fisheries Oceanography 20: 192–205.

Hoolihan JP, Luo J, Abascal FJ, Campana SE, De Metrio G, Dewar H, Domeier ML, Howey LA, Lutcavage ME, Musyl MK, Neilson JD, Orbesen ES, Prince ED and Rooker JR (2011b) Evaluating post-release behaviour modification in large pelagic fish deployed with pop-up satellite archival tags. ICES Journal of Marine Science 68: 880–889.

Humphries NE, Nuno Queiroz, Jennifer RM Dyer, Nicolas G Pade, Michael K Musyl, Kurt M Schaefer, Daniel W Fuller, Juerg M Brunnschweiler, Thomas K Doyle, Jonathen DR Houghton, Graeme C Hays, Catherine S Jones, Leslie R Noble, Victoria J Wearmouth, Emily J Southall and David W Sims (2010) Environmental context explains Levy and Brownian movement patterns of marine predators. Nature 465: 1066–1069.

Hunter JR, Argue AW, Bayliff WH, Dizon AE, Fonteneau A, Goodman D and Seckel GR (1986) The dynamics of tuna movements: an evaluation of past and future research. FAO Fisheries Technical Papers 277: 1–78.

Inagake D, Yamada H, Segawa K, Okazaki M, Nitta A and Itoh T (2001) Migration of young bluefin tuna, *Thunnus orientalis* (Temminck and Schlegel), through archival tagging experiments and its relation with oceanographic conditions in the western North Pacific. Bulletin of Far Seas Fisheries Research Lab 38: 53–81.

Itoh T, Tsuji S and Nitta A (2003) Migration of young bluefin tuna *Thunnus orientalis* in the Pacific Ocean observed with archival tags. Fishery Bulletin 101: 514–534.

Jellyman D and Tsukamoto K (2002) Swimming depths of offshore migrating longfin eels *Anguilla dieffenbachia*. Marine Ecology Progress Series 286: 261–267.

Jonsen ID, Myers RA and Flemming JM (2003) Metaanalysis of animal movement using state-space models. Ecology 84: 3055–3063.

Jonsen ID, Flemming JM and Myers RA (2005) Robust state-space modeling of animal movement data. Ecology 86: 2874–2880.

Jorgensen S, Carol A Reeb, Taylor K Chapple, Scot Anderson, Christopher Perle, Sean R Van Sommeran, Callaghan Fritz-Cope, Adam C Brown, A Peter Klimley and Barbara A Block (2009) Philopatry and migration of Pacific white sharks. Proceedings of the Royal Society of London B 277: 679–688.

Kalman RE (1960) A new approach to linear filtering and prediction problems. Trans. of the ASME Journal of Basic Engineering 82: 35–45.

Kerstetter DW, Polovina JJ and Graves JE (2004) Evidence of shark predation and scavenging on fishes equipped with pop-up satellite archival tags. Fishery Bulletin 102: 750–756.

Kitagawa T, Nakata H, Kimura S, Itoh T, Tsuji S and Nitta A (2000) Effect of ambient temperature on the vertical distribution and movement of Pacific bluefin tuna revealed with archival tags. Marine Ecology Progress Series 206: 251–260.

Kitagawa T, Boustany AM, Farwell C, Williams TD, Castleton M and Block BA (2007) Horizontal and vertical movements of juvenile Pacific bluefin tuna (*Thunnus orientalis*) in relation to seasons and oceanographic conditions. Fisheries Oceanography 16: 409–421.

Klimley A (1985) The areal distribution and autoecology of the white shark, *Carcharodon carcharias*, off the west coast of North America. Memoirs South California Academy of Sciences 9: 15–40.

Klimley AP, Prince ED, Brill RW and Holland K (1994) Archival tags 1994: present and future. NOAA Tech. Memo NMFS-SEFSC-3570. 42p.

Kurota H, McAllister MK, Lawson GL, Nogueira JI, Teo SLH and Block B (2009) A sequential Bayesian methodology to estimate movement and exploitation rates using electronic and conventional tag data: application to Atlantic bluefin tuna (*Thunnus thynnus*). Canadian Journal of Fisheries and Aquatic Sciences 66: 321–342.

Lam CH and Tsontos VM (2011) Integrated management and visualization of electronic tag data with Tagbase. PLOS ONE 6: e21810.

Lam CH, Nielsen A and Sibert J (2008) Improving light and temperature based geolocation by unscented Kalman filtering. Fisheries Research 91: 15–25.

Lam CH, Nielsen A and Sibert JR (2010) Incorporating sea-surface temperature to the light-based geolocation model TrackIt. Marine Ecology Progress Series 419: 71–84.

Loher T and Seitz A (2006) Seasonal migration and environmental conditions experienced by Pacific halibut *Hippoglossus stenolepis*, elucidated from pop-up archival transmitting (PAT) tags. Marine Ecology Progress Series 317: 259–271.

Loher T and Rensmeyer R (2011) Physiological responses by Pacific halibut, *Hippoglossus stenolepis*, to intracoelomic implantation of archival tags, with a review of tag implantation techniques employed in flatfishes. Reviews in Fish Biology and Fisheries 21: 97–115.

Lutcavage ME, Brill RW, Skomal GB, Chase BC and Howey PW (1999) Results of pop-up satellite tagging of spawning size class fish in the Gulf of Maine: do North Atlantic bluefin tuna spawn in the mid-Atlantic? Canadian Journal of Fisheries and Aquatic Sciences 56: 173–177.

Manabe R, Aoyama J, Watanabe K, Kawai M, Miller MJ and Tsukamoto K (2011) First observations of the oceanic migration of Japanese eel, from pop-up archival transmitting tags. Marine Ecology Progress Series 437: 229–240.

Maunder MN and Aires-da-Silva A (2011) Status of yellowfin tuna in the eastern Pacific Ocean in 2009 and outlook for the future. IATTC Stock Assessment Reports 11: 3–16.

Metcalfe JD (2001) Summary report of a workshop on daylight measurements for geolocation in animal telemetry. pp. 443–456. *In*: Sibert J and Nielsen J (eds.). Electronic tagging and tracking in marine fisheries. Kluwer Academic Press, Dordrecht, The Netherlands.

Methling C, Tudorache C, Skov PV and Steffensen JF (2011) Pop up satellite tags impair swimming performance and energetics of the European eel (*Anguilla anguilla*). PLOS ONE 6: e20797.

Moyes CD, Fragoso N, Brill RW and Musyl MK (2006) Predicting post release survival in large pelagic fish. Transactions of the American Fisheries Society 135: 1389–1397.

Mulcahy DM (2011) Antibiotic use during the intracoelomic implantation of electronic tags into fish. Reviews in Fish Biology and Fisheries 21: 83–96.

Musyl MK, Richard W Brill, Daniel S Curran, John S Gunn, Jason R Hartog, Roger D Hill, David W Welch, J Paige Eveson, Christofer H Boggs and Russell E Brainard (2001) Ability of electronic archival tags to provide estimates of geographical position based on light intensity. pp. 343–368. *In*: Sibert J and Nielsen J (eds.). Electronic Tagging and Tracking in Marine Fisheries. Kluwer Academic Press, Dordrecht, The Netherlands.

Musyl MK, Brill RW, Boggs CH, Curran DS, Kazama TK and Seki MP (2003) Vertical movements of bigeye tuna (*Thunnus obesus*) associated with islands, buoys, and seamounts near the main Hawaiian Islands from archival tagging data. Fisheries Oceanography 12: 152–169.

Musyl MK, Domeier ML, Nasby-Lucas N, Brill RW, McNaughton LM, Swimmer JY, Lutcavage MS, Wilson SG, Galuardi B and Liddle JB (2011a) Performance of pop-up satellite archival tags. Marine Ecology Progress Series 443: 1–28.

Musyl MK, Brill RW, Curran DS, Fragoso NM, McNaughton LM, Nielsen A, Kikawa BS and Moyes CD (2011b) Post-release survival, vertical and horizontal movements, and thermal habitats of five species of pelagic sharks in the central Pacific Ocean. Fishery Bulletin 109: 341–368.

Nakajima H and Nitta A (2001) Notes about the ecology of the ocellate puffer, *Takifugu rubripes*, using an archival tag. pp. 279–287. *In*: Sibert J and Nielsen J (eds.). Electronic Tagging and Tracking in Marine Fisheries. Kluwer Academic Press, Dordrecht, The Netherlands.

Nelson DR (1978) Telemetering techniques for the study of free-ranging sharks. pp. 419–482. *In*: Hodgson ES and Mathewson RF (eds.). Sensory Biology of Sharks, Skates, and Rays. Office of Naval Research, Department of the Navy, Arlington, Va., USA.

Nielsen A and Sibert JR (2007) State-space model for light-based tracking of marine animals. Canadian Journal of Fisheries and Aquatic Sciences 64: 1055–1068.

Nielsen A, Bigelow KA, Musyl MK and Sibert JR (2006) Improving light-based geolocation by including sea surface temperature. Fisheries Oceanography 15: 314–325.

Neilson JD, Smith S, Royer F, Paul SD, Porter JM and Lutcavage M (2009) Investigations of horizontal movements of Atlantic swordfish using pop-up satellite archival tags. pp. 145–159. *In*: Nielsen JL, Arrizabalaga H, Fragoso N, Hobday A, Lutcavage M and Sibert J (eds.). Reviews: Methods and Technologies in Gish Biology and Fisheries. Tagging and Tracking of Marine Animals with Electronic Devices. Kluwer Academic Press, Dordrecht, The Netherlands.

Patterson TA and Hartmann K (2011) Designing satellite tagging studies: estimating and optimizing data recovery. Fisheries Oceanography 20: 449–461.

Patterson TA, Thomas L, Wilcox C, Ovaskainen O and Matthiopoulos J (2008) State-space models of individual animal movement. Trends in Ecology and Evolution 23: 87–94.

Pedersen MW, Righton D, Thygesen UH, Andersen KH and Madsen H (2008) Geolocation of North Sea cod (*Gadus morhua*) using Hidden Markov Models and behavioral switching. Canadian Journal of Fisheries and Aquatic Sciences 65: 2367–2377.

Polovina JJ, Hawn D and Abecassis M (2007) Vertical movement and habitat of opah (*Lampris guttatus*) in the central North Pacific recorded with pop-up archival tags. Marine Biology 153: 257–267.

Prince ED and Goodyear CP (2006) Hypoxia-based habitat compression of tropical pelagic fishes. Fisheries Oceanography 15: 451–464.

Prince ED, Jiangang Luo, Phillip Goodyear C, John P Hoolihan, Derke Snodgrass, Eric S Orbesen, Joseph E Serafy, Mauricio Ortiz and Michael J Schirripa (2010) Ocean scale hypoxia-based habitat compression of Atlantic istiophorid billfishes. Fisheries Oceanography 19: 448–462.

Righton D, Kjesbu OS and Metcalfe J (2006) A field and experimental evaluation of the effect of data storage tags on the growth of cod (*Gadus morhua* Linné). Journal of Fish Biology 68: 385–400.

Royer F, Fromentin JM and Gaspar P (2005) A state-space model to derive bluefin tuna movement and habitat from archival tags. Oikos 109: 473–484.

Schaefer KM (2001) Reproductive biology of tunas. pp. 225–270. *In*: Block BA and Stevens ED (eds.). Tuna: Physiology, Ecology, and Evolution. Academic Press, San Diego, California.

Schaefer KM (2008) Stock structure of bigeye, yellowfin, and skipjack tunas in the eastern Pacific Ocean. IATTC Stock Assessment Reports 9: 203–221.

Schaefer KM and Fuller DW (2002) Movements, behavior, and habitat selection of bigeye tuna (*Thunnus obesus*) in the eastern equatorial Pacific, ascertained through archival tags. Fishery Bulletin 100: 765–788.

Schaefer KM and Fuller DW (2006) Comparative performance of current-generation geolocating archival tags. Journal of Marine Science and Technology 40: 15–28.

Schaefer KM and Fuller DW (2009) Horizontal movements of bigeye tuna (*Thunnus obesus*) in the eastern Pacific Ocean, as determined from conventional and archival tagging experiments initiated during 2000–2005. IATTC Bulletin 24: 191–247.

Schaefer KM, Fuller DW and Block BA (2011) Movements, behavior, and habitat utilization of yellowfin tuna (*Thunnusalbacares*) in the Pacific Ocean off Baja California, Mexico, determined from archival tag data analysis, including unscented Kalman filtering. Fisheries Research 112: 22–37.

Seitz AC, Wilson D, Norcross BL and Nielsen JL (2003) Pop-up archival transmitting (PSAT) tags: a method to investigate the migration and behavior of Pacific halibut, *Hippoglossus stenolepis*, in the Gulf of Alaska. Alaska Fisheries Research Bulletin 10: 124–136.

Sibert JR and Fournier DA (2001) Possible models for combining tracking data with conventional tagging data. pp. 443–456. *In*: Sibert J and Nielsen J (eds.). Electronic Tagging and Tracking in Marine Fisheries. Kluwer Academic Press, Dordrecht, The Netherlands.

Sibert J, Musyl MK and Brill RW (2003) Horizontal movements of bigeye tuna (*Thunnus obesus*) near Hawaii determined by Kalman filter analysis of archival tagging data. Fisheries Oceanography 12: 141–151.

Skomal GB (2007) Evaluating the physiological and physical consequences of capture on post-release survivorship in large pelagic fishes. Fisheries Management and Ecology 14: 81–89.

Smith P and Goodman D (1986) Determining fish movement from an 'archival' tag: precision of geographical positions made from a time series of swimming temperature and depth. Tech Rep NOAA-TM-NMFSSWFC-60, NOAA Tech Mem, NOAA, Silver Spring, MD.

Taillade M (1992) Animal tracking by satellite. pp. 149–160. *In*: Priede IG and Swift SM (eds.). Wildlife Telemetry: Remote Monitoring and Tracking of Animals. Ellis Horwood, New York.

Tancell C, Phillips RA, Xavier JC, Tarling GA and Sutherland WJ (2013) Comparison of methods for determining key marine areas from tracking data. Marine Biology 160: 15–26.

Taylor NG, McAllister MK, Lawson GL, Carruthers T and Block BA (2011) Atlantic bluefin tuna: a novel multistock spatial model for assessing population biomass. PLOS ONE 6: e27693.

Teo SLH, Boustany A, Blackwell S, Walli A, Weng KC and Block BA (2004) Validation of geolocation estimates based on light level and sea surface temperature from electronic tags. Marine Ecology Progress Series 283: 81–98.

Teo SLH, Sandstrom PT, Chapman ED, Null RE, Brown K, Klimley AP and Block BA (2013) Archival and acoustic tags reveal the post-spawning migrations, diving behavior, and thermal habitat of hatchery-origin Sacramento River steelhead kelts (*Onchorhynchus mykiss*). Environmental Biology of Fish 96: 175–187 .

Thorsteinsson V (2002) Tagging methods for stock assessment and research in fisheries. Report of Concerted Action FAIR CT.96.1394 (CATAG). Reykjavik. Marine Research Institute Technical Report 79: 1–79.

Thygesen UH, Pedersen MW and Madsen H (2009) Geolocating fish using hidden Markov models and data storage tags. pp. 277–293. *In*: Nielsen JL, Arrizabalaga H, Fragoso N, Hobday A, Lutcavage M and Sibert JR (eds.). Reviews: Methods and Technologies in Fish Biology and Fisheries. Tagging and Tracking of Marine Animals with Electronic Devices. Kluwer Academic Press, Dordrecht, The Netherlands.

Tremblay Y, Robinson PW and Costa DP (2009) A parsimonious approach to modeling animal movement data. PLOS ONE 4: e4711.

Tsontos VM, O'Brien FJ, Domeier ML and Lam CH (2006) Description of an improved algorithm for automated archival tag geolocational estimation based on the matching of satellite SST and *in situ* temperature data: application to striped marlin (*Tetrapturus audax*) in the north Pacific. ICES CM 2006/Q:17: 1–13.

Tsuji S, Itoh T, Nitta A and Kume S (1999) The trans-Pacific migration of a young bluefin tuna, *Thunnus thynnus*, recorded by an archival tag. Working Paper ISC2/99/15, Interim Scientific Committee for Tuna and Tuna-Like Species in the North Pacific Ocean, January 15-23, 1999, Honolulu.

Wada K and Ueno Y (1999) Homing behavior of chum salmon determined by an archival tag. North Pacific Anadromous Fish Commission Documents 425: 1–29.

Wagner GN, Cooke SJ, Brown RS and Deters KA (2011) Surgical implantation techniques for electronic tags in fish. Reviews in Fish Biology and Fisheries 21: 71–81.

Walli A, Teo SLH, Boustany A, Farwell CJ, Williams T, Dewar H, Prince E and Block BA (2009) Seasonal movements, aggregations and diving behavior of Atlantic bluefin tuna (*Thunnus thynnus*) revealed with archival tags. PLOS ONE 4: 1–18.

Welch DW and Eveson JP (1999) An assessment of light-based geoposition estimates from archival tags. Canadian Journal of Fisheries and Aquatic Sciences 56: 1317–1327.

Welch DW and Eveson JP (2001) Recent progress in estimating geoposition using daylight. pp. 369–383. *In*: Sibert J and Nielsen J (eds.). Electronic Tagging and Tracking in Marine Fisheries. Kluwer Academic Press, Dordrecht, The Netherlands.

Welch DW, Batten SD and Ward BR (2007) Growth, survival, and tag retention of steelhead trout (*O. mykiss*) surgically implanted with dummy acoustic tags. Hydrobiology 582: 289–299.

Wells RM, McIntyre GRH, Morgan AK and Davie PS (1986) Physiological stress responses in big gamefish after capture: observations on plasma chemistry and blood factors. Comparative Biochemistry and Physiology Part A: Physiology 84: 565–571.

Weng KC, Castilho PC, Morrissette JM, Landiera-Fernandez A, Holts DB, Schallert RJ, Goldman KJ and Block BA (2005) Satellite tagging and cardiac physiology reveal niche expansion in salmon sharks. Science 310: 104–106.

Weng K, Boustany A, Pyle P, Anderson S, Brown A and Block B (2007a) Migration and habitat of white sharks (*Carcharodon carcharias*) in the eastern Pacific Ocean. Marine Biology 152: 877–894.

Weng KC, O'Sullivan JB, Lowe CG, Winkler CE, Dewar H and Block BA (2007b) Movements, behavior and habitat preferences of juvenile white sharks *Carcharodon carcharias* in the eastern Pacific. Marine Ecology Progress Series 338: 211–224.

Wilson R, Ducamp J, Rees W, Culik B and Neikamp K (1992) Estimation of location: global coverage using light intensity. pp. 131–134. *In*: Priede I and Swift S (eds.). Wildlife Telemetry: Remote Monitoring and Tracking of Animals. Ellis Horwood, New York, NY.

Wilson SG, Lutcavage ME, Brill RW, Genovese MP, Cooper AB and Everly AW (2005) Movements of bluefin tuna (*Thunnus thynnus*) in the northwestern Atlantic Ocean recorded by pop-up satellite archival tags. Marine Biology 146: 409–423.

Wilson SG, Polovina JJ, Stewart BS and Meekan MG (2006) Movements of whale shark (*Rhincodon typus*) tagged at Nigaloo Reef, Western Australia. Marine Biology 148: 1157–1166.

Wilson SG, Stewart BS, Polovina JJ, Meekan MG, Stevens JD and Galuardi B (2007) Accuracy and precision of archival tag data: a multiple-tagging study conducted on a whale shark (*Rhincodon typus*) in the Indian Ocean. Fisheries Oceanography 16: 547–554.

General Index

Species Index

Printed and bound by CPI Group (UK) Ltd, Croydon, CR0 4YY

01/11/2024

01782623-0007